Lecture Notes in Electrical Engineering

For further volumes:
http://www.springer.com/series/7818

Massimo Conti · Simone Orcioni
Natividad Martínez Madrid
Ralf E. D. Seepold
Editors

Solutions on Embedded Systems

Springer

Editors
Prof. Dr. Massimo Conti
Dip. di Ingegneria Biomedica
Elettronica e Telecomunicazioni
DIBET
Università Politecnica delle Marche
Via brecce bianche 12
Ancona 60131
Italy
e-mail: m.conti@univpm.it

Prof. Dr. Simone Orcioni
Dip. di Ingegneria Biomedica
Elettronica e Telecomunicazioni
DIBET
Università Politecnica delle Marche
Via brecce bianche 12
Ancona 60131
Italy
e-mail: s.orcioni@univpm.it

Prof. Dr. Natividad Martínez Madrid
Computer Science
Reutlingen University
Alteburgstr. 150
Reutlingen 72762
Germany
e-mail: Natividad.martinez@reutlingen-university.de

Prof. Dr. Ralf E. D. Seepold
Hochschule für Technik,
Wirtschaft und Gestaltung (HTWG)
Hochschule Konstanz
Brauneggerstrasse 55
Konstanz 78462
Germany
e-mail: ralf.seepold@htwg-konstanz.de

ISSN 1876-1100 e-ISSN 1876-1119

ISBN 978-94-007-3588-0 ISBN 978-94-007-0638-5 (eBook)

DOI 10.1007/978-94-007-0638-5

Springer Dordrecht Heidelberg London New York

Softcover reprint of the hardcover 1st edition 2011

Cover design: eStudio Calamar, Berlin/Figueres

Printed on acid-free paper

Springer is part of Springer Science+Business Media (www.springer.com)

Preface

Today electronic computation is performed mainly not in personal computers, but in electronic systems integrated in devices that we use every day, like cars, mobile phones, household appliances and credit cards. Embedded computing gives a substantial added value to products. Innovation in many fields such as automotive, industrial automation, telecommunications, consumer electronics, entertainment and health equipment is mainly due to embedded computing.

Electronic systems give new features to the device, such as: energy management and power reduction, safety and security, comfort and ease to use.

The use of embedded systems in many different fields may help us to find a solution to problems that are strategic for the future of the world, such as:

- Energy production, management and delivery;
- Control and monitoring of the environment;
- Food production;
- Efficient and sustainable manufacturing;
- Traffic and mobility control and monitoring;
- Security and critical infrastructure protection;
- Home and building automation;
- Healthcare systems;
- Systems for integration of ageing and disabled people.

The book "Solutions on Embedded Systems" presents an overview on several fields of applied research, like sensor networks, network on chip and multicore systems, automotive applications, software design, system architectures, design of low power embedded systems. Each area is covered by a separate part of the book.

Contents

Part I
Sensor Networks

Chapter 1
Performance of Gossip Algorithms in Wireless Sensor Networks

Marco Baldi, Franco Chiaraluce and Elma Zanaj

1.1 Introduction

Ad-hoc wireless sensor networks are peer-to-peer systems formed by many small and simple devices, able to measure some quantities and to transmit their measured values to neighboring nodes. In such networks, nodes communicate in order to merge their single contributions into a common result. This also occurs in averaging problems, whose target is to calculate, in a distributed manner, the average value of a quantity of interest (e.g., temperature). Because of their features, these networks are suitable for many purposes, as environmental monitoring applications, allowing accurate control over large areas with favorable cost-to-benefit ratio [1]. Among these applications, however, hostile environments and scenarios of natural and man-made disasters represent great challenges, in which the network availability must be ensured, in spite of a number of possible impairments.

Among the several protocols that are available nowadays for sensors communication, an increasing attention has been devoted to simple decentralized procedures based on the gossip principle, through which the computational burden is distributed among all nodes.

M. Baldi and F. Chiaraluce (✉)
Dipartimento di Ingegneria Biomedica, Elettronica e Telecomunicazioni, Facoltà di Ingegneria, Università Politecnica delle Marche, Ancona, Italy
e-mail: f.chiaraluce@univpm.it

M. Baldi
e-mail: m.baldi@univpm.it

E. Zanaj
Departamenti i Elektronikes dhe Telekomunikacionit, Fakulteti i Teknologjise se Informacionit, Universiteti Politeknik i Tiranes, Tirana, Albania
e-mail: ezanaj@gmail.com

M. Conti et al. (eds.), *Solutions on Embedded Systems*,
Lecture Notes in Electrical Engineering, 81, DOI: 10.1007/978-94-007-0638-5_1,

The gossip algorithm was originally conceived for telephone networks [2, 3]. When gossip is applied in sensor networks, noting by x_i and x_j the local measures of the i-th and j-th nodes, an interaction among them updates one or both their values, that are then used for a subsequent interaction. The communication protocols can be managed either in a synchronous or in an asynchronous way, but the latter is more practical, because of its inherent simplicity. So, in this chapter, we will limit to consider an asynchronous time model, in which any node has a clock which ticks independently at the times of a rate 1 Poisson process. Therefore, the inter-tick times at any node are rate 1 exponentials, independent across nodes and over time.

Various implementations of gossip for averaging problems are possible; they all aim at estimating the mean value of the sensed quantity. More precisely, let us denote by N the number of nodes and by $\mathbf{x}(k) = [x_1(k), x_2(k), \ldots, x_N(k)]^T$ the vector of the estimates of all nodes after k clock ticks (superscript T denotes the transpose operation). The target of the algorithm is to find a reliable measure of the average value $x_{\text{ave}} = \sum_{i=1}^{N} x_i(0)/N$ in the shortest possible time, that is, maximizing the convergence speed.

In a first implementation, called "basic gossip" in the following, an interaction among the i-th and j-th nodes produces as output $x_i(k+1) = x_j(k+1) = x_i(k)/2 + x_j(k)/2$, that is used by both nodes for the subsequent interaction [4]. A variant of this proposal consists in the so-called "push-sum" algorithm [5]. According with such protocol, a node forwards a share of its values, properly defined, to one of its neighbors, randomly selected, while keeping the remaining part. The performance of the push-sum algorithm depends on the choice of the share, which therefore represents a degree of freedom to optimize.

Both the basic gossip and the push-sum algorithm are point-to-point protocols. However, in a wireless network, when a node transmits, all nodes in its coverage area can receive the transmitted data. This suggests implementing a "broadcast" algorithm to reduce the averaging time.

Although the fundamentals of the considered protocols are well known and a number of papers on these topics already appeared in previous literature, several issues are still open. Among them, we have mentioned above the problem of optimizing the share values in the push-sum algorithm. In [5], the authors limited to say that the choice of the shares may be deterministic or random, and may or may not depend on the time, without providing, however, a numerical evidence of the impact resulting from the different choices. The same was, at our best knowledge, in the subsequent literature. Only very recently, in [6], we presented a first set of numerical and theoretical results on this issue, focusing on ring and random geometric graph topologies.

Another relevant topic, rarely explored in the past, concerns the evaluation of the performance of gossip algorithms in the presence of link failures. Actually, when averaging algorithms are adopted in wireless sensor networks, the shadow fading or other kinds of radio impairments could prevent some links from being used, due to their poor quality in terms of signal-to-noise ratio.

The study of networks with link failures could seem not different from that of non-fully-meshed networks, where, because of a limited coverage radius, each node can reach directly only a limited set of neighbors, being linked to the others only through multiple hops (which means to pass through intermediate nodes). Really, the two situations are rather different; failures can be modeled as a stochastic phenomenon, and when a percentage x of links fail, malfunctions are generally distributed at random, without any specific correlation between distinct failures. Obviously, in some cases, failures may be due to mechanisms involving simultaneously a number of nodes that are close one each other; but this appears as a particular case, while the uncorrelation assumption seems more suitable to model practical situations. In this chapter, the convergence speed of the selected gossip algorithms, in presence of random link failures throughout the network, is investigated. Our analysis is mainly based on numerical simulations, but some theoretical issues are also discussed, particularly in regard to the share optimization when the push-sum approach is applied. We develop a number of comparisons, with the aim to show the limits and potentialities of the considered techniques.

In Sect. 1.2 we define the considered gossip versions. In Sect. 1.3 we introduce the simulation parameters and describe the graph whose performance in the presence of link failures will be investigated afterwards. In Sect. 1.4 we face, from a theoretical viewpoint, the problem of the share factor optimization in the push-sum algorithm; an analytical approach is developed, based on the computation of the potential function. In Sect. 1.5 we present a number of simulation results, first considering the various algorithms separately, and then in comparative terms. Most of the chapter contents were originally presented in [7].

1.2 The Considered Gossip Algorithms

1.2.1 Basic Gossip

The basic gossip algorithm is very simple, and has been briefly described in Sect. 1.1. Its main steps are as follows:

1. Node i chooses (at random) another node, j, inside its coverage area.
2. Nodes i and j split their information into two equal parts, $x_i(k)/2$ and $x_j(k)/2$, keeping one part and sending the other.
3. Nodes i and j calculate their new estimates by adding the received value to that already stored: $x_i(k+1) = x_j(k+1) = [x_i(k) + x_j(k)]/2$.

The choice of j is done according with a uniform distribution, conditioned on the value of the Euclidean distance D_{ij} between nodes i and j. In other words, the probability that node i contacts node j ($\neq i$) when it is selected for transmission is given by:

$$p_{ij} = \begin{cases} \frac{1}{d_i}, & D_{ij} \leq r, \\ 0, & D_{ij} > r, \end{cases} \tag{1.1}$$

where d_i is the number of nodes within its coverage area, that is delimited by a coverage radius r, assumed to be equal for all nodes. So, the coverage radius represents the maximum distance at which a node can transmit reliably. Clearly, in order to recognize the nodes inside the coverage area, a query session is required, before interaction starts. We suppose that each node performs a very simple query aimed at knowing the number of reachable neighbors, d_i, in its coverage area. More sophisticated localization strategies [8] can be adopted, that permit to implement more efficient versions of averaging algorithms. In [9], for example, a geographic gossip has been proposed, based on greedy routing, that is potentially able to provide remarkable gains. But applicability of this kind of protocols, where each node must compute and compare a large number of distances from a prefixed target, seems difficult. For this reason, we have not included these gossip versions in our study. Alternatively, the selection probabilities could be optimized with the final goal to maximize the convergence speed [4] but, once again, this would make more involved the interaction while not providing, in many cases, substantial improvements [10].

The probabilities p_{ij} can be collected in a matrix $\mathbf{P}$, with $N \times N$ entries. This matrix is stochastic, i.e., each of its rows sums to 1. On the other hand, if the link between i and j fails, the corresponding p_{ij} is set equal to zero. In this case, matrix $\mathbf{P}$ is no longer stochastic, and the i-th node has a probability to communicate $\sum_j p_{ij} < 1$. In other words, if the link between i and j fails, at some clock tick the i-th node tries transmitting to the j-th node without success, thus wasting the communication attempt. Obviously, this reflects on the averaging time, which increases in a manner dependent on the number of faulty links and their distribution.

1.2.2 Push-Sum Algorithm

The push-sum protocol proceeds as follows. At the i-th node, with $i = 1, 2, \ldots, N$, two quantities are stored and updated through the interaction with the other nodes: they are named $s_i(k)$ and $w_i(k)$, respectively. These quantities satisfy the following mass conservation properties, for any k:

$$\sum_{i=1}^{N} s_i(k) = \sum_{i=1}^{N} x_i(0) = N x_{\text{ave}}, \quad \sum_{i=1}^{N} w_i(k) = N. \tag{1.2}$$

When the protocol starts, that is, once having acquired the sensed values, we have $s_i(0) = x_i(0)$ and $w_i(0) = 1, \forall i$. Later, if the clock of the i-th node ticks at the k-th time instant (let us remind that transmission is asynchronous in the considered system), it selects randomly one of its neighbors, say j, and sends to it a fraction $(1 - \alpha)$ of its parameters, while it retains the remaining fraction α. So, the parameters at nodes i and j are modified as follows:

$$\begin{aligned} s_i(k+1) &= \alpha s_i(k), \quad w_i(k+1) = \alpha w_i(k), \\ s_j(k+1) &= s_j(k) + (1-\alpha)s_i(k), \\ w_j(k+1) &= w_j(k) + (1-\alpha)w_i(k), \end{aligned} \tag{1.3}$$

while the parameters at all the other nodes remain unchanged. This way, conditions (1.2) are certainly satisfied. A new estimate at the interacted nodes is then derived as $x_m(k+1) = s_m(k+1)/w_m(k+1)$, with $m = i, j$.

In [5], where, besides point-to-point communications, also broadcast transmissions were considered, a more general mechanism was applied, where the share factor can be different for any node and even variable in time. This model, however, seems too involved for practical applications; so, we prefer to consider a single and constant α, whose value should be optimized in order to achieve the fastest convergence speed.

On the other hand, a bidirectional version of the push-sum algorithm could also be adopted where, every time node i contacts node j, sending to it a fraction of its message, node j does the same, sending to node i a share of its own message. It is possible to demonstrate (details are omitted for saving space) that, at least for a fully-meshed network, the optimum share for this case is 1/2. So, under this choice, such modified version of the push-sum algorithm practically becomes identical to the basic gossip.

1.2.3 Broadcast Algorithm

The idea to implement a broadcast algorithm originates from the observation that, when a node transmits some information, all the other nodes in its coverage area are able to receive the transmitted data. This suggests implementing a broadcast averaging algorithm that, at the expense of a slight increase in complexity, allows reducing significantly the averaging time. This broadcast algorithm is unidirectional, as the information flows from a transmitting node to a number of receiving nodes (depending on the coverage radius and the random nodes distribution) but not in the opposite sense. Similarly to the push-sum algorithm, the i-th node maintains a sum, $s_i(k)$, and a weight, $w_i(k)$. When the algorithm starts, that is for $k = 0$, we have $w_i(0) = 1$ and $s_i(0) = x_i(0)$, that coincides with the initial sensed value at node i. When the i-th node's clock ticks, say at step k, the node splits its information into a number of parts; it may keep the first, so that $[w_i(k+1), s_i(k+1)] = \alpha_i[w_i(k), s_i(k)]$, while it sends to each neighbor j one of the remaining parts: $\alpha_{ij}[w_i(k), s_i(k)]$. Node j receives the transmission and updates its values by adding the received ones, so that $[w_j(k+1), s_j(k+1)] = [w_j(k), s_j(k)] + \alpha_{ij}[w_i(k), s_i(k)]$. As stated in the expressions, this mechanism is ruled by the share parameters, α_i and α_{ij}, that can be collected in a matrix $\mathbf{A}$, having $\alpha_{ii} = \alpha_i$ along the main diagonal. The elements of $\mathbf{A}$, that satisfy the condition $\Sigma_j \alpha_{ij} = 1$, can be chosen at random or following some suitable deterministic rule. Although different laws [11] can be adopted, in [12] we showed that good results are obtained assuming $\alpha_i = 0$ and:

$$\alpha_{ij} = \begin{cases} \frac{1}{d_i} & D_{ij} \leq r, j \neq i, \\ 0 & D_{ij} > r. \end{cases} \tag{1.4}$$

1.3 Simulation Parameters

The convergence speed of the considered protocols is evaluated through the computation of the normalized difference between the estimated average and the true average. More precisely, we determine in R simulations (with R sufficiently large) the random variable $e(k) = \|\mathbf{x}(k) - x_{\text{ave}}\mathbf{1}\|/\|\mathbf{x}(0)\|$, where $\|\mathbf{x}\|$ denotes the l_2 norm of vector $\mathbf{x}$ and $\mathbf{1}$ is the vector of all ones. A set of R curves $e^m(k)$ is obtained, $m = 1 \ldots R$, that are averaged in order to compute:

$$\langle e(k) \rangle = \frac{1}{R} \sum_{m=1}^{R} e^m(k) \tag{1.5}$$

which has the meaning of estimated mean curve. In order to average over possible different initial conditions, $\mathbf{x}(0)$ is randomly changed at the beginning of each simulation. According to the probability theory, it is known that $\lim_{R\to\infty} \langle e(k) \rangle = \widehat{e(k)}$, where $\widehat{e(k)}$ represents the true average of $e(k)$. Once having determined (1.5), the averaging time is defined as the number of clock ticks, say k^*, that permits to have a normalized difference smaller than, or equal to, a prefixed value, for example $e(k^*) \leq 10^{-10}$.

In this chapter, simulations are done over a random geometric graph (RGG), where nodes are randomly distributed in a unit square, according with a 2D homogeneous Poisson point process. The number of neighbors, d_i, the i-th node is linked to, gives its *nodal degree*. In the case of regular graphs (like the ring, for example) d_i is equal for all nodes, and it is a direct measure of the connectivity level of the network. On the other hand, in general, each node is characterized by the coverage radius r; for the RGG, even assuming that all nodes have the same r, the nodal degree is generally not unique. Moreover, for each value of r, the connectivity level can vary from graph to graph. So, an average nodal degree, $\langle d \rangle$, must be computed for the analysis purposes. The behavior of $\langle d \rangle$, as a function of r, is shown in Fig. 1.1; for each value of the coverage radius, 100 RGGs have been randomly generated, and their nodal degrees have been averaged.

As one of the objects of our study is to compare the impact of failures against that of a limited r, we suppose to start with a fully-meshed network (that implies to

Fig. 1.1 Average value of d, computed over 100 random geometric graphs

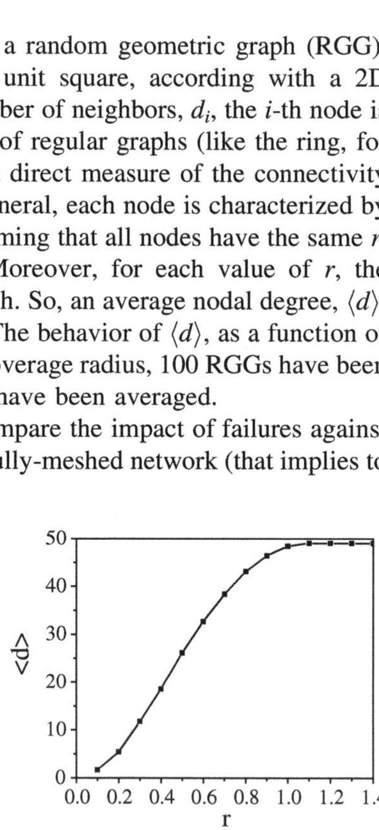

have $r \geq \sqrt{2}$ on the unit square) and to eliminate, at random, a fraction x of its links. So, while in absence of failures the network connectivity is $N - 1$ (see Fig. 1.1), in the new scenarios the average value of d becomes approximately:

$$\langle d \rangle = (N-1)(1-x). \tag{1.6}$$

The validity of (1.6) has been confirmed through simulation.

1.4 Share Factor Optimization

As mentioned in Sect. 1.2.2, for the push-sum algorithm an important issue concerns optimization of the share factor α that appears in (1.3). A useful analytical tool, in this sense, is provided by the *potential function* method.

Let us consider a vector $\mathbf{v}_i(k)$ whose components, $v_{ij}(k)$, are such that:

$$s_i(k) = \sum_{j=1}^{N} v_{ij}(k) x_j(0). \tag{1.7}$$

The following condition holds: $w_i(k) = \sum_j v_{ij}(k)$. So, if $v_i(k)$ is nearly proportional to the all-one vector, then $x_i(k) = s_i(k)/w_i(k)$ is close to the true average. The potential function at time k is defined as follows [5]:

$$\Phi(k) = \sum_{i=1}^{N} \sum_{j=1}^{N} \left[v_{ij}(k) - \frac{w_i(k)}{N} \right]^2. \tag{1.8}$$

In the limit case of all nodes perfectly aware of the true average, the potential function is null. Therefore, evaluating the mean potential function, for any k, permits us to estimate the convergence speed of the algorithm.

More precisely, assuming that, at instant k, node l is selected as the transmitter and node m as the receiver, the following difference between the potential functions at time instants k and $k+1$ can be easily derived:

$$\begin{aligned} \delta\Phi = \Phi(k) - \Phi(k+1) = 2\alpha(1-\alpha) \sum_{j=1}^{N} \left[v_{lj}(k) - \frac{w_l(k)}{N} \right]^2 \\ - 2(1-\alpha) \sum_{j=1}^{N} \left[v_{lj}(k) - \frac{w_l(k)}{N} \right] \cdot \left[v_{mj}(k) - \frac{w_m(k)}{N} \right]. \end{aligned} \tag{1.9}$$

In the following of this section we will omit, for the sake of simplicity, the argument k. We wish to compute the average of (1.9) over all possible choices, uniformly distributed, of the transmitting and receiving nodes. For a fully-meshed network, through simple algebra, it is possible to find:

$$\langle \delta\Phi \rangle = \frac{2}{N}\left[\alpha(1-\alpha) + \frac{1-\alpha}{N-1}\right]\Phi, \tag{1.10}$$

where $\Phi = \Phi(k)$. A criterion for optimizing the value of α can consist in maximizing $\langle \delta\Phi \rangle / \Phi$. According with its own meaning, in fact, to have a large $\langle \delta\Phi \rangle$, for a given Φ, should reflect in a high convergence speed. Now, from (1.10), $\langle \delta\Phi \rangle / \Phi$ is maximum for:

$$\alpha_{\text{opt}} = \frac{N-2}{2(N-1)}, \tag{1.11}$$

and, for N sufficiently large, such value can be approximated by 0.5.

In the case of non-fully-meshed network, instead, that can occur because of a limited value of r and/or the appearance of link failures, Eq. 1.10 is no longer valid, and must be replaced as follows:

$$\begin{aligned}\langle \delta\Phi \rangle = {} & \frac{2\alpha(1-\alpha)}{N}\Phi + \frac{2(1-\alpha)}{N}\sum_{j=1}^{N}\sum_{l=1}^{N}\frac{1}{d_l}\left(v_{lj} - \frac{w_l}{N}\right)^2 \\ & - \frac{2(1-\alpha)}{N}\sum_{j=1}^{N}\sum_{l=1}^{N}\frac{1}{d_l}\sum_{m\in C_l}\left(v_{lj} - \frac{w_l}{N}\right)\left(v_{mj} - \frac{w_m}{N}\right),\end{aligned} \tag{1.12}$$

where C_l is the subset of nodes that includes node l and the nodes it is linked to. The higher complexity of (1.12), with respect to (1.10), is evident. First of all, a new contribution has been added, that is null in the case of a fully-meshed network, because of the mass conservation property (1.2). Secondly, it seems not possible to evidence, at the right side, the potential function Φ, that is a necessary step toward maximization of $\langle \delta\Phi \rangle / \Phi$.

To circumvent the problem, we introduce the position $1/d_l \approx \langle 1/d \rangle, \forall l = 1 \ldots N$. Based on this approximation, Eq. 1.12 can be rewritten as:

$$\begin{aligned}\langle \delta\Phi \rangle \approx {} & \frac{2}{N}\left[\alpha(1-\alpha) + \left\langle\frac{1}{d}\right\rangle(1-\alpha)\right]\Phi \\ & - \frac{2(1-\alpha)}{N}\left\langle\frac{1}{d}\right\rangle\sum_{j=1}^{N}\sum_{l=1}^{N}\sum_{m\in C_l}\left(v_{lj} - \frac{w_l}{N}\right)\left(v_{mj} - \frac{w_m}{N}\right).\end{aligned} \tag{1.13}$$

However, the problem of evaluating the last term remains. An estimation of such term can be obtained, based on the definition of Laplacian matrix [13], as reported next. The Laplacian matrix $\mathbf{Q}(G)$ of a graph $G(V, E)$, where V is the vertex set containing the N nodes and E is the edge set, is an $N \times N$ matrix whose elements are defined as follows:

$$Q_{ij} = \begin{cases} d_i & \text{if } i = j, \\ -1 & \text{if } i \neq j \text{ and } (i,j) \in E, \\ 0 & \text{otherwise.} \end{cases} \tag{1.14}$$

The eigenvalues of $\mathbf{Q}$ are called the Laplacian eigenvalues. They are all real and non-negative, and satisfy the condition: $0 = \lambda_1 \leq \lambda_2 \leq \ldots \leq \lambda_N$. λ_2 is also known as the algebraic connectivity, and is particularly important; it is equal to zero only if G is disconnected. Other properties of matrix $\mathbf{Q}$ and its eigenvalues can be found in the literature (see [14], for example).

Let $y_{ij} = v_{ij} - w_i/N$, $i = 1\ldots N$, be the components of a vector $\mathbf{y}_j$. Through simple algebra, Eq. 1.13 can be rewritten as follows:

$$\langle \delta\Phi \rangle = -\frac{2(1-\alpha)^2}{N}\Phi + \left\langle \frac{1}{d} \right\rangle \frac{2(1-\alpha)}{N} \sum_{j=1}^{N} \mathbf{y}_j^T \mathbf{Q} \mathbf{y}_j. \tag{1.15}$$

Let us define $\mathbf{z} = (\mathbf{y}_1^T, \mathbf{y}_2^T, \ldots, \mathbf{y}_N^T)^T$; it is evident that $\mathbf{z}^T\mathbf{z} = \Phi$. Moreover, let us consider a block matrix $\mathbf{L}$, with size $N^2 \times N^2$, having N repetitions of $\mathbf{Q}$ along the main diagonal and all the other blocks equal to the null matrix. Also $\mathbf{L}$ can be interpreted as a Laplacian matrix, whose eigenvalues coincide with those of $\mathbf{Q}$, but each appears with multiplicity N. Using these further definitions, Eq. 1.15 can be rewritten as:

$$\langle \delta\Phi \rangle = \left[-\frac{2(1-\alpha)^2}{N} + \left\langle \frac{1}{d} \right\rangle \frac{2(1-\alpha)}{N} \mathrm{RQ} \right] \Phi, \tag{1.16}$$

having denoted by $\mathrm{RQ} = \mathbf{z}^T\mathbf{L}\mathbf{z}/\mathbf{z}^T\mathbf{z}$ the so-called Rayleigh quotient. So, the value of α that maximizes $\langle \delta\Phi \rangle / \Phi$ results in:

$$\alpha_{\mathrm{opt}} = 1 - \left\langle \frac{1}{d} \right\rangle \frac{\mathrm{RQ}}{2}. \tag{1.17}$$

Because of the Courant–Fischer minimax theorem [15], we have $\lambda_2 \leq \mathrm{RQ} \leq \lambda_N$. As a confirmation of the correctness of (1.17), we can observe that, in the case of a fully-meshed network, all the eigenvalues λ_i, with $i \geq 2$, are equal to N and Eq. 1.17 becomes equal to Eq. 1.11. In general, the value of RQ is not easy to determine. So, we simplify the problem by approximating RQ with the average of the non-null eigenvalues, i.e.:

$$\mathrm{RQ} \approx \frac{\sum_{i=2}^{N} \lambda_i}{N-1}. \tag{1.18}$$

In [6] we verified that the value of α_{opt} obtainable from this assumption, in case of $r < 0.6$, can be rather different from the actual optimum value. On the contrary, because of the hypothesis of randomly distributed failures, approximation (1.18) is much more acceptable when the nodal degree reduction is due to faulty links. This remark will be confirmed in Sect. 1.5.

By employing the analytical approach, for an RGG with $N = 50$ and $10 < \langle d \rangle < 49$, we have found $0.39 < \alpha_{\mathrm{opt}} < 0.49$. As, from (1.6), such range of values of $\langle d \rangle$ corresponds to $x < 0.796$, we can expect that α_{opt} does not change significantly, even for very large failure rates. This conclusion will be confirmed,

Table 1.1 Coverage radius and failure rates producing nearly identical average connectivity

Limited r ($x = 0$)	$\langle d \rangle$	Failure rate x $\left(r \geq \sqrt{2}\right)$
0.3	11.8	0.76
0.4	18.6	0.63
0.5	26.2	0.48
0.6	32.7	0.35
0.7	38.4	0.23
0.8	43.1	0.14
0.9	46.4	0.07

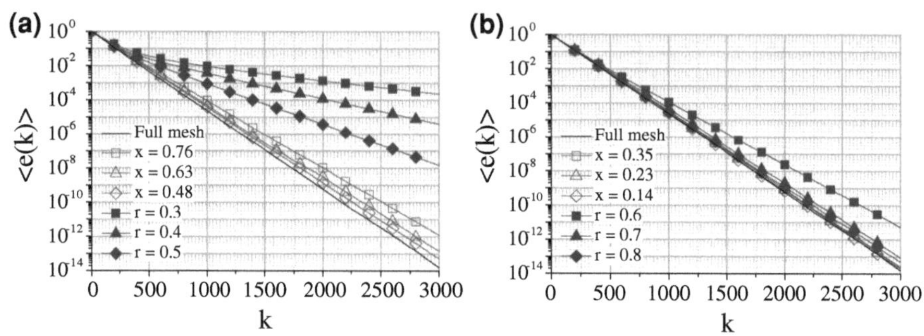

Fig. 1.2 $\langle e(k) \rangle$ for some values of radius and failure rate x (basic gossip algorithm)

in the following section, through numerical simulations, and is different from that occurring in the case of limited coverage radius, where, in the same range of $\langle d \rangle$, α_{opt} can become as low as 0.2.

1.5 Results

In this section we present a number of simulation results for the RGG with $N = 50$. Table 1.1 shows the values of $\langle d \rangle$, together with the coverage radius r (for a non-fully-meshed network with no failures) and the link failure rate x (for a fully-meshed network affected by failures). In the table, each row specifies r and x that determine (nearly) the same average connectivity level. So, we are able to compare the impact of the two different mechanisms that may be responsible for the reduction in the network connectivity. In the following, the pairs determining the same connectivity level will be denoted by r/x.

1.5.1 Basic Gossip

The simulated curves of mean normalized error are shown in Fig. 1.2. We can fix the attention on a specific error value and compare the k* needed. A couple of

Table 1.2 Number of clock ticks required to reach $\langle e(k^*) \rangle = 10^{-10}$

r/x	k^* (limited radius)	k^* (random failures)
$\geq\sqrt{2}/0$	2,165	2,165
0.6/0.35	2,661	2,210
0.4/0.63	>3,000	2,347

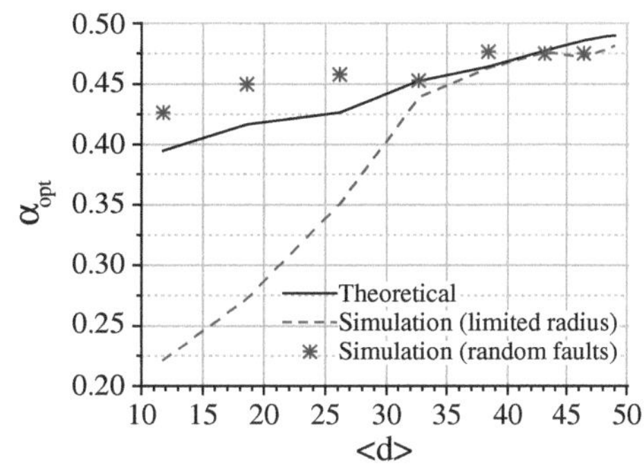

Fig. 1.3 Simulated α_{opt} for the push-sum algorithm

numerical examples are shown in Table 1.2; the k^* for the starting full-mesh network with no link failures is also reported as a benchmark. We notice that both mechanisms increase the convergence time, but the impact of the limited radius is stronger. In other words, for a given value of $\langle d \rangle$, to achieve the target requires a longer time (higher k^*) when the network is non-fully-meshed because of the limited coverage radius.

1.5.2 Push-Sum Algorithm

One problem for the push-sum algorithm is the optimization of the share factor α. The theoretical analysis developed in Sect. 1.4 gives a solid reference that, however, needs to be verified. For this purpose, we have considered $0.1 \leq x \leq 0.9$ and, for any value of the failure rate x, we have determined α_{opt} as the value of α minimizing the averaging time over a large number of repetitions of the random experiment. The result obtained is shown in Fig. 1.3, as a function of the average nodal degree. The curve is rather irregular but the optimal α is comprised between 0.43 and 0.48, which is in line with the results of the analysis in Sect. 1.4. The theoretical approach is not able to distinguish between the case of a limited radius and that of link failures. From the figure, we see that the actual network behavior is well predicted by the theory when missing links are distributed at random. When they follow from a limited coverage radius, instead, numerical simulations give results significantly different from theoretical expectations, particularly for low connectivity levels (see dashed line in Fig. 1.3).

Based on the simulation results, we can also say that the optimal value of α is close to 0.5 for the case of random faults, practically for any value of $\langle d \rangle$

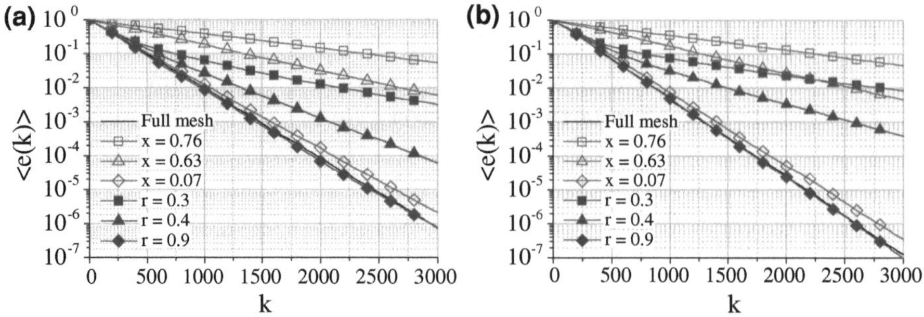

Fig. 1.4 $\langle e(k)\rangle$ for some values of coverage radius r and failure rate x by using the push-sum algorithm with **a** $\alpha = 0.3$ and **b** $\alpha = 0.5$

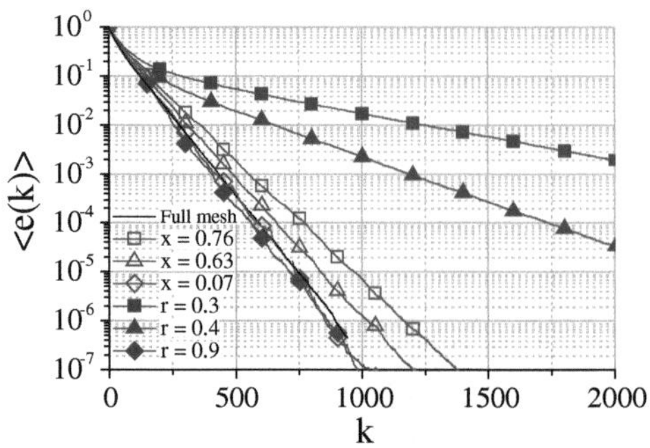

Fig. 1.5 $\langle e(k)\rangle$ for some values of radius r and failure rate x (broadcast algorithm)

(as observed, this is predicted by the theory), while smaller values should be adopted for α in the case of limited radius, particularly when $r < 0.6$. This statement is confirmed in Fig. 1.4, where $\alpha = 0.3$ and $\alpha = 0.5$ have been considered for both situations.

While in case of link failures the result for $\alpha = 0.5$ is better than that for $\alpha = 0.3$, if we focus on the results obtained with a significantly limited radius ($r < 0.6$), the opposite occurs for the non-fully-meshed network, where the smaller (although not necessarily optimum) value of α reduces the convergence time. We see that, for both values of α, the impact of faulty links is stronger than that of a limited coverage radius. This is an important difference between the behavior of the bidirectional algorithm (basic gossip) and the unidirectional one (push-sum). More will be said in Sect. 1.5.4 about the comparison between the two approaches.

1.5.3 Broadcast Algorithm

The analysis developed in the previous sections has been repeated for the broadcast algorithm. The results are shown in Fig. 1.5, for some values of $\langle d\rangle$. From the figure, we observe that the behavior of the broadcast algorithm is similar

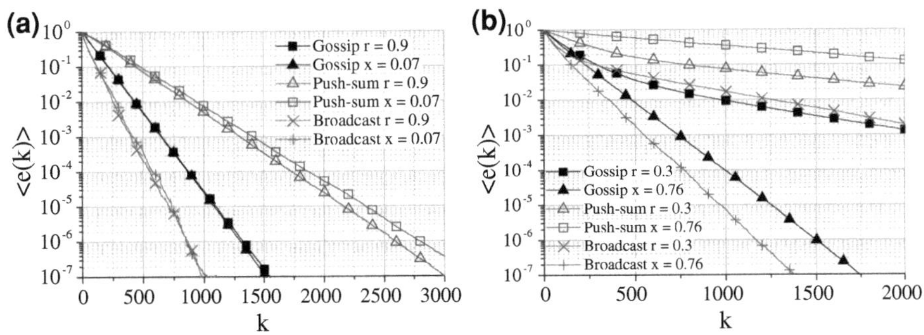

Fig. 1.6 $\langle e(k) \rangle$ for the considered averaging algorithms in the case of **a**$r = 0.9/x = 0.07$ and **b**$r = 0.3/x = 0.76$; optimum shares have been used for push-sum

to that of the basic gossip and, for the same $\langle d \rangle$, convergence of the faulty network is usually faster than that of the non-fully-meshed network.

1.5.4 Performance Comparison

For the sake of comparison, some results for the various algorithms and some pairs r/x are summarized in Fig. 1.6. As expected, the broadcast algorithm exhibits the fastest convergence to the average value, due to its point-to-multipoint nature. However, the basic gossip algorithm is also able to achieve good performance, though being simpler and requiring interaction only between couples of nodes. Its loss in terms of clock ticks, with respect to the broadcast algorithm, is usually limited within 30%. Moreover, there are situations, for very small coverage radius, where the basic gossip outperforms the broadcast algorithm (see the case $r = 0.3$ in Fig. 1.6b).

Performance of the push-sum algorithm is worse. For a network with good connectivity (see Fig. 1.6a), the push-sum algorithm requires approximately a doubled number of clock ticks with respect to the gossip algorithm to reach the same $\langle e(k) \rangle$. This can be justified by considering that, in push-sum, each interaction is unidirectional, while in the basic gossip it is bidirectional. If comparison is made on the number of transmissions, the efficiencies of basic gossip and push-sum become similar.

1.6 Conclusion

We have developed a numerical analysis of the averaging time for basic gossip, push-sum and broadcast algorithms, taking into account the impact of link failures, randomly distributed in the network. In spite of its practical importance, this topic has been rarely debated in previous literature.

Some important conclusions can be drawn from our analysis. First of all, we have demonstrated that the convergence of the algorithms is preserved, on average, regardless of the solution adopted, up to acceptably low values of the mean normalized error.

Then, we have verified that the share factor, when applicable, should be optimized for taking into account the network connectivity. However, starting from the results known for fully-meshed networks, the optimum share factor for the push-sum algorithm is less sensitive to the failure rate than to the limited coverage radius, which is another common reason for reduced connectivity. Moreover, the assumption of a random distribution for the link failures makes applicable an approximate, and inherently simple, analytical approach, based on the potential function, that instead does not provide equally accurate solutions for the case of limited coverage radius.

References

1. Barrenetxea G et al (2007) DemoAbstract: SensorScope, an urban environmental monitoring network. In: 4th European conference on wireless sensor networks (EWSN 2007), Delft, Netherlands, Jan 2007
2. Baker B, Shostak R (1972) Gossips and telephones. Discrete Math 2(3):191–193
3. Berman G (1973) The gossip problem. Discrete Math 4(1):91
4. Boyd S, Ghosh A, Prabhakar B, Shah D (2006) Randomized gossip algorithms. IEEE Trans Inf Theory 52(6):2508–2530
5. Kempe D, Dobra A, Gehrke J (2003) Gossip based computation of aggregate information. In: IEEE conference on foundation of computer science, Cambridge, MA, Oct 2003, pp 482–491
6. Zanaj E, Baldi M, Chiaraluce F (2009) Optimal share factors in the push-sum algorithm for ring and random geometric graph sensor networks. J Commun Softw Syst 5(1):9–18
7. Baldi M, Chiaraluce F, Zanaj E (2009) Fault tolerance in sensor networks: performance comparison of some gossip algorithms. In: 7th international workshop on intelligent solutions in embedded systems (WISES 2009), Ancona, Italy, June 2009, pp 11–20
8. Patwari N, Ash JN, Kyperountas S, Hero AO III, Moses RL, Correal NS (2005) Locating the nodes: cooperative localization in wireless sensor networks. IEEE Signal Process Mag 22(4):54–69
9. Dimakis AG, Sarwate AD, Wainwright MJ (2008) Geographic gossip: efficient averaging for sensor networks. IEEE Trans Signal Process 56(3):1205–1216
10. Zanaj E, Baldi M, Chiaraluce F (2007) Efficiency of the gossip algorithm for wireless sensor networks. In: 2007 international conference on software, telecommunications and computer networks (SoftCOM 2007), Split, Dubrovnik, Croatia, Sept 2007, Paper 7072
11. Zanaj E, Baldi M, Chiaraluce F (2008) Efficiency of unicast and broadcast gossip algorithms for wireless sensor networks. J Commun Softw Syst 4(2):105–112
12. Baldi M, Chiaraluce F, Zanaj E (2008) Comparison of averaging algorithms for wireless sensor networks. In: International conference on information and communication technologies (ICTTA'08), Damascus, Syria, Apr 2008, Paper TEL05_7
13. Merris R (1995) A survey of graph Laplacians. Linear Multilinear Algebra 39(1 and 2):19–31
14. Mohar B (1991) The Laplacian spectrum of graphs. In: Alavi Y, Chartrand G, Oellermann OR, Schwenk AJ (eds) Graph theory, combinatorics, and applications, vol 2. Wiley, New York, pp 871–898
15. Shawe-Taylor J, Cristianini N (2004) Kernel methods for pattern analysis. Cambridge University Press, London

Chapter 2
Using a Prioritized Medium Access Control Protocol for Incrementally Obtaining an Interpolation of Sensor Readings

Björn Andersson, Nuno Pereira, Eduardo Tovar and Ricardo Gomes

2.1 Introduction

A sensor network comprises a set of computer nodes each one equipped with a processor, memory, sensors and a transceiver for communications over a (wired or wireless) channel. The sensor network must obtain an accurate image of physical phenomena and do so with a high sampling rate in both time and space. A large number of computer nodes are needed in order to obtain a high sampling rate in space. But this generates a large number of sensor readings and since these sensor readings are located on different computer nodes, a significant amount of communication may be necessary forcing a reduction in the sampling rate in time. For systems with a very large number of computer nodes, it is therefore crucial to develop techniques that make it possible to obtain a snapshot, an approximate representation of all sensor readings, and achieve this with a time-complexity (as a function of the number of nodes) that is small.

A simple approach for obtaining an approximate representation of sensor readings would be to select a subset of the computer nodes at random and let the sensor readings at those computer nodes be used for obtaining an interpolation. Although this is fast, it has the drawback that some computer nodes with extreme sensor readings may have a significant impact on the interpolation if they would be selected but they may not be selected and this causes (as illustrated in [1]) the interpolation to be a poor representation of the physical phenomenon. And this can cause a sensor network to misperceive its physical environment.

A better approach for obtaining an approximate representation of sensor readings would be to select a subset of the computer nodes, carefully selected to be the ones that represent local extreme points and let the sensor readings at those

B. Andersson (✉) · N. Pereira · E. Tovar · R. Gomes
CISTER/IPP-Hurray Research Unit, Polytechnic Institute of Porto, Porto, Portugal
e-mail: bandersson@dei.isep.ipp.pt

M. Conti et al. (eds.), *Solutions on Embedded Systems*,
Lecture Notes in Electrical Engineering, 81, DOI: 10.1007/978-94-007-0638-5_2,

computer nodes be used for obtaining an interpolation. If one computer node had knowledge of all sensor readings then such a selection would be possible of course but in practice, a computer node only knows its own sensor reading (unless sensor readings are communicated) and therefore it has been non-obvious how to implement such an approach.

Recent work [1, 2] however have shown how to exploit a prioritized medium access control (MAC) protocol for selecting local extreme points and thereby it was shown how to quickly obtain an interpolation of sensor readings where sensor readings were taken by different computer nodes. This work assumes that the MAC protocol has a very large number of priority levels and that all sensor nodes know the priority of the node that was granted the channel. Such MAC protocols are common; the Controller Area Network (CAN) [3] bus is one such example for wired communication (with more than 300 million units sold) and a similar technology, WiDOM [2, 4] is available for wireless communication.

The algorithm [1, 2] which exploited a prioritized MAC protocol had a user-selectable parameter, k, which had the role that the k sensor nodes that contribute the most to the interpolation being a faithful representation of the physical reality are selected and the interpolation is based on those k sensor nodes. k is selected based on the number of local extrema of the signal as explained in [1]. With this approach it was possible to obtain the interpolation with a time-complexity that is $O(k)$, that is, the time-complexity is independent of the number of sensor nodes; yet the result of the interpolation was dependent on all sensor readings. The algorithm was implemented and tested both in wired systems (using CAN [3]) and in wireless systems (using WiDom [2, 4]).

The algorithm for obtaining an interpolation (i) had to run until completion and (ii) it was designed to have no prior knowledge of the physical environment. Unfortunately, these two facts bring two drawbacks:

1. There are situations where the delay from when the physical world changes until the computer system can react to this change is two times the duration required for obtaining the interpolation. (This situation occurs when the environment changed just after the algorithm for obtaining the interpolation had started; when this happens, the algorithm for obtaining the interpolation must finish execution and then take new sensor readings and finally obtain an interpolation of these new sensor readings.)
2. It is necessary that the sampling period of an application that uses the interpolation is $O(k)$ or greater. If the entire physical environment changes everywhere, it may really be necessary to obtain an interpolation from scratch. But one can expect that a change in the physical environment (such as a rapid fire, explosion or deformation) has only local effects initially (during the first milliseconds) and it changes the entire environment later. It would be desirable to use a sampling rate so high that the sampling period is independent of k and independent of the number of nodes, yet the system is able to detect extreme local changes with a duration of two sampling periods and obtain an image of the entire physical environment within k sampling periods.

Therefore, we presented, in a workshop paper [5], a new algorithm for obtaining an interpolation of sensor readings which eliminates the two above mentioned drawbacks. This chapter is an extension of that paper.

The main idea of the algorithm is that when the system starts-up, an interpolation is obtained using the previously known algorithm [1, 2]. This step of the algorithm has the time-complexity *O*(*k*), which is larger than we desire but it is done only once. Each computer node now has the *k* sensor readings that can be used to form an interpolation of all sensor readings. All computer nodes will now periodically take sensor readings with a small period; this period is independent of *k* and it is independent of the number of computer nodes. All computer nodes take their sensor readings in parallel and then each computer node computes the interpolated value at itself and compares it to its own sensor reading. The computer node whose sensor reading contributes the least to the faithfulness of the interpolation is attempted to be deselected and the computer node whose sensor reading contributes the most to the faithfulness of the interpolation is selected.

We believe this algorithm to be useful for detecting deviations from the expected behavior in the physical world very quickly, for example detecting the deformation of mechanical elements (for example in a car or aircraft) in order to enact appropriate safety actions (such as deciding which airbag to inflate or which fuel pump to be stopped or which valve to be closed).

The remainder of this paper is organized as follows. Section 2.2 gives preliminaries, that is, the main idea of how a prioritized MAC protocol can be used for computations and also the system model we will use. Section 2.3 discusses how to obtain an interpolation; this discussion leads to the new interpolation scheme. Section 2.4 gives conclusions and future work.

2.2 Preliminaries and Motivation

The basic premise for this work is the use of a prioritized MAC protocol. This implies that the MAC protocol assures that out of all nodes contending for the medium at a given moment, the one(s) with the highest priority gain access to it. This is inspired by Dominance/Binary-Countdown protocols [6]. In such protocols, messages are assigned unique priorities, and before nodes try to transmit they perform a contention resolution phase named arbitration such that the node requesting to transmit the highest-priority message succeeds.

During the arbitration (depicted in Fig. 2.1), each node sends the message priority bit-by-bit, starting with the most significant one, while simultaneously monitoring the medium. The medium must be devised in such a way that nodes will only detect a "1" value if no other node is transmitting a "0". Otherwise, every node detects a "0" value regardless of what the node itself is sending. For this reason, a "0" is said to be a dominant bit, while a "1" is said to be a recessive bit. Therefore, low numbers in the priority field of a message represent high priorities. If a node contends with a recessive bit but hears a dominant bit, then it

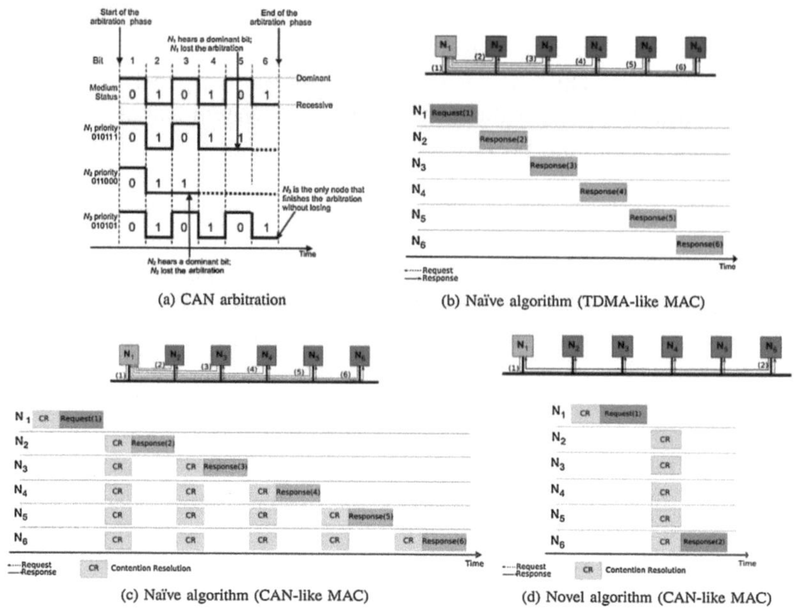

Fig. 2.1 Dominance/binary-countdown arbitration motivating examples. **a** Example of bitwise arbitration; **b** example application where N_1 needs to know the minimum (MIN) temperature reading among its neighbors (N_2–N_6); **c** possible solution for the example application using a CAN-like MAC, using fixed priorities for the messages; **d** possible solution for the example application exploiting the properties of a CAN-like MAC, where priorities are assigned at runtime according to the sensed values

will refrain from transmitting any further bits, and will proceed only monitoring the medium. Finally, exactly one node reaches the end of the arbitration phase, and this node (the winning node) proceeds with transmitting the data part of the message. As a result of the contention for the medium, all participating nodes will have knowledge of the winner's priority.

The CAN bus [3] is an example of a technology that offers such a MAC behavior. It is used in a wide range of applications, ranging from vehicles to factory automation (the reader is referred to [7] for more examples of application fields and figures about the use of CAN technologies). Its wide application fostered the development of robust error detection and fault confinement mechanisms, while at the same time maintaining its cost effectiveness. An interesting feature of CAN is that the maximum length of a bus can be traded-off for lower data rates. It is possible to have a CAN bus with a bit rate of 1 Mbit/s for a maximum bus length of 30 m, or a bus 1,000 m long (with no repeaters) using a bit rate of 50 Kbit/s. While the typical number of nodes in a CAN bus is usually smaller than 100, with careful design (selecting appropriate bus-line cross section, drop line length and quality of couplers, wires and transceivers) of the network it is possible to go well above this value. For example, CAN networks with more than a thousand nodes have been deployed and they operate in a single broadcast domain (such networks have been built; see for example [8]).

The focus of this paper is on exploiting a prioritized MAC protocol for efficiently obtaining an interpolation function which approximates the sensor readings in a geographical area. A key idea in the design of such an algorithm is the use of a prioritized MAC protocol for performing computations—this is explained next.

2.2.1 The Main Idea

The problem of obtaining aggregated quantities in a single broadcast domain can be solved with a naïve algorithm: every node broadcasts its sensor reading sequentially. Hence, all nodes know all sensor readings and then they can obtain the aggregated quantity. This has the drawback that in a broadcast domain with m nodes, at least m broadcasts are required to be performed. Considering a network designed for $m \geq 100$, the naïve approach can be inefficient; it causes a large delay.

Let us consider the simple application scenario as depicted in Fig. 2.1b, where a node (node N_1) needs to know the minimum (MIN) temperature reading among its neighbors. Let us assume that no other node attempts to access the medium before this node. A naïve approach would imply that N_1 broadcasts a request to all its neighbors and then N_1 would wait for the corresponding replies from all of them. As a simplification, assume that nodes orderly access the medium in a time division multiple access (TDMA) fashion, and that the initiator node knows the number of neighbor nodes. Then, N_1 can derive a waiting timeout for replies based on this knowledge. Clearly, with this approach, the execution time depends on the number of neighbor nodes (m). Figure 2.1c depicts another naïve approach, but using a CAN-like MAC protocol.

Assume in that case that the priorities the nodes use to access the medium are ordered according to the nodes' ID, and are statically defined prior to runtime. Note that in order to send a message, nodes have to perform arbitration before accessing the medium. When a node wins it sends its response and stops trying to access the medium. It is clear that using a naïve approach with CAN brings no timing advantages as compared to the other naïve solution (Fig. 2.1b).

Consider now that instead of using their priorities to access the medium, nodes use the value of its sensor reading as priority. Assume that the range of the analog to digital converters (ADC) on the nodes is known, and that the MAC protocol can, at least, represent as many priority levels. This assumption typically holds since ADC tend to have a data width of 8, 10, 12 or 16-bit while the CAN bus offers up to 29 priority bits. This alternative would allow an approach as depicted in Fig. 2.1d. With such an approach, to obtain the minimum temperature among its neighbors, node N_1 needs to perform a broadcast request that will trigger all its neighbors to contend for the medium using the prioritized MAC protocol. If neighbors access the medium using the value of their temperature reading as the priority, the priority winning the contention for the medium will be the minimum temperature reading. With this scheme, more than one node can win the contention for the medium. But, considering that at the end of the arbitration the priority of the winner is known to all nodes, no more information needs to be transmitted by the winning node. In this scenario, the time to obtain the minimum temperature reading only depends on the time to perform the contention for the medium, not on m. If, for example, one wishes that the winning node transmits information (such as its location) in the data packet, then one can code the priority of the nodes by adding a unique number (for example, the node ID) in the least significant bits, such that priorities will be unique.

A similar approach can be used to obtain the maximum (MAX) temperature reading. In that case, instead of directly coding the priority with the temperature reading, nodes will use the bitwise negation of the temperature reading as the priority. Upon completion of the medium access contention, given the winning priority, nodes perform bitwise negation again to know the maximum temperature value.

MIN and MAX are just two simple and pretty much obvious examples of how aggregate quantities can be obtained with a minimum message complexity (and therefore time complexity) if message priorities are dynamically assigned at runtime upon the values of the sensed quantity. In Sect. 2.3 we will show how this technique of using a prioritized MAC protocol for computations can be used for obtaining an interpolation of sensor readings.

2.2.2 System Model

The network consists of m nodes that take sensor readings where a node is given a unique identifier in the range 1… m. MAXNNODES denotes an upper bound on m and we assume that MAXNNODES is known by the designer of the system before run-time. Nodes do not have a shared memory and all data variables are local to each node.

Each node has a transceiver and is able to transmit to or receive from a single channel. Every node has an implementation of a prioritized MAC protocol with the characteristics as described earlier. Nodes perform requests to transmit, and each transmission request has an associated priority. Priorities are integers in the range [0, MAXP], where lower numbers correspond to higher priorities. Let *NPRIOBITS* denote the number of priority bits. This parameter has the same value for all nodes. Since *NPRIOBITS* is used to denote the number of bits used to represent the priority, the priority is a number in the range of $0–2^{\text{NPRIOBITS}} - 1$. Clearly, $\text{MAXP} = 2^{\text{NPRIOBITS}} - 1$.

A node can request to transmit an *empty packet*; that is, a node can request to the MAC protocol to perform the contention for the medium, but not send any data. This is clarified later in this section. All nodes share a single reliable broadcast domain.

A program on a node can access the communication system via the following interface. The `send` system call takes two parameters, one describing the priority of the packet and another one describing the data to be transmitted. If a node calling `send` wins the contention, then it transmits its packet and the program making the call unblocks. If a node calling `send` loses the contention, then it waits until the contention resolution phase has finished and the winner has transmitted its packet (assuming that the winner did not send an empty packet). Then, the node contends for the channel again. The system call `send` blocks until it has won the contention and transmitted a packet. The function `send_empty` takes only one parameter, which is a priority and causes the node only to perform the contention but not to send any data after the contention. In addition, when the contention is over (regardless of whether the node wins or loses), the function `send_empty` gives the control back to the application and returns the priority of the winner.

The system call `send_and_rcv` takes two parameters, priority and data to be transmitted. The contention is performed with the given priority and then the data is transmitted if the node wins. Regardless of whether the node wins or loses, the system call returns the priority and data transmitted by the winner and then unblocks the application.

A node N_i takes a sensor reading s_i. It is an integer in the range [0, MAXS] and it is assumed that MAXS $\leq$ MAXP.

2.3 Interpolation of Sensor Data with Location

Having seen the main idea of how to take advantage of a prioritized MAC protocol, we are now in position to present our approach for obtaining an interpolation of sensor readings. We will do so formally with pseudo-code; this pseudo-code returns an upper bound on the error of the interpolation. We will first (in Sect. 2.3.1) present the main idea of the previously known interpolation scheme.

This will lead us (in Sect. 2.3.2) to the new incremental interpolation scheme. This new incremental interpolation scheme needs to, as an intermediate result, evaluate the interpolation at certain geographical points. Therefore, we will (in Sect. 2.3.3) modify this algorithm to perform calculations at those specific points with additional speed and this results in an improvement of the new incremental interpolation scheme.

2.3.1 Previously Known Algorithm

We assume that nodes take sensor readings, but we will also assume that a node N_i knows its location given by two coordinates (x_i, y_i). With this knowledge, it is possible to obtain an interpolation of sensor data over space. This offers a compact representation of the sensor data and it can be used to compute virtually anything.

We let $f(x, y)$ denote the function that interpolates the sensor data. Also let e_j denote the magnitude of the error at node N_j; that is:

$$e_j = \left|s_j - f\left(x_j, y_j\right)\right| \tag{2.1}$$

and let e denote the global error; that is:

$$e = \max_{j=1\ldots m} e_j \tag{2.2}$$

The goal is to find $f(x, y)$ that minimizes e subject to the following constraints: (i) the time required for computing f at a specific point should be low; and (ii) the time required to obtain the function $f(x, y)$ from sensor readings should be low. The latter is motivated by the fact that it is interesting to track physical quantities that change quickly; it may be necessary to update the interpolation periodically in order to track, for example, how the concentration of hazardous gases move. For this reason, we will use weighted-average interpolation (WAI) [9–11]. WAI is defined as follows:

$$f(x,y) = \begin{cases} 0 & \text{if } S = \emptyset \\ s_j & \text{if } \exists N_j \in S : x_j = x \wedge y_j = y \\ \dfrac{\sum_{j\in S} s_j * w_j(x,y)}{\sum_{j\in S} w_j(x,y)} & \text{otherwise} \end{cases} \tag{2.3}$$

where S is a set of nodes used for interpolation. The weights $w_j(x, y)$ are given by:

$$w_j(x,y) = \frac{1}{\left(x_j - x\right)^2 + \left(y_j - y\right)^2} \tag{2.4}$$

Algorithm 1 Finding a subset of nodes to be used inWAI

```
Require: All nodes start Algorithm 1 simultaneously.
Require: k denotes the desired number of interpolation points.
Require: A node N_i knows x_i,y_i and s_i.
Require: The code below is executed by every node. A node can read the
           variable i and obtain its node index.
Require: (MAXS+1) × (MAXNNODES+1) + MAXNNODES ≤ MAXP.
function find_nodes() return a set of packets
    S ← ∅
    for q ← 1 to k do
        Calculate f(x_i,y_i) in Equation 3 and assign
          it to the variable "myinterpolatedvalue"
        error ← abs( s_i - to_integer(myinterpolatedvalue) )
        temp_prio ← error × (MAXNNODES + 1) + i
        prio ← (MAXP+1) - temp_prio
        snd_pack ←< s_i,x_i,y_i >
        <winning_prio, rcv_pack> ← send_and_rcv( prio, snd_pack)
        S ← S ⋃ { rcv_pack }
    end for
    return S
end function
```

Intuitively, Eqs. 2.3 and 2.4 state that the interpolated value is a weighted average of all data points in S and the weight is the inverse of the square of the distance. There are many possible choices on how the weight should be computed as a function of distance; the way we have selected is intended to avoid calculations of square root in order to make the execution time small on platforms that lack hardware support for floating point calculations. This is the case for typical sensor network platforms [12–14].

The original version [9] of weighted-average interpolation uses all available sensor readings for interpolation. But this would imply that computing Eq. 2.3 from sensor readings has a time complexity of $O(m)$. Fortunately, it is often the case [15] that sensor readings exhibit spatial locality; that is, nodes that are close in space give similar sensor readings. For this reason, the interpolation will offer a low error even if only a small number of carefully selected nodes are in S.

Hence, the goal is now to find those nodes that contribute to producing a low error in the interpolation as given by Eq. 2.3. We select a number of k nodes that contribute to lowering the error of the interpolation, where k is a parameter of the algorithm that will control the accuracy of the interpolation. Recall that a prioritized MAC protocol can find the maximum among sensor readings. We can exploit

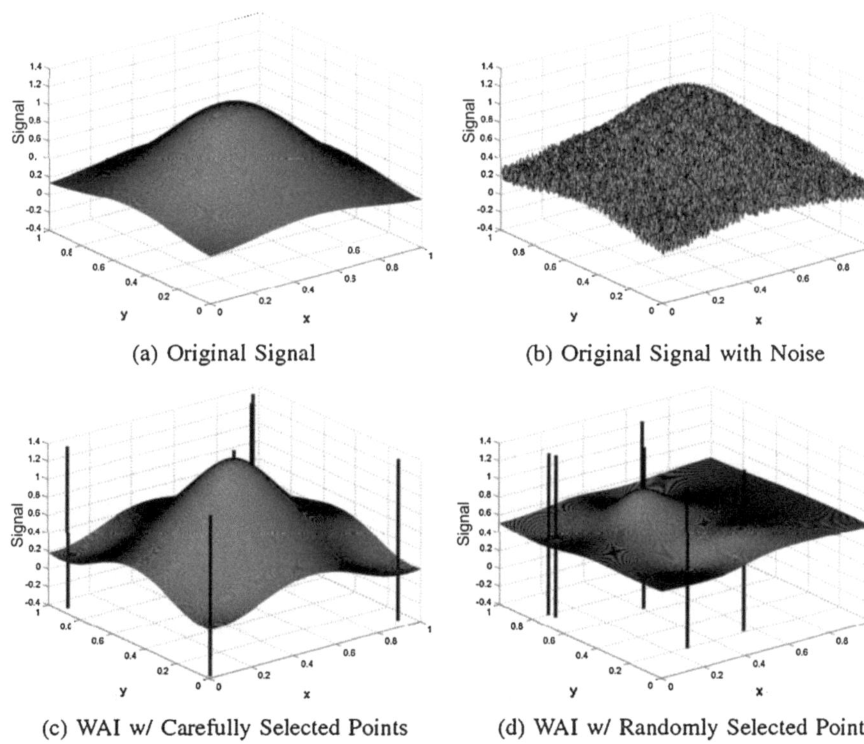

(a) Original Signal (b) Original Signal with Noise

(c) WAI w/ Carefully Selected Points (d) WAI w/ Randomly Selected Points

Fig. 2.2 Interpolation Example 1. **a** The original signal that varies over space; **b** noise added to the original signal; **c** the result of the interpolation given by our algorithm. The location of the subset of $k = 6$ nodes that were selected to be used in the interpolation is indicated with vertical lines; **d** example result of the interpolation when nodes are selected randomly this feature to find k nodes that offer a low value of the error. For this, the proposed distributed algorithm starts with an interpolation being a flat surface and then performs k iterations, where at each iteration the node with largest magnitude of the error between its sensor reading and the interpolated value will be the winner of the contention.

Algorithm 1 is designed based on this principle and it is previously published [1]. It computes (on line 5) the error. This error is concatenated with the identifier of the node (together this forms the priority of the message) ensuring that all priorities are unique. All nodes send their messages in parallel (on line 9) and exactly one will win the contention. Recall from Sect. 2.2.2 that when nodes call `send_and_rcv`, then both the priority of the winner and the data transmitted by the winner are returned to the application on every node. This packet is added (on line 10) to the set S, which keeps track of all received packets related to the problem of creating an interpolation.

Figures 2.2 and 2.3 illustrate the operation of this scheme. It can be seen that the interpolation result is smooth and that it tracks well the original signal. However, performing weighted-average interpolation with six randomly selected nodes gives poor interpolation. This is illustrated in Fig. 2.2d.

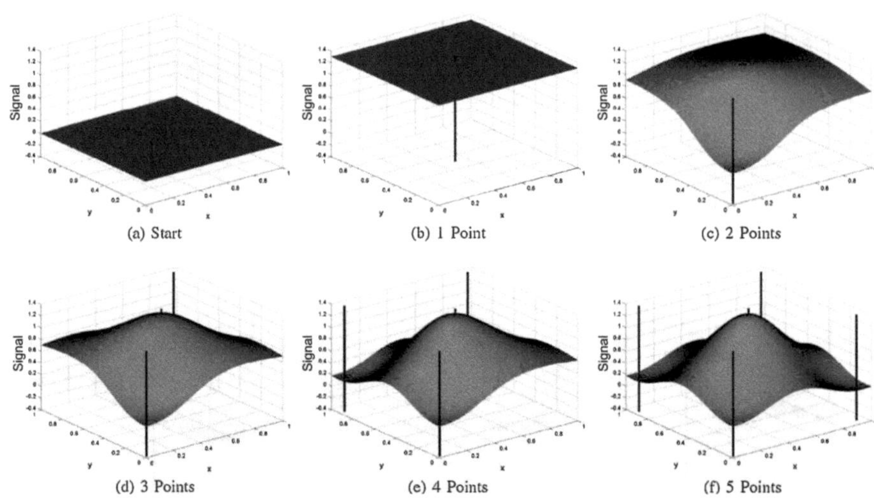

Fig. 2.3 Iterations concerning interpolation Example 1. **a** At the beginning, there is no point included in S. Therefore, from the definition of $f(x, y)$ in Eq. 2.3 it results that $f(x, y) = 0$. This gives a plane surface; **b** then, each node calculates the error between its sensor reading and the starting plane surface. This error is used in the contention of the MAC protocol, causing the node with the largest error to win, and thus, it is inserted into S; **c–e** nodes proceed similarly, calculating their error to the current interpolated surface and adding the node with the largest error to the interpolation, until the set S has k points; **f** we finally obtain the result of the final iteration, for $k = 6$

2.3.2 New Algorithm

The previously proposed algorithm [1] (which was presented in previous section and stated in Algorithm 1) obtained an interpolation from scratch every time it was run. As mentioned in the introduction of this paper, this brings two drawbacks. It is worthwhile to counter those drawbacks and therefore we need to device a new algorithm.

Algorithm 2 shows the new interpolation scheme as pseudo-code. The algorithm works as follows. First, Algorithm 1 is called and this gives us a set S with the selected data points. Then the algorithm executes lines 4–17 periodically; it is assumed that the execution of lines 4–17 is initiated periodically. The execution of lines 4–17 differs from the one in Algorithm 1 in only two respects. First, only one data point is selected instead of k data points. Second, the computation of lines 4–17 begins by removing one element in S (done at lines 5–6) and then a new element is added (done at line 17). The rationales for these adding and deleting rules are as follows. We desire to find the computer node whose sensor reading contributes the least to a faithful representation of the physical world. It would be possible to find that out using the prioritized MAC protocol but this would require some communication. Instead, we simple remove the node which was added to S least recently (done at lines 5–6). Then line 17 adds the element that contributes the most to a faithful representation of the physical world.

Algorithm 2 New Algorithm for finding a subset of nodes to be used in WAI

Require: All nodes start Algorithm 2 simultaneously.
Require: k denotes the desired number of interpolation points.
Require: A node N_i knows x_i,y_i and s_i.
Require: The code below is executed by every node. A node can read the variable `i` and obtain its node index.
Require: (MAXS+1) $\times$ (MAXNNODES+1) + MAXNNODES $\leq$ MAXP.

1: all nodes take sensor readings; the sensor reading at computer node N_j is s_j.
2: call find_nodes (in Algorithm 1) and let S denote the set that is returned
3: **while** (true) **do begin**
4: all nodes take sensor readings; the sensor reading at computer node N_j is s_j.
5: for each element in S, there is a time when it most recently became a member in S, pick the element with the earliest such time and call it OLDNODE
6: S $\leftarrow$ S $\setminus$ OLDNODE
7: **if** $N_i \in S$ **then**
8: Calculate f(x_i,y_i) in Equations 3 and 4 based on $S \setminus \{N_i\}$ and assign it to the variable "myinterpolatedvalue".
9: **else**
10: Calculate f(x_i,y_i) in Equations 3 and 4 based on S and assign it to the variable "myinterpolatedvalue".
11: **end if**
12: error $\leftarrow$ abs(s_i - to_integer(myinterpolatedvalue))
13: temp_prio $\leftarrow$ error $\times$ (MAXNNODES + 1) + i
14: prio $\leftarrow$ (MAXP+1) - temp_prio
15: snd_pack $\leftarrow < s_i,x_i,y_i >$
16: <winning_prio, rcv_pack> $\leftarrow$ `send_and_rcv(` prio, snd_pack)
17: S $\leftarrow$ S $\bigcup$ { rcv_pack }
18: **end while**

We can distinguish between two cases. One case is that the element removed (at lines 5–6) and the element added (at line 17) are the same. This occurs when the new sensor readings obtained at line 4 changed very little.

Another case is that the element removed (at lines 5–6) and the element added (at line 17) are not the same. This occurs when the new sensor readings obtained at line 4 changed a lot and therefore it is necessary that the set S is modified to reflect the changes in the physical environment.

Note that the lines 4–17 can be executed very quickly; only one transmission (line 16) is needed. Executing the lines 8 and 10 have time-complexity $O(k)$ and this is undesirable though. Therefore, the next section will present an improved version of Algorithm 2 which avoids this potential performance bottleneck.

2.3.3 An Improved Version of the New Algorithm

Evaluating $f(x_i, y_i)$ can be performed quickly and easily if $N_i \in S$. The result is simply s_i as can be seen from Eq. 2.3. Evaluating $f(x_i, y_i)$ quickly for the case $N_i \notin S$ requires additional improvements of the algorithm though. We can note that each iteration of the lines 4–17 in Algorithm 2 evaluates $f(x_i, y_i)$ based on elements in the set S. Since this set has k elements, each of these evaluations has time complexity $O(k)$. Recall (from Eq. 2.3) that for this more complex case that we are discussing now, $f(x_i, y_i)$ is defined as:

$$f(x, y) = \frac{\sum_{j \in S} s_j * w_j(x, y)}{\sum_{j \in S} w_j(x, y)} \tag{2.5}$$

Let us define num_i and $denom_i$ as:

$$num_i = \sum_{j \in S} s_j * w_j(x, y) \tag{2.6}$$

and

$$denom_i = \sum_{j \in S} w_j(x, y) \tag{2.7}$$

Hence, we can clearly (for this more complex case) write $f(x_i, y_i)$ as:

$$f(x, y) = \frac{num_i}{denom_i} \tag{2.8}$$

We can clearly evaluate $f(x_i, y_i)$ on line 8 or 10 in Algorithm 2 by calculating $f(x_i, y_i)$ from Eq. 2.8. And we can update num_i and $denom_i$ whenever the set S changes. Therefore, when we add a new node to S, we simply have to also add an extra term to num_i and $denom_i$. Analogously, removing a node from S requires that we subtract a term. Based on this observation, we can reformulate Algorithm 2.

Algorithm 3 Improved Version of the new Algorithm for finding a subset of nodes to be used in WAI

Require: All nodes start Algorithm 3 simultaneously.
Require: k denotes the desired number of interpolation points.
Require: A node N_i knows x_i,y_i and s_i.
Require: The code below is executed by every node. A node can read the variable `i` and obtain its node index.
Require: (MAXS+1) $\times$ (MAXNNODES+1) + MAXNNODES $\leq$ MAXP.

all nodes take sensor readings; the sensor reading at computer node N_j is s_j.
call find_nodes (in Algorithm 1) and let S denote the set that is returned
if $N_i \in S$ **then**
  I_am_in_S ← true
else
  I_am_in_S ← false
end if
num ← 0
denom ← 0
for each $N_j \in S \setminus \{N_i\}$ **do**
  num ← num + $s_j \cdot w_j(x_i, y_i)$
end for
for each $N_j \in S \setminus \{N_i\}$ **do**
  denom ← denom + $w_j(x_i, y_i)$
end for
while (true) **do begin**
  all nodes take sensor readings; the sensor reading at computer node N_j is s_j.
  for each element in S, there is a time when it most recently became a member in S, pick the element with the earliest such time and call it OLDNODE
  S ← S \ OLDNODE
  if $N_i = OLDNODE$ **then**
    I_am_in_S ← false
  else
    j ← OLDNODE
    num ← num + $s_j \cdot w_j(x_i, y_i)$
    denom ← denom + $w_j(x_i, y_i)$
  end if
  if I_am_in_S **then**
    myinterpolatedvalue ← s_i
  else
    myinterpolatedvalue ← $num_i/denom_i$
  end if
  error ← abs(s_i - to_integer(myinterpolatedvalue))
  temp_prio ← error $\times$ (MAXNNODES + 1) + i
  prio ← (MAXP+1) - temp_prio
  snd_pack ← $< s_i, x_i, y_i >$
  <winning_prio, rcv_pack> ← `send_and_rcv`(prio, snd_pack)
  S ← S $\bigcup$ { *rcv_pack* }
  if winning_prio=prio **then**
    I_am_in_S ← true
  else
    j ← the node in rcv_pack
    num ← num + $s_j \cdot w_j(x_i, y_i)$
    denom ← denom + $w_j(x_i, y_i)$
  end if
end while

Algorithm 3 shows this improved version of Algorithm 2. It can be seen that the execution of the lines 17–44 is independent of *m* and it is also independent of *k*. Note also that each line can execute very quickly. In particular, note that line 36, `send_and_rcv`, can be executed within 3 ms if the hardware platform from [2] is used. Because the execution of the lines 17–44 can be performed at such a high speed we see that it is possible to build a sensor network that monitors its environment by obtaining an interpolation as a representation of the physical world and obtain an update of that interpolation very quickly. Any extreme localized change in the physical environment will be detected and its position will be found within just 3 ms. Future hardware developments may make it even faster.

2.4 Conclusions

We have presented a new approach for obtaining an interpolation based on sensor readings. We left open the problem of analyzing the error of the interpolation and also finding out whether smaller errors can be achieved if knowledge about the physical phenomenon is known.

Acknowledgements This work was partially funded by CONET, the Cooperating Objects Network of Excellence, funded by the European Commission under FP7 with contract number FP7-2007-2-224053, the ARTISTDesign Network of Excellence on Embedded Systems Design ICT-NoE- 214373 and by the Portuguese Science and Technology Foundation (Fundção para Ciência e Tecnologia—FCT) and the project SmartSkin supported by ISEP.

References

1. Andersson B, Pereira N, Elmenreich W, Tovar E, Pacheco F, Cruz N (2008) A scalable and efficient approach to obtain measurements in CAN-based control systems. IEEE Trans Ind Inform 4(2):80–91
2. Pereira N, Gomes R, Andersson B, Tovar E (2009) Efficient aggregate computations in large-scale dense WSN. In: 15th IEEE real-tune and embedded technology and applications symposium (RTAS'09), San Francisco, CA, USA
3. CAN Specification, ver: 2.0 (1991) Bosch GmbH, Stuttgart
4. Pereira N, Andersson B, Tovar E (2007) Widom: a dominance protocol for wireless medium access. IEEE Trans Ind Inform 3(2):120–130
5. Andersson B, Pereira N, Tovar E, Gomes R (2009) Using a prioritized medium access control protocol for incrementally obtaining an interpolation of sensor readings. In: Proceedings of the 7th workshop on intelligent solutions in embedded systems (WISES'09), Ancona, Italy
6. Mok AK, Ward S (1979) Distributed broadcast channel access. Comput Netw 3:327–335
7. (CiA), CAN in automation website [Online]. http://www.can-cia.org
8. Kimaldi, network of readers website section [Online]. http://www.kimaldi.com/kimaldi_eng/productos/lectores_de_tarjetas/red_de_lectores_can/red_de_lectores_ampliacion_de_informacion
9. Shepard D (1968) A two-dimensional interpolation function for irregularly-spaced data. In: Proceedings of the/96823rd ACM national conference, pp 517–524

10. Tynan R, O'Hare G, Marsh D, O'Kane D (2005) Interpolation for wireless sensor network coverage. In: Proceedings of the second IEEE workshop on embedded networked sensors, pp 123–131
11. Sharifzadeh M, Shahabi C (2004) Supporting spatial aggregation in sensor network databases. In: Proceedings of the 12th annual ACM international workshop on geographic information, pp 166–175
12. Crossbow, MICA2—wireless measurement system product datasheet. http://www.xbow.com/products/Product_pdf_files/Wireless_pdf/MICA2_Datasheet.pdf
13. Crossbow, MicaZ—wireless measurement system product datasheet. http://www.xbow.com/products/product_pdf_files/Wireless_pdf/MICAz_Datasheet.pdf
14. Polastre J, Szewczyk R, Culler D (2005) Telos: enabling ultra-low power wireless research. In: Proceedings of the fourth international conference on information processing in sensor networks: special track on platform tools and design methods for network embedded sensors (IPSN/SPOTS'05). IEEE Computer Society, New York, pp 364–369
15. Guestrin C, Bodik P, Thibaux R, Paskin M, Madden S (2004) Distributed regression: an efficient framework for modeling sensor network data. In: Proceedings of the third international conference on information processing in sensor networks (IPSNO4)

Chapter 3
Embedded Systems in the Poseidon MK6 Rebreather

Microcontroller Network in a Life Supporting System

Arne Sieber, Nigel A. Jones, Bill Stone, Richard Pyle, Bernhard Koss and Kurt Sjöblom

3.1 Introduction

In the past 30 years, underwater activities have registered a steep increase across Europe, going from a few thousand people in the 1980s, when diving was prevalently an elite activity, to about 5 million today. The typical recreational diver uses a self contained underwater breathing apparatus (SCUBA), where breathing gas is stored in a tank at a pressure up to 300 bar. A one or two stage pressure regulator reduces that pressure to ambient pressure allowing the diver to breathe. Exhaled gas is then vented into the environment (open circuit). The first stage reduces the tank's pressure to an intermediate pressure around 8–10 bar higher than ambient pressure. The second stage, also known as the regulator, reduces the intermediate pressure to ambient pressure thus allowing the diver to breath underwater. Exhaled air is then vented through an exhaust valve into the water. The maximum time a diver can stay under water is mainly determined by the amount of gas he is carrying, the depth, and the breathing volume per minute. In normal conditions during a relaxed dive a diver only metabolizes about 0.8–2.5 bar l/min, but the overall gas consumption is

A. Sieber (✉) · K. Sjöblom
Poseidon Diving Systems AB, Goeteborg, Sweden
e-mail: asieber@gmx.at

N. A. Jones
R.M.B. Consulting Inc, New Market, MD, USA

B. Stone
Stone Aerospace, Del Valle, TX, USA

R. Pyle
Bishop Museum, Honolulu, HI, USA

B. Koss
Scuola Superiore Sant'Anna, Pisa, Italy

M. Conti et al. (eds.), *Solutions on Embedded Systems*,
Lecture Notes in Electrical Engineering, 81, DOI: 10.1007/978-94-007-0638-5_3,

much higher (circa 20 bar l/min on the surface to for example 100 bar l/min at 40 m), the gas efficiency of such open circuit systems is very low. An alternative with a much higher gas efficiency is so called rebreather or closed circuit diving systems [1, 2].

The main idea of a rebreather is that exhaled gas is collected in a flexible bag, the so called counterlung, CO_2 is chemically filtered in a scrubber and metabolized oxygen is substituted with fresh oxygen from a small tank. As the breathing gas is recycled, only small tanks have to be carried for sufficient gas support. During descent, the gas inside the loop is compressed, thus additional gas needs to be added. Breathing pure oxygen is only safe to a maximum depth of 6 m, thus a diluent gas is needed to keep the partial pressure of O_2 (pO_2) inside the loop below toxic values (maximum pO_2: 1.4–1.6 bar). In addition to high gas efficiency that allows long diving times, rebreathers also offer advantages like silence (no exhaled gas creating "loud" bubbles) and warm and humid breathing gas. Rebreathers are classified into either semi-closed circuit rebreathers (SCRs) or manually or electronically controlled closed-circuit rebreathers (mCCR or eCCR; CCR). In closed circuit rebreathers the pO_2 inside the loop is measured with one or for redundancy purposes several pO_2 sensors (Fig. 3.1). Dependent on the design of the system, the pO_2 is then controlled by either manual O_2 injection or electronically controlled O_2 injection, typically by using a solenoid. The electronic control usually includes one or two microcontrollers. As well as pO_2 control, the microcontrollers usually also perform decompression calculations. In terms of safety requirements, rebreather diving systems fall under the Personal Protective Equipment Directive [3]. CE marking according to EN14143 (normative for rebreather diving systems [4]) also requires a notified body.

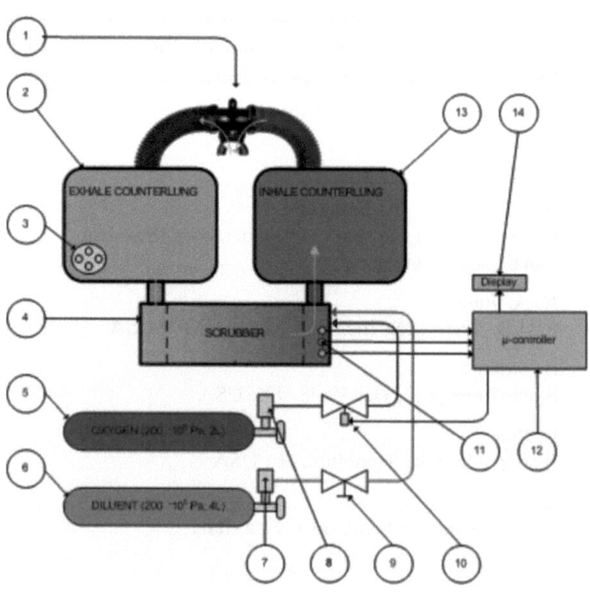

Fig. 3.1 Standard components of an electronically controlled closed circuit rebreather: *1* mouthpiece, *2* exhale counterlung, *3* overpressure valve, *4* carbon dioxide scrubber, *5* oxygen cylinder, *6* diluent cylinder, *7, 8* pressure regulators, *9* manual diluent valve, *10* solenoid, *11* pO_2 sensors, *12* microprocessor, *13* inhale counterlung, *14* display

Traditionally CCR diving requires continuous high level training and technical understanding of the equipment so that the diver is able to safely dive and second to detect and safely handle malfunctions of the system.

Thus most rebreathers, available on the market today, address very advanced (mostly technical) divers. A recreational diver, with limited diving experience and low technical understanding is most likely not able to handle such systems.

The Poseidon MK6 closed circuit rebreather is especially designed for the recreational diver. A network of microcontrollers is the basis to assure maximum safety of the diver and to facilitate operation. Sophisticated software monitors and controls the system, and in the case of a failure, alarms are initiated. This allows the user to enjoy his (recreational) dive without continuously checking instruments. In the case of a problem, the user will be warned and eventually, in the case of a severe problem, the user has to abort the dive by using the system in open circuit mode. While the previous article in Popular Science [5] and in IEEE spectrum [6] were presenting the idea of the Poseidon MK6 from the end users point of view, the present paper is focusing on the embedded systems and the algorithms incorporated in the design. The Poseidon MK6 has been briefly presented at the WISES 2009 workshop [7].

3.2 Methods

3.2.1 Mechanical Design of the System

Figure 3.2a shows the back view of the MK6. The scrubber is placed in the center. Fast scrubber change is assured by using a pre packed scrubber instead of a user packable one. Moreover this helps to avoid accidents that happen due to incorrect packing techniques. Next to the scrubber the diluent tank and the O_2 tank are mounted. On top of the scrubber there sits the backpack electronics. The battery (Fig. 3.2c) is plugged into the backpack.

Figure 3.2b shows the handset. The IrDA port is located to the right of the LCD. Moreover there is a light sensor incorporated for automatic adjustment of the LCD's backlight. On the back of the handset two contacts are used as a wet-switch; as soon as the contacts are in water the system is activated. Figure 3.2d shows the mouthpiece with the open circuit (OC)/closed circuit (CC) switch. On top of the mouthpiece there is the so called head up display (HUD). It includes a red warning LED, a vibrator motor and 2 magnetic (Hall) sensors, which can detect the position of the OC/CC switch.

3.2.2 Embedded Systems in the MK6

It is clear that malfunctions of such a life supporting system have to be seen as potentially life threatening. The components have to be carefully selected to

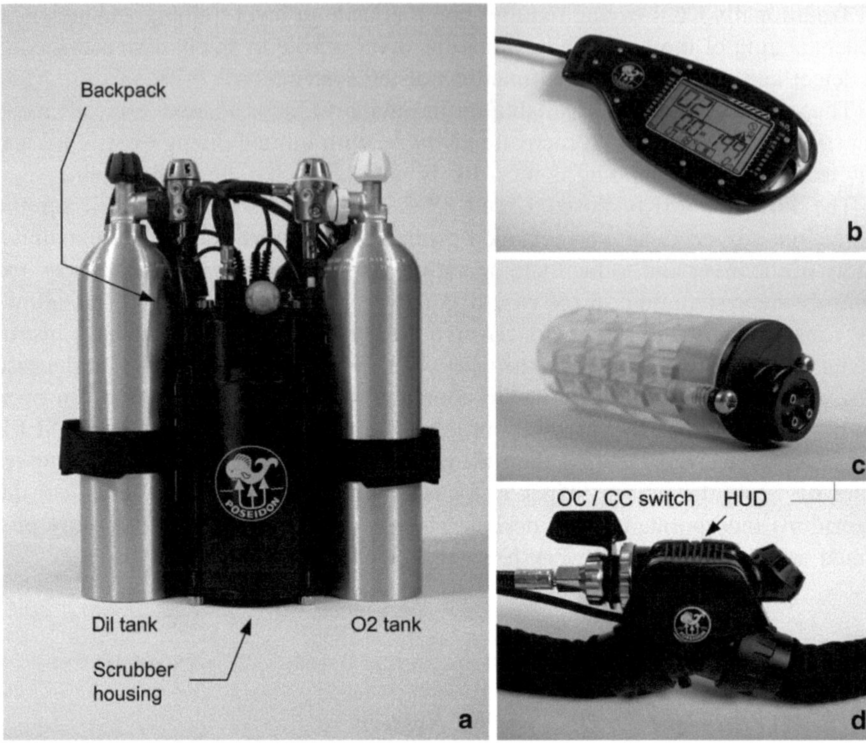

Fig. 3.2 **a** Scrubber housing with backpack on top and tanks; **b** display; **c** battery; **d** mouthpiece

achieve maximum reliability. As this is an autonomous battery powered system, special attention has to be directed to the power consumption of each component.

Until now most of the commercially available rebreathers use at least two separate sets of electronics to monitor and control the pO_2 inside the breathing loop. In the case of a failure of the main electronics, the second electronics is used for manual control of the system using manual gas injection buttons.

The MK6 is designed in a different way. Instead of having two complete separate sets of electronics, the system is based on a network of microcontrollers (Fig. 3.3). All of them can communicate via a network. In the case of a failure, three nodes in the network are able to warn the diver and depending on the severity of the failure, encourage him to abort the dive using the system in the alternative open circuit mode.

The ATmega 8 bit RISC microprocessor family from Atmel was chosen for the design of the system as they are offered in a wide variety of package styles, have low power consumption over a wide operation voltage range and allow excellent code density.

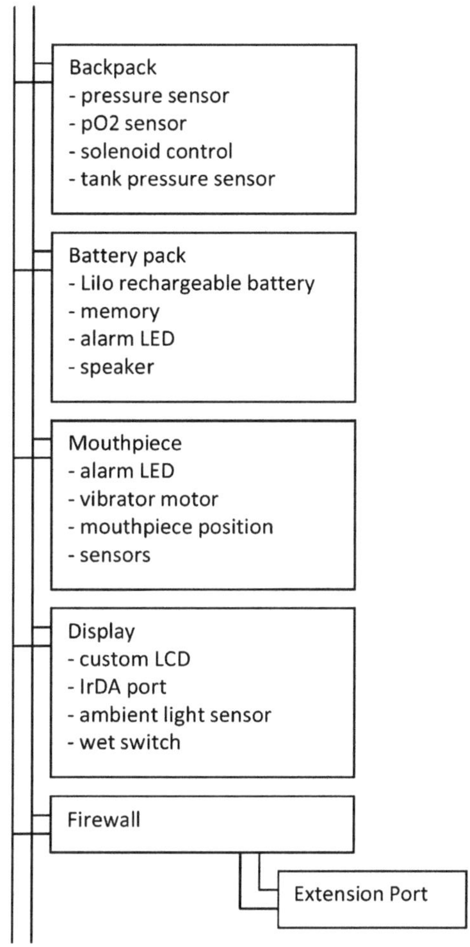

Fig. 3.3 Electronic components of the system

3.2.2.1 Backpack

The backback (Fig. 3.2a) processor's main function is to control the pO_2 inside the loop. The core component is an ATmega processor. A digital ambient pressure sensor (Intersema, Switzerland) is read out via a derivative of the serial peripheral interface (SPI). The pO_2 sensors are amplified and connected to the inputs of the analog to digital converter of the microprocessor. Two high pressure sensors are mounted on the pressure regulators on the O_2 and the diluent tank. The analog outputs of these are already temperature compensated and amplified. These signals can be sampled directly by the analog digital converter. The electronics is encapsulated in polyurethane.

3.2.2.2 Handset

The handset contains a custom LCD. Below this, an ATmega processor is situated. A light intensity sensor allows automatic regulation of the backlight. An IrDA transceiver is situated on the right of the display. An IrDA encoder is used for interfacing the ATmega's USART to the IrDA transceiver. A digital interface to a PC is necessary for two reasons. Firstly it allows dive data download and post dive analysis. Dive specific data like depth, time, pO_2, and decompression data are stored in an onboard serial memory. Secondly the interface permits reconfiguration of the rig and firmware update. In principle also a wired connection is possible, which requires an expensive watertight connector. Data download to a mobile device is also desirable, thus it was decided to use a wireless link. To establish an IrDA connection the two communication devices have to be pointed at each other. Having several units next to each other (for example in a diving school) it is quite unlikely that a link would be established with the wrong unit. Using Bluetooth, where communication within a 10 m range is feasible, it is theoretically possible, that one user connects to the wrong machine. Especially when performing firmware updates or configuring the system, communication with the wrong system could be life threatening. IrDA was chosen mainly for safety reasons. To achieve water pressure resistance the Handset is encapsulated in a silicon gel.

3.2.2.3 Battery

The air filled watertight and pressure resistant battery pack houses one rechargeable LiIon cell. It also incorporates an ATmega processor and a 4 MBit external Flash memory chip. All relevant dive data are stored in that location. The battery pack also houses a speaker and a pair of orthogonally mounted bright red warning LEDs. The battery pack can be inserted in an external charger module. This one also offers a USB port for dive data download.

3.2.2.4 HUD

The HUD is mounted above the mouthpiece. It has 2 main functions:

- it detects with 2 magnetic sensors if the mouthpiece switch is in OC or CC position (the rotating cylinder in the mouthpiece is equipped with 2 magnets). The information about the position is broadcasted on the bus;
- it alarms with a bright red LED and a vibrator motor like for example in the case of malfunctions of the rebreather or if depth limits are exceeded.

3.2.2.5 Communication

Four processors are incorporated in the design. Each one of the processors has to be able to communicate with the others. As detailed above, the backpack is responsible for controlling the pO_2 inside the loop and sensor signal readout. All the other components receive data from the backpack, and in the case of a problem, alarm the user. The communications bus allows multi-mastering and thus allows any processor to talk to any other.

3.2.2.6 Firewall

For interfacing and upgrading the MK6 with additional hardware, a "firewall" can be optionally integrated. The firewall is based on another ATmega microcontroller. The firewall acts as an interface between the MK6 and the external hardware. Communication between the firewall and the external hardware takes place via a USART.

3.2.3 Algorithms

The software of the system was developed using Embedded Workbench and visualSTATE (both from IAR systems).

3.2.3.1 State Machine

The architecture of the system is a state machine. With sufficient power, the possible machine states can be either Running, Start up or Power Down. During start up multiple pre dive tests (Sect. 3.2.3.2) are carried out. Running includes "ready to dive", meaning that the system has successfully passed the pre-dive tests, 'diving' (system is under water) and 'post dive' (back on the surface). In the power down state the system's microcontrollers are in a standby mode, which while dramatically lowering power consumption, still allows them to respond to certain inputs.

3.2.3.2 Pre-dive Checks

For a life supporting system it is essential that all the components are correctly working. The 35 pre-dive checks of the MK6 include firstly checks of the basic electronics and then checks of the sensors and the overall machine. The electronic tests include:

- current consumption of electronics (parts are switched on one after the other);
  - all four microcontrollers,
  - solenoids,
  - pressure sensors,
  - alarm LED's,

  - vibrator,
  - speaker:

- memory of the microcontrollers (checksums).

Most of the dive relevant data is stored twice in the internal EEPROMs, thus in the case of a failed checksum test, information can be recovered from the second data set. Even if the electronic check of the solenoids or the pressure sensors are passed, correct function is not guaranteed (it was just an electronic check). In the second part of the pre dive checks their correct function is validated:

- the tank pressures are measured;
- the battery's capacity is checked;
- the correct detection of the OC/CC mouthpiece switch is validated;
- a positive loop pressure test is carried out (the loop is inflated to a certain pressure) this allows checking:

  - if the solenoids can pass gas,
  - if the solenoids are not leaking,
  - if the depth sensor can measure micro-pressure changes,
  - the over pressure valve.

In the next step the pO_2 sensors are calibrated. Here the oxygen sensor membranes are flushed with pure oxygen. In the second part of that calibration the pO_2 sensors are flushed with diluent for 20 s. This allows checking the sensors for linearity and the correct time constant (typically pO_2 sensors have a t90 time constant of about 6 s at 20°C).

3.2.3.3 pO_2 Control Algorithm

The major task of the running state is the control of the pO_2 inside the loop. As detailed earlier, the pO_2 inside the loop is measured with two pO_2 sensors. In the case of a too low pO_2, O_2 from the O_2 tank is injected via a solenoid into the loop. The pO_2 setpoint is dependent on several parameters:

- depth of the diver;
- configurable limits (pO_2 min 0.35–0.9 bar, pO_2 max: 0.5–1.4 bar);
- decompression ceiling (during a dive the body is exposed to an increased pressure, thus the human body is saturated with inert gas such as nitrogen (that is a part of the breathing gas). If the tissue saturation tensions are too high, an immediate return to the surface might be dangerous, thus decompression stops have to be included. A "ceiling" is calculated, representing the minimum depth to which a diver can ascend safely.)

Figure 3.4 details the control of the pO_2 setpoint over depth. Traditional electronically controlled rebreathers use normally 2 setpoints, one for shallow and the second for deeper parts of the dive. Operating such a system includes setting

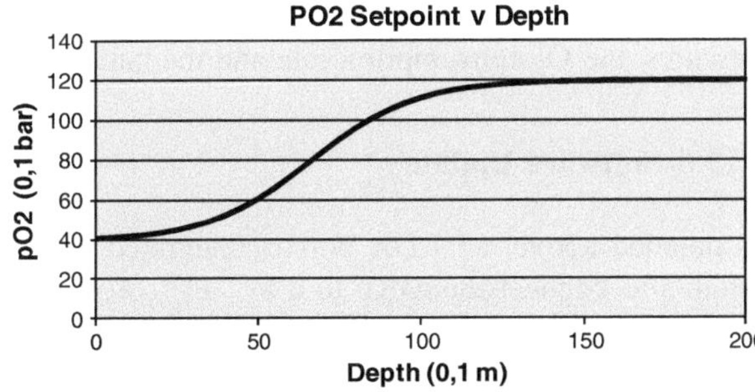

Fig. 3.4 Setpoint control

the setpoints and eventual manually selecting them underwater. As the partial pressure of a gas component in a mixture depends on its fraction and the ambient pressure, control of the pO_2 in shallow water, where large relative pressure changes occur often, is difficult. Typical problems are that a lot of O_2 is injected, resulting in too much positive buoyancy. In order to relieve the recreational diver from manually selecting the pO_2 setpoint and difficult buoyancy control in shallow water, instead of using fixed setpoints, the pO_2 is controlled in the MK6 according to a function detailed in Fig. 3.4.

3.2.3.4 Sensor Signal Validation

Prior to the advent of the MK6, the norm in measuring the pO_2 in rebreathers was to use three oxygen sensors and to use some form of voting logic to determine the correct pO_2. However such a scheme implicitly assumes that sensor failures are statistically independent, or, to put it in another way, that two sensors would not fail at the same time for the same cause. This is a very dubious assumption. The approach taken in the MK6 is to validate the performance of the primary sensor throughout the dive. This is done by periodically exposing the primary sensor to a known pO_2 and determining that the sensor responds correctly. This not only allows Poseidon to determine that the sensor is behaving correctly, it also allows them to show that the sensor is linear in the hyperoxic region. Conversely, other rebreathers merely 'hope' that the sensor is operating correctly in this region. The use of a second (truly redundant) pO_2 sensor allows for detection of other classes of failures, including leaking solenoids. More detailed information about the principle of sensor signal validation principle can be found in [8, 9].

3.2.3.5 Controlling Resource Algorithm

Open circuit divers usually have to check their diving computers for their calculated no decompression time (the time remaining for an ascent to the surface without decompression stops) and the tank pressure gauge to decide when they have to start their ascent. In the MK6 the controlling resource algorithm attempts to predict the remaining dive time. It takes into account the decompression status, remaining tank pressures, the O_2 consumption rate and the battery consumption rate.

3.2.3.6 Software Update

As detailed above, an IrDA port is integrated in the handset. Via this port it is possible to connect the MK6 to a PC. PC software was developed to allow both easy dive data download and firmware update too. To perform a firmware update, an encrypted data package is uploaded to the handset's processor and then dispatched to the other processors in the network.

3.3 Results

The Poseidon MK6 was launched and presented to the public on the major diving trade shows in the world. It only weights 18 kg, less than a standard 12l SCUBA set, but allows at least 3–4 times longer diving times. A test team was using the first units under various conditions including also decompression dives. The MK6 was certified according to EN14143.

3.4 Discussion

The MK6 is an innovative electronically controlled rebreather designed to address the needs of a recreational diver. Automatic pre-dive checks facilitate the operation and shorten preparation times. Continuous sensor signal validation ensures maximum safety of the system. Multiple networked microprocessors provide a distributed alarm function. All of the processors are monitoring the bus and the system status, thus in the case of a failure, the user will be warned and then can switch the OC/CC lever on the mouthpiece to open circuit mode and safely abort the dive.

The safe and easy operation of a rebreather is the key element for wide acceptance on the recreational market. The MK6 with all its safety functions and automated tests is designed to fulfill these needs and will allow the recreational diver to enjoy silent and bubble free diving. This will allow the recreational diver much better integration into the underwater environment compared to using open circuit systems.

3.5 Outlook/Ongoing Work

As stated above, the MK6 was developed especially for the needs of recreational divers. As the system is smaller, lighter and easier to prepare than other (technical) rebreathers on the market, it also attracts the attention of technical

Fig. 3.5 Gasblock

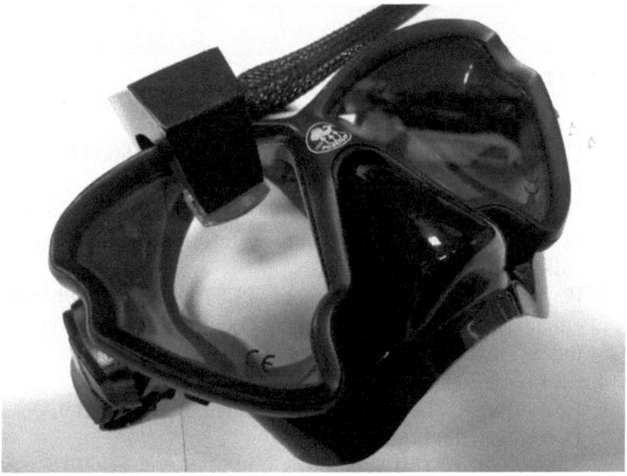

Fig. 3.6 Innovative graphical head up display mounted on a commercial diving mask

and professional divers. In order to safely allow deeper and longer diver dives beyond recreational limits, an extension to the MK6 was developed together with the Scuola Superiore Sant'Anna in Pisa, Italy and Divesystem in Massa Marittima, Italy. This extension consists of a completely independent secondary set of electronics, also based on an Atmel ATmega processor. It has its own decompression algorithms and includes readout of a third pO_2 sensor. Moreover via the above described firewall it can read out the main parameters of the MK6, such as depth, pO_2, ceiling, etc.

Figure 3.5 shows this upgrade. Adjacent to the electronics is a gas switch, which offers additional redundancy:

- in the cases of the primary MK6 electronic failure, the unit pO_2 in the loop can be controlled manually (therefore the gas block is equipped with 2 manual gas addition buttons for O_2 and diluents gas);
- in the case of a failure of all pO_2 sensors, the unit can be used in a self mixing semi closed mode, where O_2 and diluents gas are mixed with 2 sonic orifices.

For depths greater than 20 m the diver uses semi closed mode 2 (SCR2), between 20 and 6 m the diver uses semi closed mode 1, and shallower than 6 m the diver uses pure O_2 for accelerated decompression. The orifices can be changed by the user and adapted to different depths and gas mixtures.

Figure 3.6 shows an innovative display that can be mounted directly on a diving mask. It consists of a 128 $\times$ 64 pixels micro OLED display. A first prototype was discussed in [10].

This enhanced system is designed to allow diving the Poseidon MK6 with He mixtures. In that way, dives deeper than 50 m are possible. Moreover the ability to manually control the system (2 manual injection buttons for O_2 and diluent) or to use the unit as a semi closed rebreather offers maximum redundancy. This increases the safety of the diver, which is especially necessary, when the diver cannot return to the surface due to decompression obligations or because he simply cannot due to mechanical reasons—for example during a cave or wreck dive.

References

1. U.S. Navy Diving Manual (2008) Revision 6. SS521-AG-PRO-010. Direction of commander. Naval sea systems command, USA, vol 4
2. NOAA Diving Manual (2001) Diving for Science and Technology, 4th edn. US Department of Commerce, National Technical Information Service, Springfield USA. Chapter 3, pp 7–8, ISBN: 0-941332-70-5
3. Personal Protective Equipment (PPE) Directive 89/686/EEC. Available at http://ec.europa.eu/enterprise/sectors/mechanical/documents/legislation/personal-protective-equipment/index_en.htm. Accessed 26 Apr 2010
4. EN14143:2003. ISBN 058042738-2; 2003
5. Best of what's new 2008 (2008) Recreation, poseidon discovery rebreather. Popular Science, December 2008, p 94
6. Shreeves K (2009) Winner: poseidon discovery. IEEE Spectrum, 1.2009, p 3, pp 20–21
7. Sieber A, Jones NA, Stone B, Pyle R, Koss B, Sjoeblom K (2009) Embedded systems in the poseidon MK6 rebreather. WISES 2009, Italy
8. Sieber A, Koss B, Bedini R, L'Abbate A, Dario P (2008) Novel controller for rebreather diving systems—true sensor signal validation and safe oxygen injection. Biodevices 2008, Funchal
9. Sieber A, L'Abbate A, Bedini R (2008) Oxygen sensor signal validation for the safety of the rebreather diver. Diving Hyperb Med 38:38–45
10. Koss B, Bedini R, Woegerer C, L'Abbate A, Dario P, Sieber A (2008) Rebreather diving safety: a novel graphic head up diving computer. EUBS Conference 2008, Graz, Austria

Chapter 4
Embedded Data Logging Platform for Research in Diving Physiology

Monitoring ECG and Blood Oxygenation of Apnea Divers

Benjamin Kuch, Remo Bedini, Antonio L'Abbate, Matthias Wagner, Giorgio Buttazzo and Arne Sieber

4.1 Introduction

Medical concerns about professional (commercial, scientific, rescue, etc.) as well as recreational diving safety derive from two major shortcomings: scanty knowledge of diving physiology and lack of monitoring of vital parameters during diving. Both deficiencies are virtually related to the total absence of instrumentation suitable for underwater measurements of simple but crucial physiological parameters such as heart rate, blood pressure, cardiac function, blood oxygen saturation, etc. Actually, none of the available clinical devices used in everyday clinical practice for assessing health status can be used underwater because of a variety of problems related to the liquid environment, especially its salinity and the high hydrostatic pressure. Thus, with regard to performance of physiological measurements, underwater medicine is still at the level it was centuries ago.

With regards to the lack of direct measurements, the results of series of 'models' of underwater diving are considered as valid surrogates and inferences from the clinical world are commonly adopted. Unfortunately, both processes are intrinsically

B. Kuch (✉) · G. Buttazzo
Real-Time Systems Laboratory, Scuola Superiore Sant'Anna, Pisa, Italy
e-mail: benjamin.kuch@gmx.de

R. Bedini
CNR-Institute of Clinical Physiology, Pisa, Italy

A. L'Abbate
Extreme Center, Scuola Superiore Sant'Anna, Pisa, Italy

M. Wagner
University of Applied Science Frankfurt, Frankfurt, Germany

A. Sieber
Institute of Micro and Nanotechnology, Gothenburg, Sweden

M. Conti et al. (eds.), *Solutions on Embedded Systems*, Lecture Notes in Electrical Engineering, 81, DOI: 10.1007/978-94-007-0638-5_4, uncertain and scientifically incorrect. Thus, the transfer to the underwater environment of routine clinical instruments would represent a great advancement, both in terms of knowledge and safety, just as it has already occurred in space medicine. This task requires designing novel underwater diagnostic and monitoring instrumentation and developing ad hoc support infrastructure. Beyond the design of waterproof instruments, special attention must be paid to selecting, placing and protecting the sensors and transducers especially for long term monitoring.

Measurement of underwater blood pressure was detailed somewhere else [1]. The idea of the present work was to develop a platform that is able to record physiological parameters, but also physical parameters like water temperature and water pressure and store them on suitable memory. Minimum needed recording time is 8 h [2].

With regards to diving, O_2 is of special interest. Under normal conditions, more than 98% of the O_2 in the blood normally binds to Hemoglobin (Hb) in the red blood cells. The rest is dissolved in plasma. In arterial blood, at a normal pO_2 of 13 kPa, the arterial O_2 saturation of Hemoglobin (s_aO_2) is about 97.5%. In the venous blood the pO_2 drops to 5 kPa which corresponds to a venous O_2 saturation of Hemoglobin (s_vO_2) of approximately 75%. Thus s_aO_2 reflects the amount of O_2 that the blood can deliver to the tissues. The maximum possible arterial oxygen saturation depends on the partial pressure of O_2 in the gas inside the lungs. The critical value of pO_2 inside the lungs should not fall below this value. Otherwise the sO_2 will rapidly decrease and may lead to vasoconstriction [3].

During breathhold dives in shallow waters after excessive hyperventilation, the breathing reflex occurs with a time delay. This may help to increase the overall possible diving time, but at the end of the dive the pO_2 inside the lungs can reach dangerously low levels that may result in a blackout. In deep breathhold dives the pO_2 inside the lungs drops rapidly during ascend. Many accidents occur during the last meters before surfacing, because of the quick drop of the relative pressure [4].

Electrocardiography (ECG) recording underwater requires special attention to both the recorder and the electrodes. For electrical bio-potential processing, an amplifier with high input impedance is necessary. Water (especially sea (salt) water) represents a good electrical (ionic) conductor means with respect to the typical ECG electrode–skin impedance. For example the impedance between two metal electrodes (with each a surface of about 1 cm^2) may drop to some hundred Ω in fresh water and to values below 10 Ω in salt water. From the electrical point of view, the low resistance of water is in parallel with the amplifiers input impedance—i.e., it acts like a shortcut for poorly insulated electrodes, and, as a consequence, the recording of significant bio potentials becomes impossible. Thus, without a suitable electrical insulation, ECG signal recording on a water immersed body ranges from difficult (fresh water) to impossible (sea water), even if the diver uses a typical humid neoprene suit. Moreover, putative field tests on large diver cohorts require cheap electrodes and effective insulation.

Authors' first attempts were based on the use of adhesive electrodes, clinically used for electro-stimulation, which were well insulated with silicon grease. Signal quality was adequate, however, several limitations manifested: the electrodes' high cost, the low acceptance by the divers, and the difficulty of removing silicon grease from diving suits. Thus, we used standard ECG (self-) adhesive electrodes together with a suitable insulation technique. The two component impression material (Elite H–D+, Zhermack Hydrophilic Vinyl Polysiloxane) was chosen for its suitable characteristics like fast curing abilities, biocompatibility, and waterproofness [5].

4.2 Methods

4.2.1 Hardware

The main idea for the novel physiological data recorder was to combine a 2-channel ECG with a pulseoximeter and a pressure sensor (see Fig. 4.1). Suitable bases for in field physiological recordings are PDAs together with exiting interfaces. Unfortunately PDAs are neither water nor pressure resistant and underwater housings are not available. Additionally PDAs are bulky and have only a short operation time. An alternative is the development of a dedicated embedded system, which is as small as possible, so that a diver can carry it easily, without getting disturbed during a dive. Since some field tests require long term recording, overall power consumption is an issue. Thus a low power 8-bit microcontroller was chosen as the core element.

4.2.2 Data Logging Module

The core component is a hardware module with an Atmel ATmega644p 8-bit RISC microcontroller (Atmel) with the following specifications:

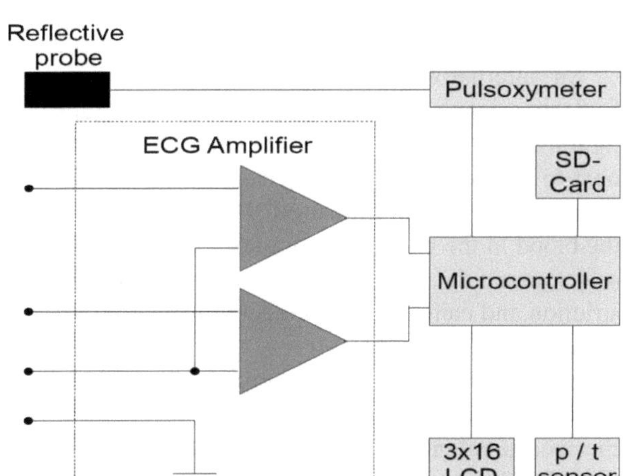

Fig. 4.1 Principle design of the module

- 4 Kbytes SRAM
- 2 Kbytes EEPROM
- 64 Kbytes Flash
- Up to 20 MIPS Throughput at 20 MHz
- Power Consumption at 1 MHz–0.4 mA in Active Mode

A secure digital memory card (SD-Card) connector is connected to the serial peripheral interface (SPI) of the microprocessor. Low-drop, linear regulators are used to provide 3.3 V. For visualization of active data, a 16×3 alphanumerical display is integrated on the board and interfaced via software SPI to dedicate the microprocessor's inbuilt SPI solely to the SD-Card. For depth and temperature measurement, a digital sensor was integrated (Intersema MS5541B, Switzerland). It is specified for a depth of 150 m, but unofficially depth measurements down to 330 m are possible. The overall low power consumption allows powering the whole circuit via a single LiIon battery. The whole device is housed in a lexan tube.

4.2.3 sO_2 Measurement

Clinical O_2 saturation (sO_2) meters (pulseoximeters) are normally based on transmissive light absorbance measurements with red and near infra-red light. Probes are usually attached on the ear lobe or on a finger [6, 7]. sO_2 while breath holding was already investigated in [8]. There a standard transmissive pulsoxymeter was used.

Underwater tests were carried out using standard transmissive pulseoximeter probes attached to the finger but did not produce significant data. Physiological adoptions to immersion in cold water are understood under the term "diving response". The vasoconstriction phenomenon, where peripheral parts are most affected, reduces the peripheral blood flow thus inhibits sO_2 measurements on a finger. Immersions in cold water may further intensify vasoconstriction.

An alternative to transmissive pulseoximetry is reflective pulseoximetry. Reflectance pulseoximetry is not based on transmissive absorbance anymore. A light transmitter and receiver are situated in a probe at a short distance (like 8 mm) next to each other. Light is transmitted into the underlying tissue and the reflected light is received and measured. The intensity depends then on the O_2 saturation of the blood in the underlying tissues. For placing the reflective probe it is possible to select particular parts of the head that are certainly less affected by vasoconstriction and can be easily protected from cold water temperature (i.e. the glabellar or temple artery zones).

For the prototype a commercial pulseoximeter module (OEM III, Nonin) is chosen. It is interfaced to the microcontroller via USART at 9,600 bit/s. To avoid the measurement problem caused by the "diving response" a reflectance probe (8000R, Nonin) was chosen, which can be placed on the forehead or the temple.

Due to its small size, integration in a commercial diving mask is possible. Alternatively it may also be glued directly to the skin with adhesive tape.

4.2.4 ECG Measurement

A 2-channel ECG was integrated into the design (see Fig. 4.2). The ECG signal is picked up with three electrodes and a reference electrode. Two precision instrumentation amplifiers (AD620, Analog Devices) were selected as first input stage. Advantages of these amplifiers which are especially interesting for our battery operated devices are their low power consumption and that they can be single supply operated starting from 3 V.

The amplification of the Instrumentation amplifiers is set to 11 (defined with R6 and R7). R8 and C5 and, respectively, R9 and C6 are high pass filters with a corner frequency of approximately 2 Hz.

C7 and R10 and, respectively, C8 and R11 are low pass filters (100 Hz). A quad singly supplies rail to rail operational amplifier (OP491) servers for further amplification of the ECG signals by a factor of 32. Moreover it is also used to

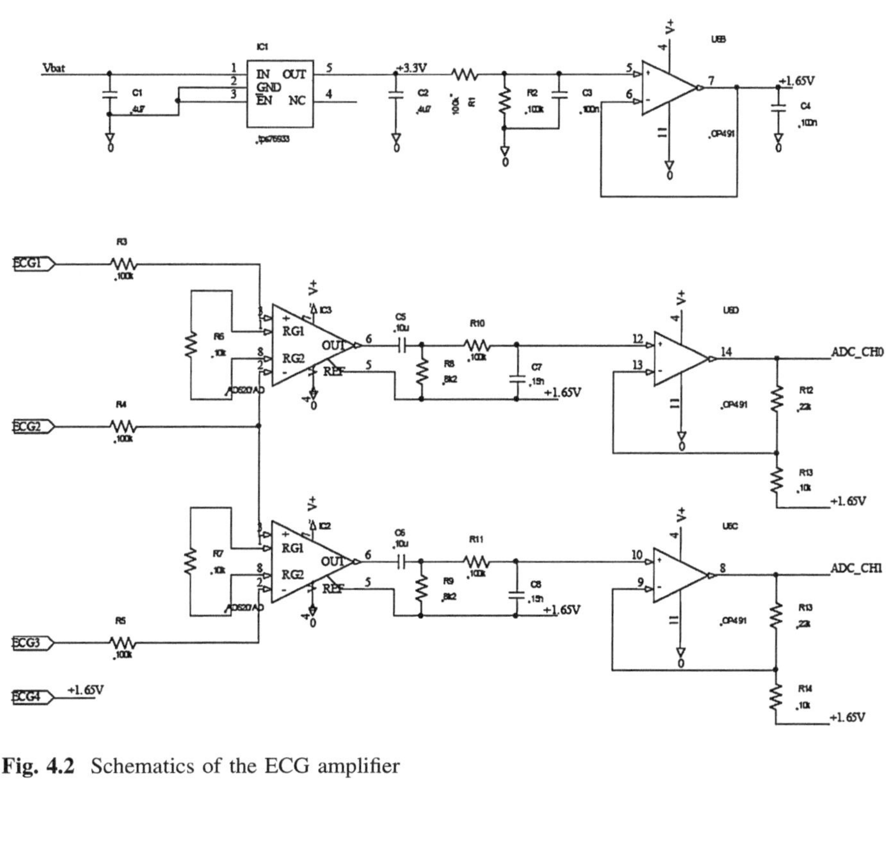

Fig. 4.2 Schematics of the ECG amplifier

generate a 1.65 V reference voltage, to which the forth (reference) ECG electrode (ECG4) is connected. The output of the amplifiers is then directly sampled with the AD converter of the microprocessor (analogue channel 0 and 1). The operational amplifiers, as well as the instrumentation amplifiers, are both supplied with 3.3 V by a low dropout linear voltage regulator.

4.2.5 Software

4.2.5.1 Firmware

The main functions of the data logging module are

- To start and read out a raw temperature measurement in 1 s intervals.
- To start and read out a raw pressure measurement.
- To convert the raw temperature values into grad Celsius.
- To convert the raw pressure values into mbar.
- To read out ECG channel 1.
- To read out ECG channel 2.
- To read out heart rate and oxygen saturation.
- To store all measurements in a FAT16 Text file on SD-Card.

The firmware of the module is developed in C with the GNU C compiler under AVR Studio 4.13 (Atmel). ECG channels are sampled with 500 Hz. Parallel read out of several channels of the microcontroller is not possible, thus the channels are read out one after the other in 1 ms intervals. The pressure/temperature sensor has to be read out in 1 s intervals. Additionally every second the pulsoximeter sends the measured heart rate and oxygen saturation via USART to the microcontroller. Quasi in parallel to that, the data has to be visualized on the display and stored on a SD-Card.

Data storage on a SD-Card can be done in three different ways. One way is to simply write the data into the memory blocks. This method is fast, but since this method does not use a FAT file system, data processing afterwards is complicated. Using a FAT file system to store measurements is more efficient, because the data is stored in a file and the data access via PC-applications is simple. Data storage in a FAT file system is possible in two ways—either in binary or in ASCII form. For the prototype the latter method was chosen as it allows for easy processing. Data are stored in a FAT16 filesystem in ASCII format in hexadecimal values and the data are written in 512 bytes blocks. Measurements are stored in a buffer. As soon as the buffer size reaches 512 bytes, the block is written to the SD-Card. Storage of 1 block takes up to 18 ms.

Advanced scheduling is necessary, in order to handle all these tasks within an 8 bit RISC microcontroller operating at 8 MHz. To achieve a precise timing of analogue sampling of the ECG, the AD conversions are controlled by interrupt. The internal Timer2 of the ATmega644p is triggered every 1 ms and creates an interrupt. The last converted ADC value of 1-channel is then read out and stored in a FIFO buffer and the conversion of the other channel is initiated. As soon as there are 2 × 10 entries in the FIFO buffer, the data are stored on the SD-Card.

With a sampling frequency of 500 Hz, data storage occurs in intervals of 20 ms. Writing a data block on the SD-Card takes up to 18 ms. To avoid resource conflicts between the complex data management and storage, it is important to schedule all other procedures carefully. For each task only a small time window can be dedicated, as in the worst case only 2 ms (20 ms intervals—18 ms for data storage) are left.

Fortunately the other procedures do not need such a high sampling frequency like the ECG processing. The pulseoximeter sends heart rate and oxygen saturation every second via USART. That is the reason why all other procedures are triggered in 1 s intervals, too. Since the time window is 2 ms and the interval time for everything which does not correspond to ECG recording is 1 s, all other processes are distributed into <2 ms long tasks and triggered by a scheduler within the 1 s interval (see Fig. 4.3).

First the temperature and pressure are measured. The MS5541B needs up to 40 μs to measure the raw temperature/pressure. To increase software efficiency, the measurement procedure is split into starting, reading and converting the measured value into a usable value. Afterwards the display is updated in four tasks. And last but not least, the received heart rate and oxygen saturation is appended together with temperature and pressure to the actual ADC data block and written on the SD-Card.

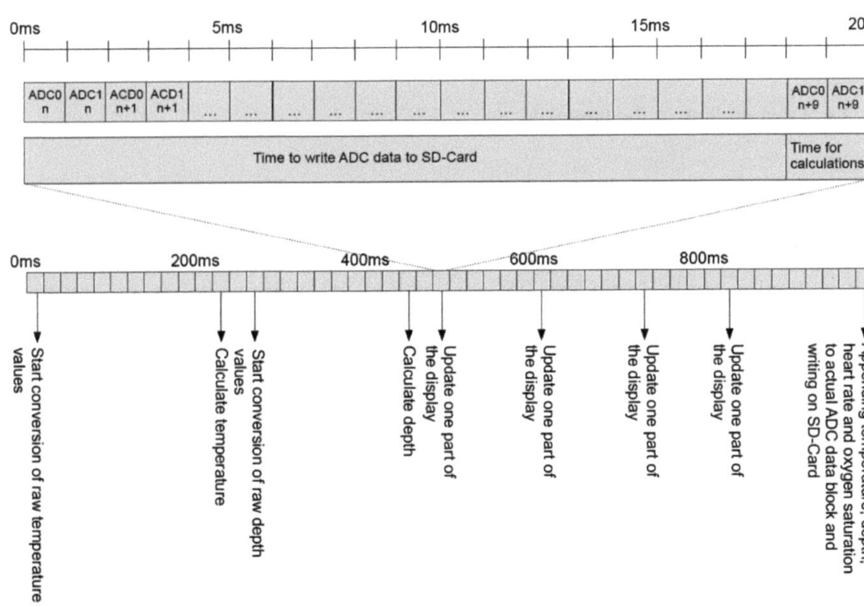

Fig. 4.3 Software scheduler to trigger time critical tasks

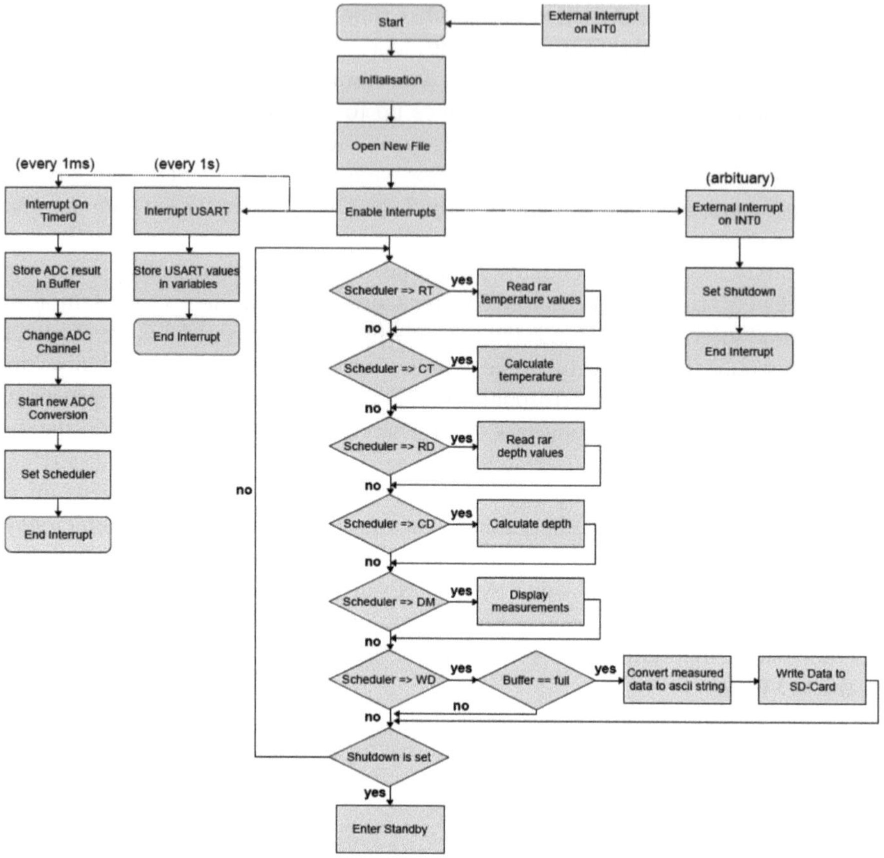

Fig. 4.4 Program flow of the prototype

The system runs as long as it is switched into standby mode. This is done by an external interrupt on INT0. A reed contact together with a magnet is used to initiate standby mode. In underwater applications magnetic switches are preferred as they do not require any mechanical connection to a switch thus avoiding o-rings. To wake up the system, only an additional external interrupt via INT0 is required. The whole program flow is shown in Fig. 4.4.

4.3 Data Processing

For post dive data visualization and processing a graphical user interface was developed under National Instruments LabVIEW 8.5. The depth, temperature, sO_2, heart rate and the two ECG channels are detailed in six graphs.

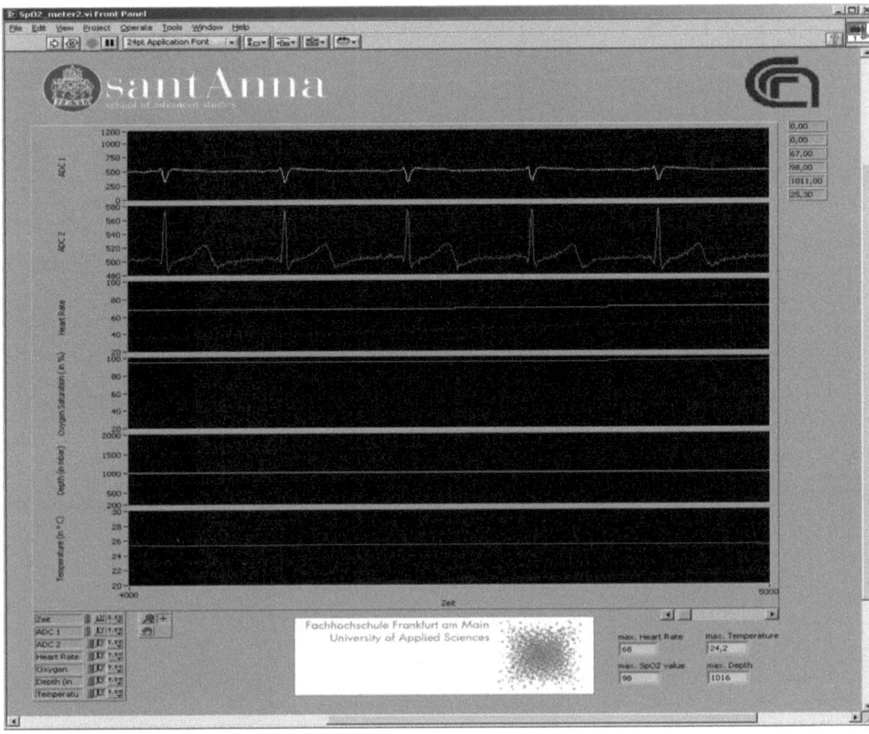

Fig. 4.5 LabView software for data processing

The software reads line by line the data originally stored on the SD-Card. This data are loaded into system memory and converted from hexadecimal to decimal values. Finally, these values are presented in six graphs.

Figure 4.5 shows a screen shot of the graphical user interface. The six graphs visualize the measured values, which were stored on SD-Card. The first and second graphs show both ECG channels. The third graph shows the heart rate. The fourth and fifth graphs visualize oxygen saturation and depth. The last graph presents the environmental temperature. The graphs are connected to each other. By scrolling through one graph all other graphs are scrolled, too. Thus, on an arbitrary point during a recorded dive, all measured values are visible.

4.3.1 Validation

Figure 4.6 shows the validation of the device in the public swimming pool of Pisa, Italy. The prototype has been used by volunteer apnea divers of the Italia Apnea Diving Academy. Measurement results were compared to processed results of a commercial available pulseoximeter (ChipOx, Weinman Medical Technology).

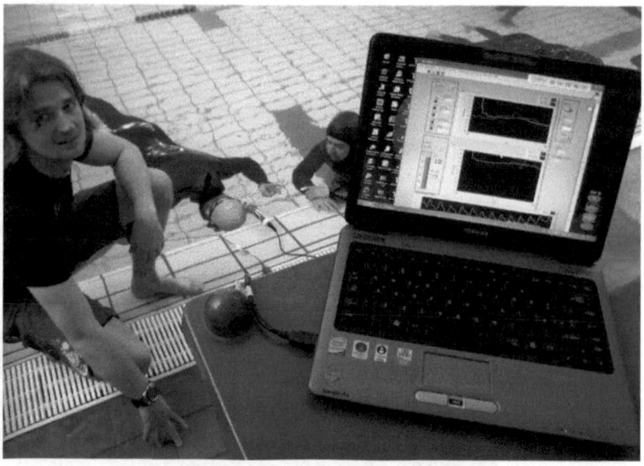

Fig. 4.6 Validation of the pulsoximeter part of the prototype

Fig. 4.7 First prototype in a Lexan housing

Since a ChipOx pulseoximeter measures sO_2 via a transmissive sensor at the fingertip, which is affected by vasoconstriction, comparative measurements could only be performed in warm water. Moreover the reference pulseoximeter was connected via cable on a laptop and the fingertip sensor was pressure/water resistant only up to 1 m. Thus test were only performed in static apnea.

4.4 Results

A prototype (see Fig. 4.7) was build and housed in a Lexan tube with 42 mm outer diameter and a length of 100 mm. The tube is rated to 200 m. The device is powered by 3 $\times$ 1.5 V rechargeable AAA batteries (900 mAh). The overall power consumption is approximately 30 mA which allows 30 h continuous recording.

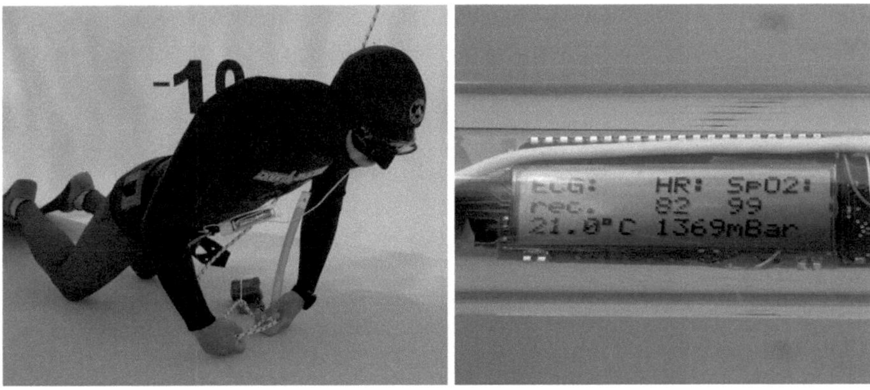

Fig. 4.8 Field tests in the 10.5 m deep research pool

ECG sampling frequency is 250 Hz per channel. This results in a data storage rate of 4 kbytes/s. sO_2, heart rate, depth and temperature are recorded at 1Hz.

First field tests were carried out in the 10.5 m deep research pool at Divesystem, Massa Marittima, Italy (Fig. 4.8). Four electrodes were placed and carefully sealed.

To get a clear signal, the reflectance forehead pulseoximeter sensor is fixed with tape on the temple of the apnea diver and kept on the right position by the hood of the diving suit.

The next generation of prototype will have a specially prepared scuba mask to place the reflectance sensor in the right position. A series of breath hold dives were conducted to 10.5 m for a maximum duration of 3.5 min. During all the dives the quality of the recorded data was good, and the reflectance probe showed significant values.

4.5 Discussion

Figure 4.9 shows a typical dive profile. After approximately 2 min of apnea at 10.5 m the sO_2 starts to drop. At the beginning of the ascent the sO_2 has reached a value of approximately 90%. On the surface the values dropped to nearly 70%. The reason for the fast decrease is that by returning from 10 m to the surface, the pressure drops from 2 to 1 bar. Thus also the partial pressure of O_2 inside the lung decreases rapidly.

During deep breath-hold diving the alveolar pO_2 increases in proportion to the increase of environmental pressure.

So even if the fraction of O_2 is low, due to the increased ambient pressure at depth the alveolar pO_2 might still be high enough. When surfacing it then drops rapidly, particularly in the last meters before surfacing, as the relative pressure drop per meter of depth reaches the highest values. This is one major reason for

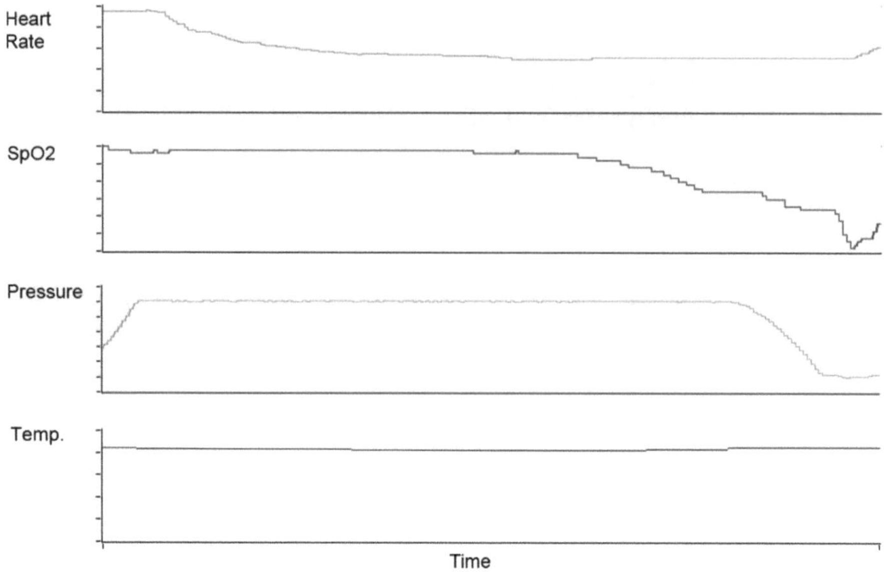

Fig. 4.9 Analysis of measured data

fatalities in deep breath-hold diving. Thus, it is obvious that the measurement of sO_2 can definitely contribute to a better understanding of diving physiology as well as be an important warning factor for safety assurance.

4.6 Conclusions

The presented prototype allows simultaneous recording of 2-channel ECG, sO_2, heart rate, depth and temperature. Additional AD inputs allow easy expendability and extension of the device with additional measurement parameters. The device was successfully validated against a transmissive pulseoximeter. Several long time tests were done in the laboratory and several test dives were successfully conducted in the 10.5 m research pool (Dive System, Massa Maritima, Italy). We envisage that this novel device will lead, first, to a better understanding of the human diving physiology and, second, will provide a novel tool for adding the physiological information necessary for a safer and more effective training.

4.7 Future Work

Future work will include the development of a dedicated apnea diving computer. In addition to displaying and recording sO_2, heart rate and depth, it will also be capable of storing the complete plethysmogram with a sample rate of 75 Hz.

This then will also allow a RR-analysis and estimation of the heart rate variability even without having ECG derivations, which is of great advantage, especially when performing measurements in salt water.

References

1. Sieber A, Kuch B, Wagner M, L'Abbate A, Dario P, Bedini R (2008) Underwater blood pressure measurement device. Diving Hyperb Med 38(3):128–134
2. Kuch B, Bedini R, L'Abbate A, Wagner M, Buttazzo G, Sieber A (2009) Embedded data logging platform for research in diving physiology. In: Proceedings of the 7th workshop on intelligent solutions in embedded systems, Ancona, Italy, pp 43–48
3. Silbernagl S, Despopoulos A (2007) Taschenatlas der Physiologie. Georg Thieme Verlag, Stuttgart
4. Ehm OF, Hahn M, Hoffmann U, Wenzel J (2003) Tauchen noch sichere—Tauchmedizin fuer Freizeittaucher, Berufstaucher und Aerzte. Mueller Rueschlikon Verlag AG, Cham
5. Sieber A, Bedini R, Yong X, Navarri A, Dalle Luche M, L'Abbate A, Dario P (2007) High resolution ECG and depth data logger. Eur J Underw Hyperb Med 8(3):56–57
6. Zonios G, Shankar U, Iyer VK (2004) Pulse oximetry theory and calibration for low saturations. IEEE Trans Biomed Eng 51(5):818–822
7. Dresher RP, Mendelson Y (2006) A new reflectance pulse oximeter housing to reduce contact pressure effects. In: Proceedings of the IEEE 32nd Annual Northeast, Easton, USA, pp 49–50
8. Schagatay E, Lodin A, Richardson M (2007) Diving response and arterial oxygen saturation during apnea in apneists and untrained subjects of both genders [abstract]. 33rd annual scientific meeting of the European underwater and baromedical society, Sharm el Sheikh, Egypt 8(3):43–44

Chapter 5
IEEE 1451 Sensor Interfacing and Data Fusion in Embedded Systems

Gas Leak Detection Case Study in H_2 Vehicles

Sergio Saponara, Luca Fanucci and Bruno Neri

5.1 Introduction

A smart transducer is the integration of an analog or digital sensor plus an embedded system including a mixed-signal conditioning and data conversion unit, an application processor with associated software for digital signal processing and data fusion, calibration and diagnostics, and a communication controller. The latter passes the measured or calculated parameters to a host or monitoring system in a network by using of a network communication protocol [1–3]. To address the need for standardized interfaces and architectures for networked transducers, this work illustrates in Sect. 5.2 the architecture of a system of networked transducers and reviews the emerging IEEE 1451 family of standards. In Sect. 5.3 an application case study is proposed: a safety system based on hydrogen/methane sensors to monitor the gas leak in multi-fuel and zero emissions vehicles. Section 5.4 deals with sensor selection, modelling and compensation through data fusion. Section 5.5 presents the design of the mixed-signal embedded system for intelligent sensor interface and data fusion and show experimental results. Conclusions are drawn in Sect. 5.6. This work extends the WISES2009 paper [3].

5.2 IEEE 1451 Smart Network Sensor System

The IEEE Instrumentation and Measurement Society's Technical Committee on Sensor Technology sponsored the development of a suite of smart transducer interface standards, known as IEEE 1451, to enable Plug and Play transducers

---

S. Saponara (✉), L. Fanucci and B. Neri
Department of Information Engineering, Università di Pisa, Pisa, Italy
e-mail: sergio.saponara@iet.unipi.it

M. Conti et al. (eds.), *Solutions on Embedded Systems*,
Lecture Notes in Electrical Engineering, 81, DOI: 10.1007/978-94-007-0638-5_5,

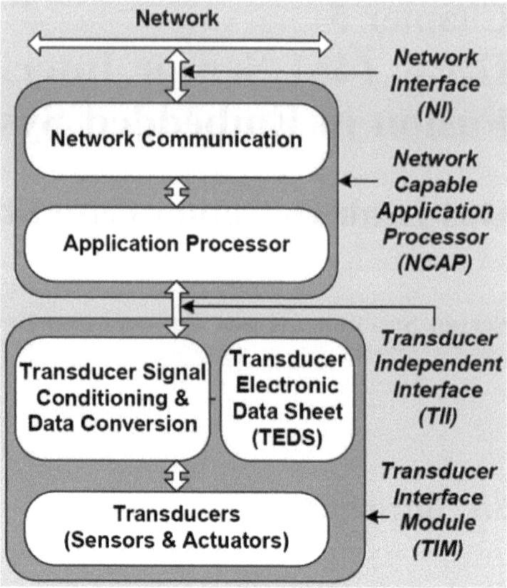

Fig. 5.1 IEEE 1451 smart transducer architecture

networking and capabilities of self-identification, self-diagnosis, self-calibration, location and time awareness, data processing and fusion, alert notification [2, 3]. Figure 5.1 illustrates the architecture of an IEEE 1451 smart transducer characterized by the partitioning into two main components: a Network Capable Application Processor (NCAP) and a Transducer Interface Module (TIM), connected through a Transducer Independent Interface (TII). The NCAP, a network node, performs application processing and network communication functions, while the TIM consists of a number of sensors and actuators with the relevant signal conditioning and data conversion circuits. The TII defines a communication medium and a protocol for transferring sensor information. This interface provides a set of operations, such as read and write messages, commands and responses. The network interface (NI) defines a network communication protocol for NCAP transactions over the network. Another key feature of the IEEE 1451 is the introduction of the Transducer Electronic Data Sheet (TEDS) which stores information such those provided by the manufacturer in the data sheet. The TEDS can be read and reprogrammed depending if they are stored in a ROM, if the content never changes, or an EEPROM, if its content can be updated. As defined in the IEEE 1451.0 standard, four kinds of TEDS are mandatory, while the others are optional. Among the required TEDS, the Meta TEDS gives some worst-case timing parameters used by the NCAP to set time-out values in the communication software to determine when the TIM is not responding. The Transducer Channel TEDS provides detailed information about the specific transducer (measured physical parameter, operating range, I/O characteristics, operational modes, timing information). In the User's Transducer Name TEDS, the user can store the "nickname" by which the system will know the transducer. The PHY TEDS stores information dependent on the physical communications media connecting the TIM to the NCAP. The format of a TED foresees an header field specifying TEDS length and type, a variable length field containing TEDS data, a final checksum for transfer robustness. The IEEE 1451 standards family provides a suite of protocols for wired and wireless distributed transducers applications [4–7].

The IEEE 1451.0 standard defines a set of common functionalities, commands, and TEDS for the family of IEEE 1451 smart transducer interface standards. These functionalities are independent from the physical communication media (1451.X) between the transducer and the NCAP. They include the basic functions to read from and write to the transducers, to read and write TEDS, and to send configuration, control, and operation commands to the TIM. The IEEE 1451.0 helps achieving data-level interoperability for the IEEE 1451 family when multiple wired and wireless sensor networks are connected together. The IEEE 1451.1 defines the possible ways to access sensors and actuators in the TIM from a network and its focus is mainly on the communications between NCAPs and between NCAPs and other nodes in the system. The physical interface between the NCAP and TIM is defined by the other IEEE 1451.X standard and can be (i) a point-to-point interface that meets the IEEE Standard 1451.2 (with communication layer based on serial interfaces such as SPI, UART or USB), (ii) a distributed multi-drop interface that meets the IEEE Standard 1451.3, (iii) a wireless interface that meets the IEEE Standard 1451.5 (WiFi, Bluetooth, ZigBee or Wireless Personal Area Networks), (iv) a CANopen interface that meets the IEEE Standard 1451.6, (v) an RFID interface that meets the IEEE Standard 1451.7. The interfaces between transducers and signal conditioning and conversion circuits inside each TIM unit are not standardized, with the exception of the IEEE 1451.4, which specifies a low-level, mixed-mode interface for transducers. The use of TEDS and the NCAP-TIM partitioning approach provide many benefits: a transducers can identify and describe itself to the host or network by sending the TEDS information; the TEDS can be updated and it can store information, such as the location of the sensor, recalibration date, repair records and maintenance-related data thus enabling long-term self-documentation of the transducer; the automatic transfer of the TEDS data to the network eliminates the need of manually entering the sensor parameters thus reducing human errors; plug-and-play capability is provided since TIM and NCAP can be connected with a standardized physical communication media and are able to operate without change of the host software. TIMs from different sensor providers can interoperate with NCAPs from different network operators through the same IEEE 1451 communication module. IEEE 1451 smart transducers represent a step forward towards the Internet of Things paradigm.

5.3 Networked Gas Leak Sensing System

As an application case study of the above concepts of smart transducers interfacing and IEEE 1451 sensor networks, this work presents a monitoring system for gas leak in vehicles using hydrogen as energy vector (multi-fuel internal combustion

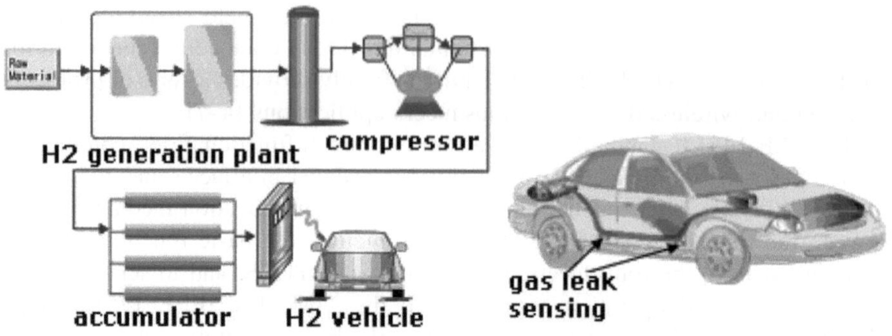

Fig. 5.2 Processing chain of the future hydrogen-based transport system

engine or fuel cells), or in hydrogen storage and distribution facilities. Figure 5.2 presents the processing chain of hydrogen-based transport system (covering all aspects of the future hydrogen economy: generation, distribution and storage, use as propulsion gas) and a detailed view of hydrogen monitoring inside a vehicle [8–10]. The threshold for the concentration of hydrogen in air to generate an explosion (defined as LEL, Lower Explosivity Limit) is 40,000 ppm, i.e. 4×10^{-2}. This value is similar to that of methane (CH_4), also used in internal combustion engines, whose LEL is 50,000 ppm, i.e. 5×10^{-2}. For safety reasons, a LEL-warning system able to detect the presence of hydrogen (and methane) at concentrations as low as 10^{-3} is needed. Multi-fuel vehicles are also announced using both CH_4 and H_2. In future storage and distribution systems for propulsion gas both CH_4 and H_2 may be present. Therefore, for a warning system it is important to independently detect a leak of H_2 or CH_4. The risk of undetected H_2 leaks is higher since this gas is odorless and not visible.

The system architecture for the monitoring unit foresees two scenarios. First, for gas leak monitoring inside the car few smart sensing units should be placed near the tank, near the fuel-cells or internal combustion engine, and inside the passenger compartment. Such smart sensing units will be connected to the already available digital in-vehicle networks: typically a CAN backbone connecting the main vehicle Electronic Control Units (ECU) from which there are local connections to smart transducers based on low-rate serial links such as LIN or UART or SPI [11]. As further discussed in Sect. 5.4, COTS (commercial off the shelf) hydrogen sensors have an analog output and their response should be compensated since they are sensitive also to methane and ambient conditions (temperature, humidity). Therefore the smart sensing unit we have designed, see Fig. 5.3, is composed by one hydrogen sensor plus methane and temperature compensating sensors plus a multi-channel embedded system integrating a mixed-signal front-end for analog signal conditioning and digital conversion, a signal processing and sensor fusion core in the digital domain and a communication circuitry for connections to the in-vehicle network through a serial link. This module is an IEEE 1451.2 compliant TIM connected to an ECU which will act as an NCAP towards

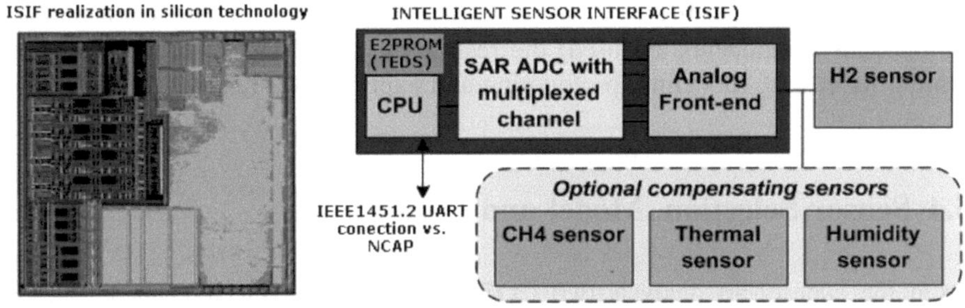

Fig. 5.3 Architecture of the smart sensing node and chip realization of the ISIF

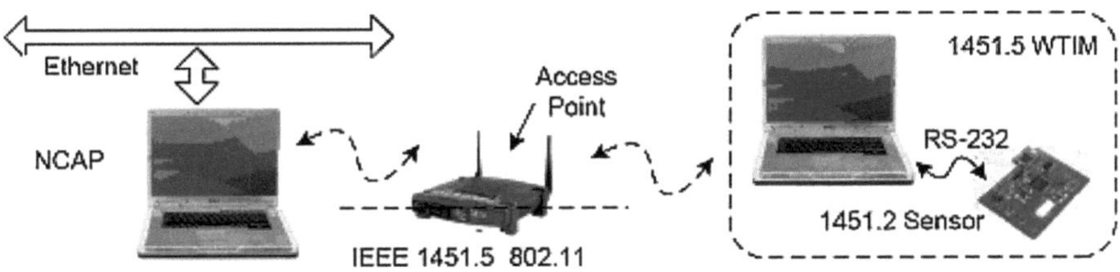

Fig. 5.4 Smart IEEE 1451-based wireless sensing network configuration

the main CAN-based vehicle network. The HW/SW design, prototyping and experimental results on the IEEE 1451.2 compliant TIM are reported in Sects. 5.4 and 5.5. This TIM is already realized using technologies compliant with a migration to the large volume low-cost automotive market.

Secondly, in hydrogen distribution/storage stations a wireless link has been selected as most suited to interconnect the smart gas leak sensing nodes to the host system. In this case the overall networked monitoring system consists of a NCAP node connected to the main network through an Ethernet link and to the wireless TIMs through wireless links. For prototyping purpose, see Fig. 5.4, in our work we adopted a configuration with the NCAP realized on commercial mobile computers using Java language (to ease the software testing) connected to a single IEEE 1451.5 wireless TIM by means of a WI-FI 802.11 protocol. For the higher networking layers classic client–server communication models and TCP/IP protocols are used.

The wireless TIM is realized by connecting the IEEE 1451.2 TIM described in the first scenario through the UART to a commercial mobile computer which implements the 802.11 wireless connection towards the NCAP. Direct point-to-point Wireless connections between the NCAP and the Wireless TIM and network connections mediated by a commercial access point have been implemented. The proposed prototyping configuration allows assessing the full functionality of the system but obviously, in case of a real production scenario, the tasks implemented on the mobile computers may be easily ported in embedded HW/SW nodes equipped with 802.11 transceivers. Since the main innovation of this work is the development of the IEEE 1451 TIM (for the NCAP and for the wireless links commercial components are used) the rest of the work will be focused on presenting its design, modeling, prototyping and experimental results.

5.4 Sensor Selection, Model and Data Fusion

From the state-of-art analysis two COTS sensors were selected whose performances are good representatives of the available classes of H_2 sensors: the FIGARO TGS6812, which can be used for the detection of large leaks of hydrogen in explosiveness warning systems and the TGS821, more suited for small H_2 concentrations [12]. Commercial sensors from other suppliers, e.g. Synkera, Kebaili, E2V are available with similar performances; hence the proposed approach keeps its validity also if applied to other COTS sensors.

The TGS6812 is a catalytic resistive H_2 sensor for the detection of concentrations up to 100% of LEL. Like other COTS sensors, the TGS6812 is sensitive also to hydrocarbon gases, mainly CH_4 which will be present in future gas-based vehicles or storage/distribution systems using both methane and hydrogen. The front-end reading requires a Wheatstone bridge with dummy resistors of 1 kΩ. A potentiometer can be used to adjust the offset. The sensor has a linear output (with sensitivity to H_2 of about 15/4,000 mV/ppm) and is insensitive to variations of humidity and temperature. The TGS821 is a tin dioxide (SnO_2) semiconductor sensor which has low conductivity (i.e. high resistance R_0) in clean air. In the presence of a detectable gas, the sensor's conductivity increases (the sensor resistance R_S decreases) depending on the gas concentration in the air. Figure 5.5 shows the functions that determine the variation of output resistance (R_S/R_0) in relation to the gas concentration (mainly H_2 and CH_4) and to environmental factors such as relative humidity (%RH) and temperature (T). The functions in Fig. 5.5 have been obtained by curve fitting of experimental measurements on the sensor. Such functions are used to build an accurate model that is the basis for the

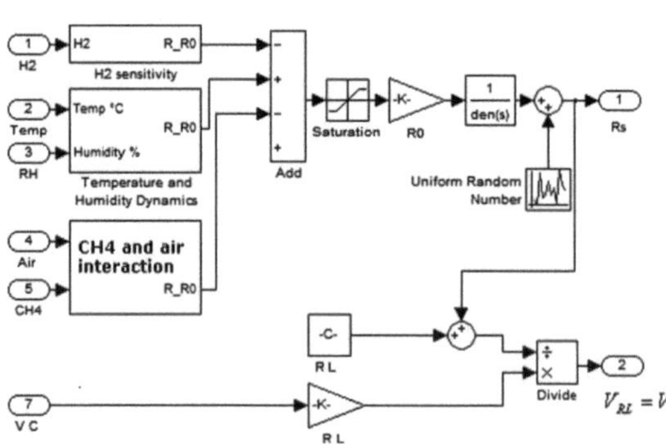

$$V_{RL} = V_C \cdot \frac{R_L}{R_S + R_L}$$

Sensitivity to $[R_S/R_0]$	Function
H_2	$5^{2-\log(H_2 ppm)}$
CH_4	$1.8^{9.4-\log(CH_4 ppm)}$
Temperature	$1.0188^{-T+12}-1$
Relative Humidity	$-\frac{0.3}{100}\cdot(\%RH)+95$

Fig. 5.5 Model and sensitivity functions for the TGS821 sensor

development and prototyping of data fusion techniques. As example, Fig. 5.5 shows the block diagram of the proposed TGS821 Simulink model which takes into account: (i) the dependence on environment conditions (temperature, humidity); (ii) error sources such as saturation, noise, offset and sensitivity errors; (iii) timing (and frequency) response; (iv) main dependencies on other gases. The response of the proposed model takes into account also the statistical uncertainty of the sensor offset and sensitivity due to technology spreading. The uncertainties are calculated each time by an initial script, using a function that extracts a random Gaussian number with mean equal to zero and variance equal to 1/3 of the range of uncertainty, so the whole range of uncertainty is in $\pm 3\sigma$. To be noted that for the considered sensors the declared technology spreading is not negligible, up to 10%, and hence in practical application a pre-calibration phase is always applied. The developed model is then integrated in a test environment simulating external conditions change and interactions with non idealities of other components: external temperature and humidity; concentration of other gases; power supply ripple, tolerance and errors of the circuitry needed to produce a voltage output (V_{RL} in Fig. 5.5, where V_C is a supply voltage of 10 V and R_L is 4 kΩ) proportional to the sensor resistance R_S and hence to the H_2 concentration. The obtained output signal V_{RL} is then processed by analog circuitry before being converted in a digital form by an ADC: for the specific TGS821 device in the analog domain the amplitude of V_{RL} is scaled by a factor $K = 0.2$ and an offset $V_{OFF} = 1$ V is subtracted. The signal is then low-pass filtered and the output V^*_{RL} is sent to the ADC.

Such top-down modeling approach annotated bottom-up with physical device characteristics leads to a better trade-off between simulation accuracy and time vs. state-of-art models which are computational intensive physical models or fast but inaccurate linearized models [13].

The same modeling approach has been followed for the TGS6812 H_2 sensor and the relevant circuitry for output voltage reading (based on Wheatstone bridge). Being linear in response and with negligible dependence on temperature and humidity, the TGS6812 model is simpler than the TGS821 one. As example Fig. 5.6 shows Simulink diagrams obtained with (a) the TGS6812 model and (b) the TGS821 model and compared to experimental measured data (different H_2 concentrations at ambient temperature conditions). As discussed above, the TGS821 sensor was selected and modeled to be used in small leakage detection warning systems, targeting a concentration range up to 1,000 ppm while the TGS6812 is selected to cover larger H_2 concentrations up to several thousands of ppm. In the target ranges the modeled sensor responses show a good match with both data sheet values and experimental data (e.g. differences within 1% between simulated and measured results in Fig. 5.6). Although the developed sensor models take into account all relevant interferences, the fusion algorithms proposed in this work aim at removing the dependence of the target sensor response only on the main interfering variables for the considered application, as a trade-off between complexity/cost and accuracy of the measure. In fact, the correction of the measured concentration w.r.t. all possible sources of interference, including the ones

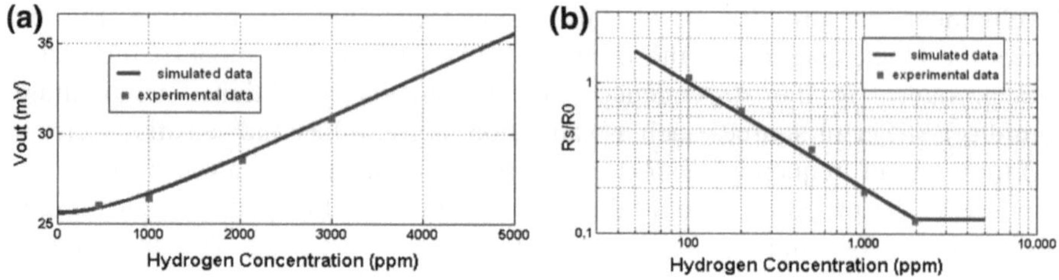

Fig. 5.6 (**a**) TGS6812 and (**b**) TGS821: model response vs. experimental data

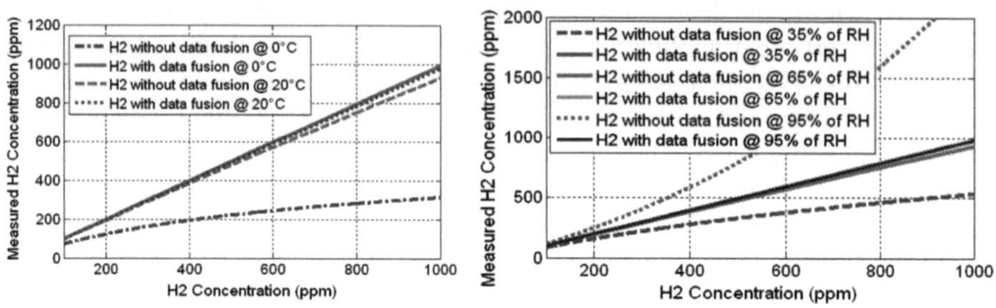

Fig. 5.7 Data fusion for TGS821, temperature and humidity variation

giving just a minor contribution, would result in a too complex and too costly fusion technique not suitable for automotive applications. Since the high sensitivity of TGS821 to changes in temperature and humidity (see Fig. 5.7) can corrupt the measured H_2 concentration, compensating sensors have to be used.

Furthermore, if the vehicle or the storage/distribution system is using also methane it is necessary to correct the measure obtained with the TGS821 and TGS6812 sensors by using a methane-specific sensor. Moreover, given the wide variation of sensitivity for the TGS6812 device, it is necessary to calibrate this sensor, determining by experimental data the value of offset and sensitivity (otherwise calibration measuring errors in the range of thousands of ppm can occur). For TGS821 compensation a humidity sensor able to measure relative humidity levels in a wide range and preferably with a linear output characteristic is required. The selected commercial sensor is the SY HC1 by RHOPOINT, a linear capacitive sensor able to measure %RH ranging from 5 to 95% with sensitivity of 0.6 pF/[%RH]. The effect of noise and the dependence on temperature and frequency have been considered when modeling the humidity sensor. The used temperature sensor for thermal compensation of the TGS821 H_2 and SY HC1 sensors is a simple PT100 resistor featuring a good linearity and whose response (following the classic Callendar–Van Dusen equation) does not depend on humidity or gas concentration. This feature is important to reduce cross-correlation between the sensors involved in the compensating process. The PT100 sensor has been modeled considering the Callendar–Van Dusen law linking the output resistance to the temperature and taking into account errors due to offset, noise, sensitivity gain. The last sensor used for compensation of both TGS6812 and TGS821 H_2 sensors is the FIGARO TGS6810, a methane sensor that can detect concentrations up to 100% of LEL (50,000 ppm for methane in air). The TGS6810 CH_4 sensor is similar to the TGS6812 H_2 one in terms of structure and model, except that TGS6810 is not influenced by H_2 concentration. The dependence on temperature and humidity of this device is zero thus avoiding cross-dependency with the other compensating sensors.

To compensate by data fusion the TGS821 sensor the outputs of temperature, humidity and methane sensors are first converted into voltage signals. Then, for the three sensors, an electronic circuit adjusts the dynamics, eliminates the offsets and applies the anti-aliasing filtering before sampling. The resulting signals are then passed on to the A/D converter. The bandwidths of the sensors are low, and consequently, it is possible to employ only a single A/D converter multiplexing the signals provided by the three sensors. After sensor acquisition the following equation is used, linking the change in resistance of the sensor output to the parameters measured by the other three sensors (temperature, methane and humidity; the value of the latter is first thermally compensated using the PT100 sensor):

$$R_S = R_0 \cdot \left[\frac{R_S}{R_{0\text{Air}}} + \frac{R_S}{R_{0\text{H}_2}}(\text{H}_2) + \frac{R_S}{R_{0\text{RH}}}(\text{RH}) + \frac{R_S}{R_{0\text{CH}_4}}(\text{CH}_4) + \frac{R_S}{R_{0\text{Temp}}}(T) \right] \quad (5.1)$$

where $R_S/R_{0\ \text{Air}}$ is the resistance ratio in air, while R_S/R_0 (Eq. 5.1) represents the resistance variation caused by the interfering XX compound, by relative humidity or by temperature (see Fig. 5.5). Figure 5.7 shows the simulation results obtained with this algorithm, after sensor pre-calibration, for different H_2 concentrations and different temperature and humidity operating points: measurements not taking into account the effect of temperature or humidity can be extremely inaccurate. The same data fusion approach is also followed for the TGS6812 sensor for which compensation is required only if the vehicle (or the distribution/storage system) uses also methane. In this case the TGS6810 sensor has to be used to correct the measure of hydrogen.

5.5 Embedded ISIF Design and Realization

To fast identify, trim and verify at experimental level an architecture to interface and compensate a given sensor, a mixed-signal embedded hardware platform for Intelligent Sensor Interface (ISIF) has been developed by University of Pisa in collaboration with SensorDynamics AG [14]. Realized in 0.35 μm BCD (Bipolar CMOS DMOS) technology with 3.3 V supply for the digital part and 5 V for the analog one, the IC has been developed according to a platform based design strategy [15], by assembling a set of analog, digital and software intellectual property (IP) modules in the same multi-channel sensor interfacing chip.

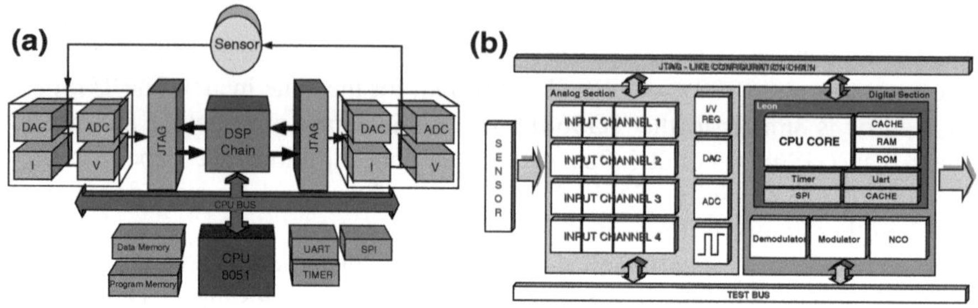

Fig. 5.8 Architectures of the **a** 8051-based and **b** of the LEON-based ISIF

The architecture of the mixed-signal embedded system is sketched in Fig. 5.8. It is composed of an analog front end and a digital processing section with a JTAG standard interface between the two signal domains. The basic idea behind the architecture in Fig. 5.8 is using a low-cost sensor and reducing to a minimum the analog signal processing, while compensating non-ideality through digital signal conditioning, since digital circuitry can be easily designed and scaled in micro-electronics technologies. The analog front-end in Fig. 5.8 mainly accomplishes tasks of driving sensor's electrodes (in case of sensor requiring external excitations), through couples of thermometer-type DACs and performing signal acquisition by means of SAR-type ADCs and programmable-gain operational amplifiers. It also provides a regulated power supply to the digital section. All modules are digitally controlled since gain coefficients, offset values and reference voltages are set by means of dedicated registers accessed via the JTAG bridge by the digital processor. Hence, also the analog part is configurable while the digital part is both HW-configurable and SW-programmable. The default HW configuration plane and the TEDS can be stored in on-chip PROM. All non-trivial signal processing required for sensor conditioning, i.e. filtering, function generation and demodulation, is performed by the digital section which also monitors system activity and handles communication with external devices. Both dedicated and general purpose computing resources are available in the digital part for a good trade-off between power consumption and flexibility. The HW DSP chain in Fig. 5.8 contains dedicated circuits for digital signal processing: FIR/IIR filters to remove noise/interference sources and a digital Phase Locked Loop (based on numerically controlled oscillators) for demodulating the sensor response and for function generation (e.g. sensor stimuli) based on the direct digital synthesis concept. General purpose tasks are managed by CPU core provided in a configuration with on-chip program/data memories and standard parallel I/O plus UART and SPI interfaces for communication. In this system the CPU core is in charge of monitoring the DSP chain and managing communication/control flows among the mixed-signal part via JTAG, the DSP unit, the internal memories and the external devices. This basic architecture has been implemented in silicon in two configurations where the main differences are the used CPU core and the amount of on-chip memory resources. The first configuration in Fig. 5.8a is based on a power-optimized 8051-compliant core we presented in [16, 17]. A prototype chip has been realized in 0.35 μm BCD technology.

The architecture in Fig. 5.8a with a configuration of 32-kbit PROM and two 4-kbit SRAM memories and an 8-bit 8051 core with timer/counter, UART and SPI has an overall area of roughly 20 mm^2 and works at 20 MHz clock frequency. The ADCs are sized for 10 bits and 100 kS/s max sample rate. The second ISIF generation targets more powerful sensor conditioning systems; it is based on a 32 bit SPARCV8 LEON2 CPU core and is more suited for fast sensor conditioning prototyping, or for applications requiring computation-intensive signal processing but with less bounded limits in terms of chip size and cost. Still realized in 0.35 μm BCD technology, this embedded platform (Fig. 5.8b) has an area of roughly 70 mm^2 and enhances the first generation in Fig. 5.8a with: four multi-stage configurable analog (instrumentation amp + filter) acquisition channels; four 12-bit SAR ADCs (max. 150 kS/s) and two 16-bit sigma-delta ADCs (max. 15 kS/s); six high-precision 12-bit and six high-speed 10-bit on chip DACs; a 32 bit 20 MHz SPARC V8 fixed-point core (LEON2); on-chip 32 kbytes EEPROM and 32 kbytes RAM; UART/SPI interface plus 2 timer peripherals and 16-bit GPIO. For the target sensors of this work the 8051-based ISIF embedded device is enough and ensures lower cost and size vs. the second architecture chip. Other mixed-signal platforms for sensor interfacing are available on the market, such as the Actel Fusion (AFS600) based on an ARM7 core and a single 12-bit ADC [18] or the Cypress PSOC CY8C featuring 4 ADCs and 4 DACs configurable from 6 to 14 bits and based on 8-bit CPU. With respect to the above COTS platforms the proposed ISIF device is preferable in the 8051-based version when the application is more dedicated to the conditioning of sensors with limited size and cost budgets, while the LEON[19]-based version is preferable if high-precision ADCs or a powerful CPU are required. From a SW and algorithmic point of view, the development flow of an application on the ISIF platform is integrated with a system-level Simulink/Matlab flow. The starting point is the realization of a Matlab/Simulink model of the whole system, which is made of a set of functional blocks allowing co-simulation of the sensor model with the analog/digital conditioning circuitry. To this aim to each sub-block in Fig. 5.8 a simulating model of the relevant circuitry is associated (detailing input–output transfer characteristic plus main error sources such as saturation, offset, noise, temperature dependence, frequency response). Using such configurable models at an early stage of the design it is possible to run a number of simulations for identifying the critical parameters for the overall system performance, and hence for correctly sizing the sensor signal processing circuit. A system exploration phase, based on simulations, design iterations and functional blocks refinements leads to a first partitioning of the system in analog and SW-programmable digital building blocks. After architecture definition and HW/SW partitioning in the Matlab/Simulink environment, each block is modeled with the most appropriate description language and EDA tools or proper pre-designed IPs to be reused are selected, configured and assembled. Conventional flows are used for the lower level design phases (VHDL-based for digital HW, VHDL-AMS and Spice for analog circuitry and C/C ++ for SW routines). The top-down platform-based design flow ends up with the prototyping phase, through which the whole system can be tested under practical operating conditions. This methodology enables a rapid managing also of complex designs thanks to high reuse of concepts, architectures, and IPs among different projects.

As proved in Sect. 5.4 data fusion algorithms are very important for measuring H_2 concentration both in explosion warning systems (targeting with the TGS6812 sensor a dynamic range of several thousands of ppm with a measuring resolution of 100 ppm) and in H_2 leak detection systems for early warning, targeting with the TGS821 sensor a dynamic range up to 1,000 ppm, but with a fine grained resolution. However, it must be noted that the data fusion algorithm, especially the one for TGS821, is computationally complex. Beyond simple additions, subtractions and products, some exponential, logarithm and division operations must be performed to calculate the formulas described in Fig. 5.5 and Eq. 5.1. The use of a DSP with Floating Point Unit is required, but this device is expensive compared to a traditional microcontroller and it is not suited for automotive applications, which are intended for a high-volume market. A solution relies on the use of a tabulated form of the needed non linear analog functions, based on pre-calculated Look-Up Tables (LUT) in the digital domain. Naturally, it is necessary to find a trade-off between the number of levels of the tabulated functions and the related memory cost. A coarse function approximation can lead to unacceptable measuring errors; on the contrary, an approximation with a large number of levels and bits for their encoding can lead to a memory resource requirement too high for the limited budgets of embedded automotive HW. When choosing the LUT-based approach it is necessary to use a processing platform, such as the ISIF, with a PROM that contains the entire array describing the tabulated functions. When sizing the HW platform, first of all the number of ADC's bits has to be defined. For the H_2 data fusion algorithm extensive simulations in the Matlab environment have been carried out trying to find the lower number of ADC data size and LUT levels that ensure at system level a quantization error below a 'quality target' of 10 ppm for leak detection in the range 0–1,000 ppm and below 100 ppm for explosivity warning in a wider measuring range up to 10,000 ppm. As a result of this design exploration activity, a 10-bit ADC was selected. It has to be noted that from our analysis for humidity and for temperature sensors an 8-bit converter would be sufficient. Therefore, the SAR-type 10-bit ADC of the ISIF platform can be used to acquire the required signals: since the bandwidths of the sensors are in the order of Hz and the sampling rate of the ADC is up to 100 kS/s, all sensor inputs can be multiplexed on the same ADC. To reduce glitch noise when switching between different inputs and to reduce quantization noise (thus having an effective number of bits equal to the nominal 10 bits), the digital conversion can be performed at a higher frequency w.r.t. the required Nyquist-rate and then the digital signal can be cleaned by decimating the samples. This way the resolution obtained using the SAR ADC of the ISIF platform is 2 ppm for TGS821 acquisitions and about 40 ppm for TGS6812 acquisitions.

After sizing the ADC the next choice to be made is related to the size of the tables that will describe the various non-linear functions used for data fusion.

As example, for the TGS821 the functions in Fig. 5.5 have to be tabulated. Hereafter we report, for each formula, the criteria for sizing the corresponding table, emerged after extensive simulations in the Matlab environment targeting an overall quantization error below 10 ppm. The table that represents the CH_4 sensitivity function is sufficiently informative if it is implemented with 256 lines of input, divided in the range of concentration of CH_4 from 0 to 4,000 ppm, with outputs represented over 8 bits, for a total memory usage of 256 bytes. In order to get a satisfactory approximation of the temperature and humidity sensitivity functions it is necessary a table with 256 entries, where outputs have an 8-bit representation. This way the memory requirement amounts to 256 bytes for each function. The H_2 sensitivity function can be calculated in tabular form, using a table with 2,048 input lines and outputs with 8-bit precision. This table requires 2 kbytes of memory. Finally the function that extracts the hydrogen measure from the read output voltage requires a table with 256 entries encoded on 12 bits. The low-cost implementation of the sensor fusion compensation for the TGS6812 device is easier since its dependence on temperature and humidity is negligible and the correlation with methane measurement is based on a linear law. Following the above considerations, the fusion technique for both H_2 sensors, TGS8612 and TGS821, can be implemented on a mixed-signal embedded device, such as the 8051-based ISIF platform. Indeed, the required HW resources after the LUT-based strategy and bit sizing reported above are the following: 10-bit ADC with 4 multiplexed channels (one for the hydrogen sensor, TGS6812 or TGS821 depending on the target dynamic range, while the others are for methane compensation and in case of the TGS821 for temperature and humidity compensation); on-chip memory of roughly 3 kbytes for LUT-based implementation of complex processing functions; 8-bit CPU of few MIPS which implements only digital samples decimation and signal control tasks at low repetition frequencies (the sensors have bandwidth of few Hz); IEEE 1451.2 UART serial interface towards a NCAP host controller. Figure 5.9 shows the results obtained implementing data fusion for TGS821 on the ISIF platform. Both simulated data of the overall Simulink model and experimental data (different H_2 concentration points) are illustrated. From Fig. 5.9 it can be noted the effect of the representation of values on a limited number of bits and of the LUT-based realization of non-linear equations. The use of LUTs and of a fixed point arithmetic allows the algorithm

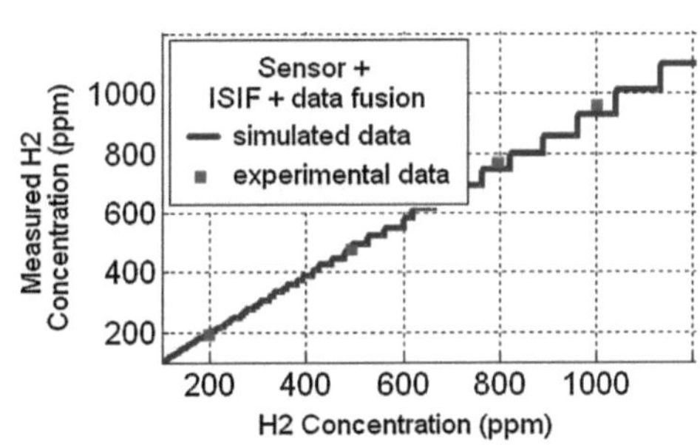

Fig. 5.9 TGS821: LUT-based implementation of data fusion algorithm

implementation in the low-complexity 8051-based ISIF device, instead of using a floating point DSP, and the introduced error with respect to the ideal response curve amounts to tens of ppm. Similar simulation and experimental analysis have been repeated after implementing data fusion for TGS6812 on the ISIF platform. In this case the maximum error amounts to few hundreds of ppm in a range up to 10,000 ppm.

5.6 Conclusions

The design of an IEEE 1451-compliant embedded system for sensor network interfacing and processing is presented. The achieved results on the case study of gas leak detection for H_2-based vehicles prove that intelligent sensor interface IC can be designed integrating on-chip the mixed-signal processing chain plus data fusion and communication digital resources. Beside the detailed design of the smart sensing module (IEEE 1451.2 compliant) the whole monitoring system architecture is discussed, including network capable application processors (NCAP) for both wired (CAN in-vehicle network) and wireless (Wi-Fi networking) scenarios. This work is supported by the projects *Filiera* H_2 (Tuscany region) and *Pollux* (EU).

References

1. Elmenreich W, Pitzek S (2003) Smart transducers—principles, communications, and configuration. In: IEEE international conference on intelligent engineering systems, vol 2, pp 510–515
2. IEEE Instrumentation and Measurement Society's Technical Committee on Sensor Technology. Standards IEEE 1451.0-2007, IEEE 1451.1-1999, IEEE 1451.2-1997, IEEE 1451.3-2003, IEEE 1451.4-2004, IEEE 1451.5-2007. http://ieee1451.nist.gov/
3. Saponara S, Petri E, Fanucci L, Terreni P (2009) Smart transducer interface in embedded systems for networked sensors based on the emerging IEEE 1451 standard: H_2 detection case study. In: IEEE WISES 2009, June, pp 49–56
4. Lee K et al (2004) IEEE-1451-based smart module for in-vehicle networking systems of intelligent vehicles. IEEE Trans Ind Electron 51(6):1150–1158
5. Song E, Lee K (2007) Smart transducer web services based on IEEE 1451.0 standard. In: IEEE instrumentation and measurement technology conference
6. Wobschall D (2008) Networked sensor monitoring using the universal IEEE 1451 Standard. IEEE Instrum Meas Mag 11(2):18–22
7. Song E, Song EY (2005) Object-oriented application framework for IEEE 1451.1 standard. IEEE Trans Instrum Meas 54:1527–1533
8. Petrecca G, Decarli M (2008) A review of hydrogen applications: technical and economic aspects. In: IEEE MELECON 2008, May, pp 658–662
9. Saponara S et al (2011) Sensor modeling, low-complexity fusion algorithms and mixed-signal IC prototyping for gas measures in low-emission vehicles. IEEE Trans Instrum Meas 60(2):372–384. doi: 10.1109/TIM.2010.2084230

10. Sun L, Liang R, Wang Q (2008) A serial hybrid bus with methanol-hydrogen engine. In: IEEE VPPC2008, September, pp 1–4
11. Navet N et al (2005) Trends in automotive communication systems. Proc IEEE 93(6):1204–1223
12. Figaro TGS6812 data sheet, rev 09/06, Figaro TGS821 data sheet, rev 10/04
13. Saponara S et al (2011) Modeling, sensitivity-analysis and prototyping of low-g acceleration acquisition systems for spacecraft testing and environmental-noise measurements. IEEE Trans Instrum Meas 60(2):385–397. doi: 10.1109/TIM.2010.2084231
14. Volpi E et al (2010) A mixed-signal embedded platform for automotive sensor conditioning. J Embedded Syst 2010:1–15
15. Sangiovanni-Vincentelli A, Martin G (2001) Platform-based design and software design methodology for embedded systems. IEEE Design Test Comput 18:23–33
16. Saponara S et al (2007) Architectural-level power optimization of microcontroller cores in embedded systems. IEEE Trans Ind Electron 54(1):680–683
17. Fanucci L, Saponara S, Morello A (2005) Power optimization of an 8051-compliant microcontroller. IEICE Trans Electron E88-C(4):597–600
18. Tanurhan Y (2006) Processors and FPGAs Quo Vadis? IEEE Comput 39:108–110
19. Saponara S, Fancci L, Tonarelli M, Petri E (2007) Radiation tolerant space wire router for satellite on-board networking. IEEE Trans Aerosp Eletron Syst Mag 22(5):3–12

Part II
Network on Chip and Multicore Systems

Chapter 6
Cost-Based Deflection Routing for Intelligent NoC Switches

Martin Radetzki and Adán Kohler

6.1 Introduction

The continuing reduction in feature sizes of digital VLSI circuits enables the integration of dozens, and in the future hundreds of processing elements (cores, resources) on a single chip. Traditional on-chip buses can no longer sustain the increasing demand for communication between these cores. In order to overcome the performance gap, interconnection networks-on-chip (NoC) with packet-switched multi-hop communication and high-bandwidth point-to-point links between switches are being researched.

The two-dimensional nature of current chip layout mandates the use of NoC topologies, e.g. the two-dimensional mesh or the Spidergon [1], that have a mapping to a planar physical implementation. While the Spidergon, based on a ternary ring topology with cross-section shortcuts, primarily targets heterogeneous multi-core chips, the quarternary 2d mesh (Fig. 6.1) is more suitable for homogeneous manycore systems. However, traffic hotspots tend to appear in the center of 2d mesh structures. To avoid or reduce such congestion, load balancing mechanisms have to be introduced, and network congestion has to be taken into account when making routing decisions.

In addition to congestion, NoC communication is also distorted by faults. Furber in [2] projects that in 2016, up to 20% of a chip's transistors will have production defects, and that further 10% may fail in the first year of the chip's lifetime. Beyond such permanent faults, the diminishing electrical charges used in storage and transmission will be highly susceptible to soft errors, e.g. from radiation impact. In order to avoid packet corruption and loss due to such faults,

M. Radetzki (✉) · A. Kohler
Institut für Technische Informatik, Pfaffenwaldring 47, Stuttgart 70569, Germany
e-mail: martin.radetzki@informatik.uni-stuttgart.de

M. Conti et al. (eds.), *Solutions on Embedded Systems*,
Lecture Notes in Electrical Engineering, 81, DOI: 10.1007/978-94-007-0638-5_6,

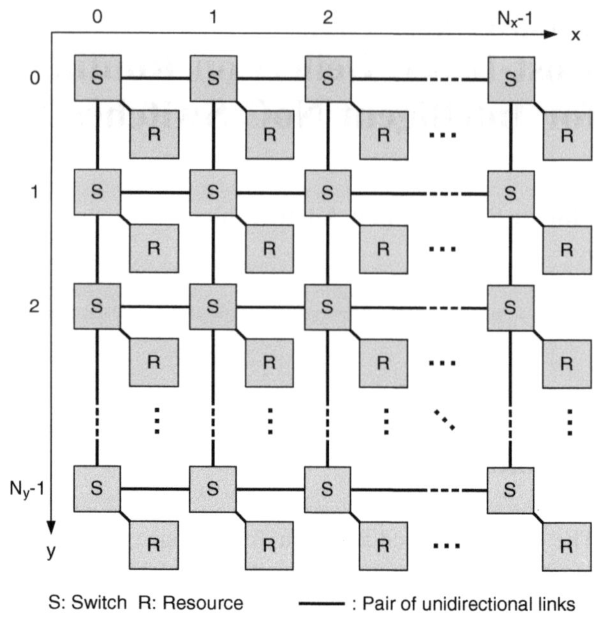

Fig. 6.1 Two-dimensional mesh topology

NoCs will have to perform self-diagnosis and route packets around faulty parts of the NoC infrastructure.

Both, load balancing and fault tolerance, are active research areas. In Sect. 6.2, we give an overview of some relevant related work in these fields. The remainder of this chapter is devoted to a novel approach which for the first time integrates load balancing and fault tolerance in a common framework based on an algorithm that minimizes routing cost at run-time. Section 6.3 outlines the fundamental concepts of cost based routing, and Sect. 6.4 provides details on the design of a cost function that takes basic routing functionality, faults, and congestion into account. In Sect. 6.5, we present an efficient implementation of cost-based routing. Experimental results on achieved communication performance are provided in Sect. 6.6, Sect. 6.7 concludes our contribution.

6.2 Related Work

Due to the many feasible combinations of topologies, routing algorithms, switching and flow control mechanisms, the NoC design space is large. Since the subject emerged (e.g. [3]), a multitude of NoC architectures have been developed. Of particular importance to our work is the Nostrum architecture [4], which includes deflection routing in a 2d mesh topology with 128 bit wide links.

The network is synchronously clocked, transmits a complete 128 bit packet over a link in a single cycle, and performs routing and switching in a combinational way. We adopt these principles as the basis for our work. The concept of deflection—redirecting a packet if its preferred direction is already occupied by another packet—can also be employed for load balancing and fault tolerance, but the latter has not yet been investigated by other authors. The performance of alternative deflection routing policies, without consideration of load or faults, has been investigated in depth in [5].

The field of network diagnosis is not covered in this chapter; however, diagnosis is an essential activity that yields the network fault status information which is an input to our routing method. In order to react to faults that emerge during system operation, diagnosis must be performed on-chip, and possibly concurrent with regular network operation (online). The on-chip diagnosis algorithm presented in [6] injects directed test packets at the network boundaries and locates faults based on lost packets. An online diagnosis method that checks error-detecting codes at every switch to gain information on faulty links and switches has been presented in [7]. The model of stuck-at port faults has been used [8] to further discriminate between different faults inside a switch. In [9], faults are pinpointed to individual connections within a crossbar switch, which yields the information required by our routing algorithm.

Fault-tolerant routing can either employ informational redundancy—sending replicated packets over different routes [10]—or exploit a network's structural redundancy to route around known faults. We concentrate on the latter in order to avoid the performance penalty of transmitting multiple packet copies. In source routed networks, with knowledge of the global network status, alternative routes can be devised by the packet source, which is often a programmable processor capable of executing complex algorithms. Fault-tolerant source routing methods have been investigated in [11]. Distributed fault-tolerant routing has the advantage of better scalability for large networks. Wu and Wang [12] identify convex regions enclosing faulty network components. Given such regions, deadlock-free distributed fault-adaptive routing can be implemented, e.g. with the odd-even turn model (e.g. [13]) or with cycle-free contours [14].

Load balancing in NoCs is usually based on distributing stress information off-band, via dedicated signals, in the proximity of a congested switch. Adaptive routers use this information to bypass congested areas, but need to take special care in order to avoid deadlock situations [15]. In the Nostrum deflection router, the load (stress) of a switch is measured by averaging the number of packets routed in a given amount of cycles with a sliding window method [16]. A switch is considered congested if its load is above 70% of the maximum load. Neighbor switches avoid routing packets to the congested switch, employing packet deflection instead. Deadlocks cannot appear since packets never wait, and livelocks are prevented by giving older packets a higher priority so that they can take their preferred direction instead of being deflected indefinitely.

6.3 Cost Based Routing

In his analysis of deflection routing, Lu [5] distinguishes between two phases. In the first phase, named *routing algorithm*, a switch computes the preferred route of each packet, i.e., the output to which it should be switched in the ideal case. This may lead to conflicts as multiple packets may require the same resources on their preferred route. To resolve these conflicts, packets are served in order of their *priority* in a second phase. If a lower priority packet would use an occupied resource on its preferred route, a deflection is computed according to a *deflection policy*.

For phase one, Lu [5] classifies routing algorithms as *random*, *dimension XY* (x direction preferred over y), and *delta XY* (routing minimizes distance to target without preferring x over y). For phase two, the following deflection policies are distinguished:

- *non-priority* with random deflection,
- *straight-through*, where straight-through packets have higher priority than packets that take a turn, and
- *weighted priority*, where priority is a weighted sum of packet age, distance from target, deflection count, and packet type.

The two-phase approach has a disadvantage: A route selected in phase one constrains the choices available in phase two, and may therefore lead to a suboptimal overall routing. Another routing algorithm, *minimum deflection*, combines both phases by selecting packet routes so that the total number of packet deflections is minimized. It has, however, not been implemented in NoC so far, possibly due to a suspected implementation overhead.

To specify minimum deflection routing, we formalize the routing done within a switch as a permutation, a bijective mapping

$$\pi : I \rightarrow O \quad \text{where } I = O = \{N, E, S, W\} \tag{6.1}$$

represent the input and output directions of a 2d mesh switch. The definition can easily be extended to include a local input and output, L. This is not needed in our approach since we immediately route all incoming packets destined to the local processing element, and since packets from L are routed only if an output is left after serving the packets that are already in the network.

The simplest case of minimum deflection routing can be specified with a routing cost function as follows: The cost of routing a packet from input $i \in I$ to output $j \in O$ is

$$c_{ij} = \begin{cases} 0 & \text{if } j \text{ is preferred output for packet from } i \\ 1 & \text{else} \end{cases} \tag{6.2}$$

according to a given routing algorithm and deflection policy. The resulting matrix is referred to as the *cost matrix*. We obtain a routing cost function by summing up the cost of a given routing permutation:

$$c(\pi) = \sum_{i \in I} c_{i\pi(i)} \tag{6.3}$$

The cost-optimal (deflection-minimal) routing is determined by solving the following optimization problem:

$$\pi_{\text{opt}} = \arg\min\{c(\pi) | \pi \in \text{Sym}(\{N, E, S, W\})\} \tag{6.4}$$

where Sym(*X*) represents the symmetric group of all permutations on the set *X*.

The solution can be computed by enumerating all permutations and evaluating their cost function or by employing specialized optimization algorithms such as the Hungarian method [17]. An efficient parallel hardware implementation is devised in Sect. 6.6.

6.4 Cost Function Design

In the following, we develop a cost function for fault-tolerant and load-balancing deflection routing step by step. We first show how a particular deflection routing algorithm and policy are encoded in the cost function. Second, we extend the cost function to take faults into account. Finally, we add a cost component to implement load balancing.

6.4.1 Deflection Routing

This section specifies a cost function that emulates *delta XY routing* with *weighted priority* deflection policy. It relies on a cost matrix with entries defined as

$$c_{ij} = \begin{cases} 2 & \text{if} \quad i = j \\ 0 & \text{if} \quad i \neq j \wedge d(i,j) \\ 1 & \text{else} \end{cases} \tag{6.5}$$

where the predicate $d(i, j)$ is *true* if and only if routing the packet from input i to output j brings it closer to its destination. Note that the value of 2 penalizes the reflection of a packet back to the direction from where it arrived, avoiding packet oscillations between adjacent switches.

Packet priorities are accounted for by a weight vector $w \in \{1, 2, 3, 4\}^{|I|}$ where the element w_i represents the priority of the packet arriving via input i and the value 4 stands for the highest priority.

In our experiments, priorities are assigned only on the basis of packet age, determined by a hop count field that is incremented in each cycle. More complex weighted priorities and a larger value range can be specified.

We further define an $I \times O$ *routing matrix*, R, with entries

$$r_{ij} = w_i c_{ij} \tag{6.6}$$

and modify the cost function (cf. Eq. 6.3) as follows:

$$c(\pi) = \sum_{i \in I} r_{i\pi(i)} \tag{6.7}$$

Solving the optimization problem (Eq. 6.4) now reduces packet reflection and the choice of non-minimal paths, giving priority to older packets. The resulting routing performance is analyzed in Sect. 6.6.2; it is also significantly better than minimal deflection (cf. Sect. 6.3), for which performance analysis is available in [5].

6.4.2 Fault-Tolerant Routing

To modify the cost function so that it takes faults into account, we assume knowledge of the fault status of a switch and its adjacent links, which can be obtained using the online diagnosis method detailed in [9]. Let F denote a *fault matrix* with entries

$$f_{ij} = \begin{cases} 0 & \text{if switching from } i \text{ to } j \text{ involves no faults} \\ 15 & \text{else} \end{cases} \tag{6.8}$$

The routing matrix, R, is modified as follows:

$$r_{ij} = \max\{w_i c_{ij}, f_{ij}\} \tag{6.9}$$

This results in a penalty of 15 for each packet that is routed via a faulty connection. The four bit value of 15 is larger than any product of cost and weight (8 at maximum), giving highest priority to avoiding packet corruption or loss due to faults. The headroom between 8 and 15 is used to express congestion cost in the next section.

While a single entry f_{ij} of the fault matrix represents the fault status of a crossbar connection from input i to output j, a faulty link can be represented by a row or column filled with the value 15: if the outgoing link j is unavailable, $f_{ij} = 15 \ \forall i \in I$. A faulty incoming link i is captured with $f_{ij} = 15 \ \forall j \in O$.

With these definitions, solving the optimization problem (Eq. 6.4) gives highest priority to avoiding faults, in addition to following the deflection routing algorithm and policy specified in Sect. 6.4.1.

6.4.3 Load Balancing

To perform load balancing, we count the number of packets, p_t, routed by a switch in the past t cycles, $t \in \{1, ..., n\}$. To account for the reduced number of available resources, an unavailable link (faulty or tied off at the mesh boundary) is counted as one packet. The stress level of the switch is defined as:

$$s = \frac{1}{4}\sum_{t=1}^{n} p_t \tag{6.10}$$

If $s = n$, all four network outputs (not counting the output to the local processing element) of the switch have been utilized, in the worst case, during the past n cycles, meaning full switch utilization.

Let s_j denote the stress levels of the switches neighboring a given switch in direction $j \in O$. To penalize the routing of a packet to a congested switch, we add a cost component to the routing matrix as follows:

$$r_{ij} = \max\{\max\{\min\{m + s_j - n, m\}, 0\} + w_i c_{ij}, f_{ij}\} \tag{6.11}$$

where m is the largest congestion penalty and the inner maximum and minimum operations limit the penalty to the range of [0, m].

In our experiments, we have obtained good results with a sliding window of size $n = 12$ and maximum penalty of $m = 4$, resulting in penalties shown in Table 6.1.

6.5 Efficient Implementation

When implementing cost-based routing, we must solve the optimization problem (Eq. 6.4) with acceptable overhead in terms of delay and area. As far as delay is concerned, Nostrum deflection routing requires routing decisions to be taken in a single clock cycle. This mandates a combinational implementation, evaluating all possible routings in parallel. This section explains the design of a corresponding datapath in which area efficiency is achieved by sharing of common subexpressions and early pruning of sub-optimal routings.

The fundamental datapath structure is derived from the tree shown in Fig. 6.2. All possible permutations of the directions {*N*, *E*, *S*, *W*}, i.e. all possible routings, are constructed from right to left. In level 3, all routing options for the *N* input are enumerated as mappings to one of the outputs {*N*, *E*, *S*, *W*}. In levels 2, 1, and 0, the *E*, *S*, and *W* inputs, respectively, are mapped to the remaining outputs.

The datapath (Fig. 6.3) sums up the costs of routings (i, j) from inputs i to outputs j, given by routing matrix entries r_{ij}. This is done starting from the leaves of the tree from Fig. 6.2 and progressing towards its root. In each stage of the circuit, corresponding to the levels of the tree, only the minimal cost value is

Table 6.1 Penalties for routing to congested neighbors

Stress level s_j	<9	9	10	11	>11
$\max\{\min\{m + s_j - n, m\}, 0\}$	0	1	2	3	4

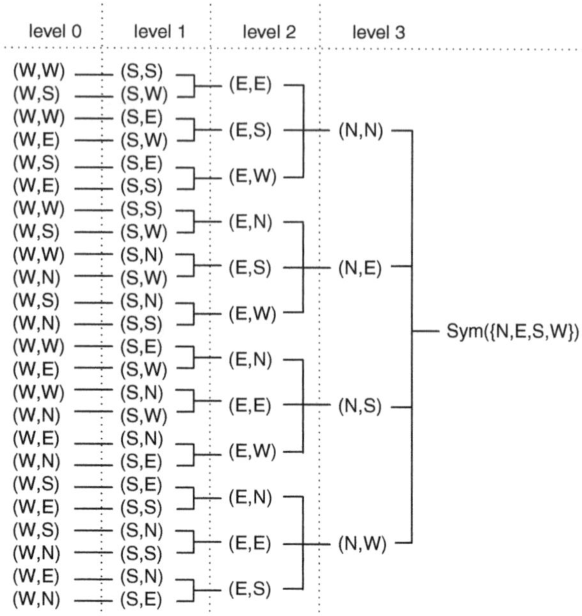

Fig. 6.2 All possible routings as permutations of {*N*, *E*, *S*, *W*}

selected by the min units, and all other branches are pruned. In addition, each min unit encodes with its sel output the (*i*, *j*) mapping that corresponds to the cost-minimal input. In stage 1 of the circuit, sel is defined as

$$sel_{ab}(s_{ab}, s_{ba}) = \begin{cases} \{(S,a),(W,b)\} & \text{if} \quad s_{ab} \leq s_{ba} \\ \{(S,b),(W,a)\} & \text{if} \quad s_{ab} > s_{ba} \end{cases}, \tag{6.12}$$

where the two alternatives can be encoded with 0 and 1, respectively. With each further stage, the number of alternatives increases by one. In stages 2 and 3, only the mapping represented by the cost-minimal branch is selected and forwarded to the next stage. This is done by multiplexers that have the additional functionality of prepending the (*i*, *j*) decision made in their own stage, given by the respective *sel* signal, to the selections received as inputs. Their functionality is specified in Eq. 6.13 (for mux_N in stage 2) and Eq. 6.14 (*mux* in stage 3). Like *sel*, the output of *mux* employs a binary encoding. The rightmost *mux* output encodes the cost-optimal routing permutation π_{opt}. It controls the crossbar switch of the NoC router.

$$mux_N(sel) = \begin{cases} sel_{SW} \cup \{(E,E)\} & \text{if} \quad sel_N = 0 \\ sel_{EW} \cup \{(E,S)\} & \text{if} \quad sel_N = 1 \\ sel_{ES} \cup \{(E,W)\} & \text{if} \quad sel_N = 2 \end{cases} \tag{6.13}$$

$$mux_\pi(sel) = \begin{cases} mux_N \cup \{(N,N)\} & \text{if} \quad sel_\pi = 0 \\ mux_E \cup \{(N,E)\} & \text{if} \quad sel_\pi = 1 \\ mux_S \cup \{(N,S)\} & \text{if} \quad sel_\pi = 2 \\ mux_W \cup \{(N,W)\} & \text{if} \quad sel_\pi = 3 \end{cases} \tag{6.14}$$

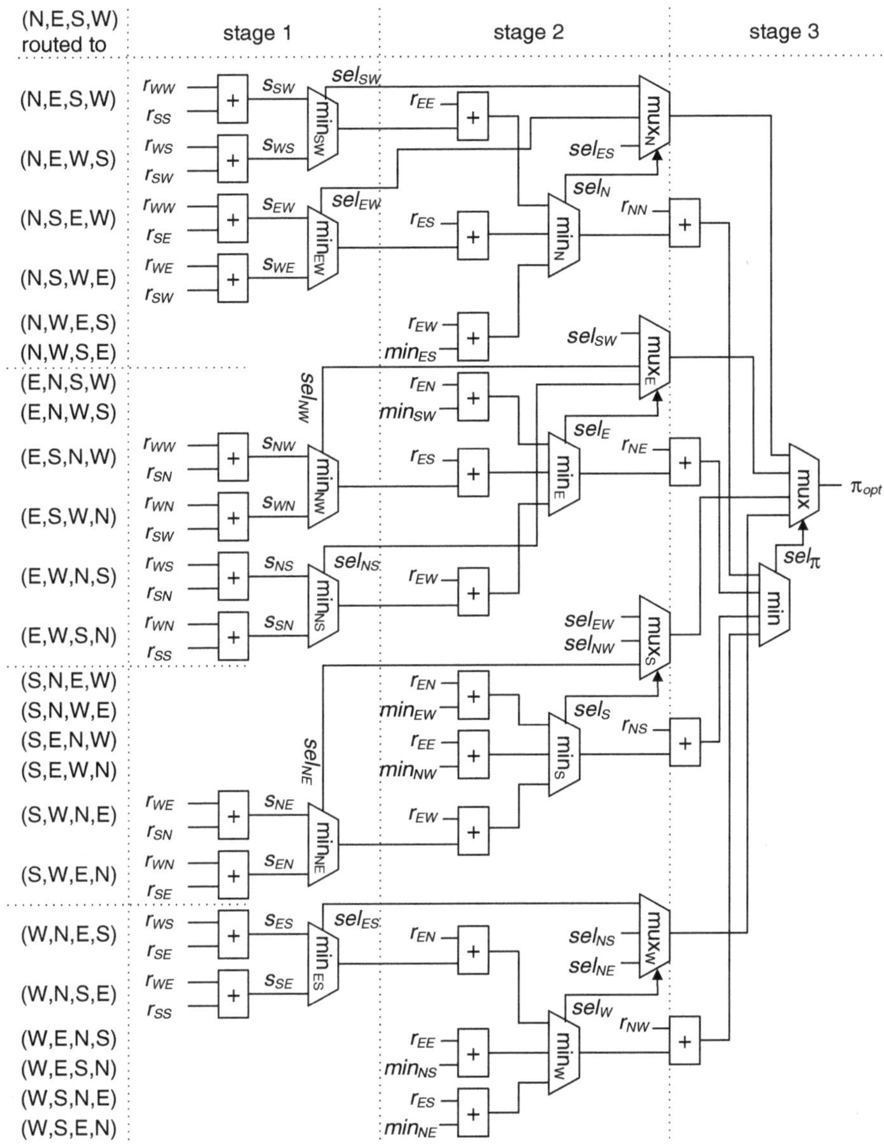

Fig. 6.3 Cost minimizing datapath

Table 6.2 compares area cost and performance of different variants of cost-based routing, corresponding to Sects. 6.4.1, 6.4.2, and 6.4.3 of this chapter, with the original implementation of a Nostrum router. All designs contain not only the routing logic, but also any error or congestion detection logic necessary for the full functionality, but not shown here. They have been designed and synthesized with

Table 6.2 Cost and timing performance

Design	Area (μm^2)	Equivalent gate count (NAND2)	Critical path (ns)	Maximum operating frequency (MHz)
Nostrum	13,570	21,695	9.36	106.8
Cost-based (Sect. 6.4.1)	12,320	19,739	10.67	93.7
Fault-tolerant (Sect. 6.4.2)	16,790	26,961	12.92	77.4
Retimed version	19,129	28,546	9.25	108.1
Pipelined version	21,580	34,738	5.45	183.5
Load balancing (Sect. 6.4.3)	17,299	27,776	13.13	76.2

RTL VHDL. Since r_{ij} inputs are restricted to 4 bits, arithmetic results can be represented with 5 bits in stage 1 and 6 bits in stages 2 and 3.

A cost-based implementation of deflection routing (Sect. 6.4.1) requires less area than the original Nostrum design, which employs costly sorting and selection operations on packets. The area overhead of the designs with fault tolerance (Sect. 6.4.2) and additional load balancing (Sect. 6.4.3) stems from the detection logic.

Performance-wise, our cost-based routing implementation suffers from increased logic depth, compared to the original Nostrum router. Retiming during synthesis and design of a two-stage pipelined router, as exemplified with the fault-tolerant design, are effective in avoiding a performance penalty.

6.6 Communication Performance

We have performed extensive simulation studies to validate the intelligent, cost-based routing approach. The most significant results are presented subsequently.

6.6.1 Experimental Setup

Our simulation experiments use a two-dimensional mesh topology (cf. Fig. 6.1) of size $N_x \times N_y = 8 \times 8$. Links are 128 bit wide and transmit a complete packet in one cycle. Switches implement deflection routing without flow control as each incoming packet is forwarded to an output in the same cycle. Input ports are registered for synchronous operation. Switches at the mesh boundaries are of the same type as all other ones, but their unused inputs and outputs are tied off. The only design parameter varied between simulations is the routing policy.

Resources in our simulation generate and receive network traffic. We have used synthetic traffic patterns, *uniform traffic* with pseudo-random destination addresses uniformly distributed over the address range, and *complement traffic* from resource positioned at (x, y) to destination $(N_x - 1 - x, N_y - 1 - y)$ for all $x \in \{0, ..., N_x - 1\}$, $y \in \{0, ..., N_y - 1\}$. Packets are generated with a configurable, constant rate, and are stored in an unbounded packet FIFO per resource. A packet is injected into the NoC when a packet FIFO is non-empty and the attached switch has an unused output after routing all other packets.

Simulation models have been implemented in SystemC [18] based on the Transaction Level Modelling (TLM) extension version 2.0 [19] and the object-oriented approach from [20]. While simulation allows switching between modes of different accuracy in an adaptive way [21], we employ a strictly cycle-accurate simulation here in order to obtain best possible accuracy. For each individual parameter set, 2,000 cycles have been simulated.

6.6.2 Latency and Throughput

We evaluate the performance impact of cost-based intelligent routing by comparing its throughput and hop count against a traditional router with weighted priority deflection routing. Simulation has been performed under varying load conditions, shown here for complement traffic, with and without load balancing. At rates below saturation (linear zone in Fig. 6.4, left), all routing methods provide similar throughput. The cost-based intelligent deflection router reaches a higher saturation throughput than the priority based two-stage deflection routing mechanism. Load balancing has small impact in case of intelligent routing, but reduces saturation throughput significantly in conjunction with two-stage deflection. This is because packets are unnecessarily deflected in a network that is fully congested anyway. Cost-based intelligent routing avoids this pitfall by not favoring deflection if stress is equally high in all directions.

Another advantage of the cost-based approach can be seen in Fig. 6.4 (right): it reduces the average hop count significantly, compared to the priority based

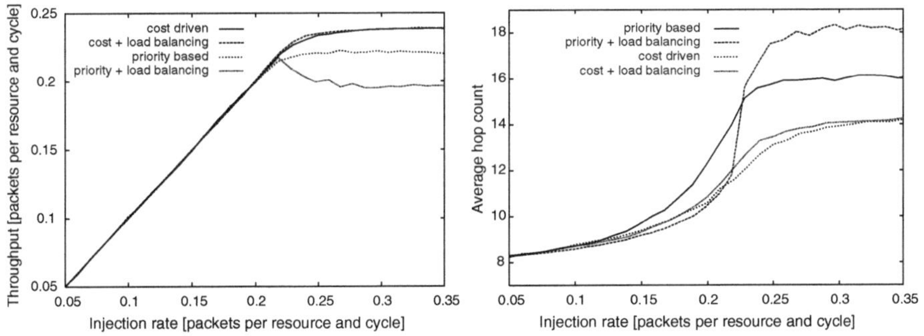

Fig. 6.4 Packet throughput (*left*) and average hop count (*right*)

variant. In the sub-saturation zone, adding load balancing to priority based routing also provides significant improvement, but it fails under saturation due to unnecessary deflections in a fully congested network. Adding load balancing to cost-based routing generally increases hop count slightly; however its advantage is in the reduction of FIFO backlog (cf. Sect. 6.6.4).

6.6.3 Fault-Tolerance

Here we measure NoC performance by means of its saturation throughput achievable in presence of faults and under uniform traffic. Faults have been simulated as permanent or transient with duration t = 1, 4, ..., 256. Figure 6.5 (left) shows maximum packet throughput over varying failure rate, under the assumption that a failure makes a switch fully unavailable. This is the best assumption that can be made (and has been made by previous work) if no intelligence about the internal switch status is available. It results in a significant negative performance impact already at small failure rates. With increasing fault duration, performance degrades more rapidly. In the case of permanent faults, even low failure rates can reduce throughput to near zero.

Figure 6.5 (right) shows achievable throughput using our concept of fault matrix. For bidirectional link faults, i.e. complete rows and columns indicating faults, performance degrades significantly less compared to the previous case. When assuming single crossbar connection faults, i.e. singular fault entries in the fault matrix, performance is reduced just slightly, even at the highest failure rates. Of course, this is because a single connection fault is much less severe than the complete breakdown of a switch. We argue that in practice, most faults would affect only part of a switch. In this case, using intelligence on the switch's internal status for making routing decisions vastly increases the performance of fault-tolerant routing mechanisms.

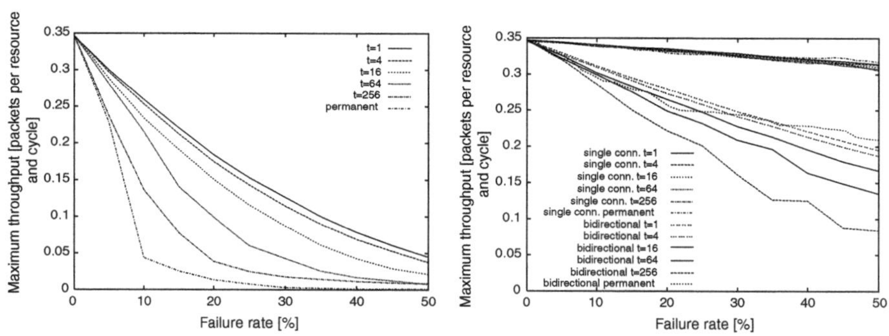

Fig. 6.5 Throughput under switch faults (*left*), link and crossbar faults (*right*)

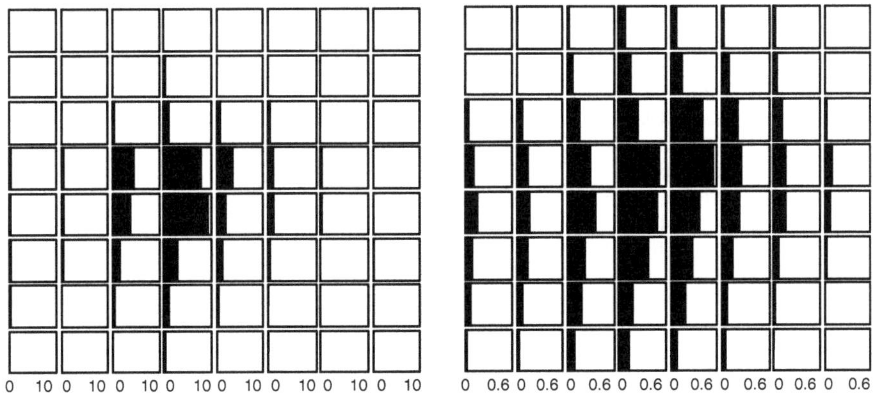

Fig. 6.6 Average packet FIFO filling level without (*left*) and with load balancing (*right*)

6.6.4 Benefits from Load Balancing

Figure 6.6 shows the average packet FIFO filling levels (backlog) at the different positions in the mesh, without (left) and with (right) load balancing. These filling levels have been obtained with complement traffic at a packet generation rate of 0.21, just slightly below network saturation.

Without load balancing, FIFOs in the center of the network are filled, on average over all simulated cycles, with up to almost 10 packets. Load balancing according to Sect. 6.4.3 significantly improves the situation at the given packet generation rate (right). Note the different scale: the largest FIFO has an average filling level of less than 0.6. Moreover, FIFO backlog is much better distributed over the network than without load balancing, where it is centered in the middle.

Above results have been obtained with the parameters $m = 4$ and $n = 12$ (cf. Sect. 6.4.3). Shorter window length, e.g. $n = 4$, and less differentiation through penalty steps, e.g. $m = 1$, both yield inferior results.

6.7 Conclusion

We have presented an intelligent NoC routing algorithm that uses information on the router's and its environment's status. This information is weighted and combined in a cost function which enables the computation of locally optimized routing permutations. The technique has been employed to combine, for the first time in NoC, fault-avoidance and congestion avoidance as criteria for selecting deflections. Models of the intelligent cost-based deflection router show performance improvements over previous deflection routing variants employed in NoC. A VHDL implementation shows that cost-based routing causes no area overhead.

References

1. Coppola M, Grammatikakis MD, Locatelli R, Maruccia G, Pieralisi L (2008) Design of cost-efficient interconnect processing units—Spidergon STNoC. CRC Press, Boca Raton
2. Furber S (2006) Living with failure: lessons from nature. In: Proceedings of the European test symposium (ETS), pp 1–4
3. Dally WJ, Towles B (2001) Route packets, not wires: on-chip interconnection networks. In: Proceedings of the design automation conference (DAC), pp 684–689
4. Penolazzi S, Jantsch A (2006) A high level power model for the Nostrum NoC. In: Proceedings of the Euromicro conference on digital system design (DSD), pp 673–676
5. Lu Z, Zhong M, Jantsch A (2006) Evaluation of on-chip networks using deflection routing. In: Proceedings of the great lakes symposium on VLSI (GLSVLSI), pp 296–301
6. Raik J, Ubar R, Govind V (2007) Test configurations for diagnosing faulty links in NoC switches. In: Proceedings of the European test symposium (ETS), pp 29–34
7. Grecu C, Ivanov A, Saleh R, Sogomonyan ES, Pande PP (2006) On-line fault detection and location for NoC interconnects. In: Proceedings of the international on-line testing symposium (IOLTS), pp 145–150
8. Alaghi A, Karimi N, Sedghi M, Navabi Z (2007) Online NoC switch fault detection and diagnosis using a high level fault model. In: Proceedings of the international symposium on defect and fault-tolerance in VLSI systems (DFT), pp 21–29
9. Kohler A, Radetzki M (2009) Fault-tolerant architecture and deflection routing for degradable NoC switches. In: Proceedings of the 3rd ACM/IEEE international symposium on networks-on-chip (NOCS), pp 22–31
10. Bogdan P, Dumitras T, Marculescu R (2007) Stochastic communication: a new paradigm for fault-tolerant networks-on-chip. Hindawi VLSI design, p 17
11. Mediratta SD, Draper J (2007) Performance evaluation of probe-send fault-tolerant network-on-chip router. In: Proceedings of the conference on application-specific systems, architectures and processors (ASAP), pp 69–75
12. Wu J, Wang D (2002) Fault-tolerant and deadlock-free routing in 2-d meshes using rectilinear-monotone polygonal fault blocks. In: Proceedings of the international conference on parallel processing, pp 247–254
13. Hu J, Marculescu R (2004) Dyad—smart routing for networks-on-chip. In: Proceedings of the design automation conference (DAC), pp 260–263
14. Zhang Z, Greiner A, Taktak S (2008) A reconfigurable routing algorithm for a fault-tolerant 2d-mesh network-on-chip. In: Proceedings of the design automation conference (DAC), pp 441–446
15. Li M, Zeng Q-A, Jone W-B (2006) DyXY: a proximity congestion-aware deadlock-free dynamic routing method for network on chip. In: Proceedings of the design automation conference (DAC), pp 849–852
16. Nilsson E, Millberg M, Öberg J, Jantsch A (2003) Load distribution with the proximity congestion awareness in a network on chip. In: Proceedings of the design, automation and test in Europe (DATE), pp 1126–1127
17. Kuhn HW (1955) The Hungarian method for the assignment problem. Nav Res Logist Quart 2:83–97
18. IEEE Standard 1666 (2005) SystemC 2.1 language reference manual. IEEE Standards Association, Piscataway
19. Open SystemC Initiative (2008) OSCI TLM-2.0 user manual. Software version TLM-2.0. Document version JA22, http://www.systemc.org
20. Radetzki M (2006) SystemC TLM transaction modelling and dispatch for active objects. In: Proceedings of the forum on design languages (FDL), pp 203–209
21. Radetzki M, Salimi Khaligh R (2008) Accuracy-adaptive simulation of transaction level models. In: Proceedings of the design automation and test in Europe (DATE), pp 788–791

Chapter 7
NOCEXplore

A SystemC Platform for NoC Analysis

Stefano Gigli and Massimo Conti

7.1 Introduction

Thanks to continuous advances in semiconductor technology, a modern System-on-Chip (SoC) can contain a great number of modules implementing different functionalities; more complex systems need more complex and careful design, verification and testing, but aggressive time-to-market requirements need faster resolution of these phases.

The re-use methodology speeds up the three design phases by re-using parts of projects previously designed, verified and tested. The reusable modules must be easily and efficiently tunable, in order to get the best performances in terms of computation, communication, cost, power consumption and reuse degree.

Some issues arise on communication architectures when the number of Intellectual Properties (IP) sending and receiving information increases: bandwidth requirements increases because the number of modules increases; additional services associated to the communication protocol can be needed, because transactions of the more complex system have different priorities; the clock domain partitioning too could be a very critical design.

The more traditional communication architecture, the bus, has an intrinsic limit on bandwidth, so the Network-on-Chip (NoC) paradigm [1, 2] tries to overcome this limit. NoC is the communication architecture that comes from computer and processor networks. It is composed by three types of modules: routers, links and interfaces. The IPs do not communicate between them directly: the messages are sent from an IP source to a router and forwarded by this to other routers until they will arrive to the router connected to the destination IP. Routers are connected each other by links forming a net of chosen topology, size and connection degree.

S. Gigli · M. Conti (✉)
DIBET, Universita Politecnica delle Marche, Ancona, Italy
e-mail: m.conti@univpm.it

M. Conti et al. (eds.), *Solutions on Embedded Systems*,
Lecture Notes in Electrical Engineering, 81, DOI: 10.1007/978-94-007-0638-5_7,

NoC architecture has many degrees of freedom. Topology for regular network can be chosen between a wide variety: the most common ones are two dimensional mesh and torus, but examples of other topologies can be hypercubes, spidergon [3], hexagonal [4, 5], binary tree and variants [6–8], butterfly and Benes networks [9]. Topologies affect performances like cost (router and link number), maximum and average distance between nodes and fault-tolerance through alternative path. A network can use circuit switching and/or packet switching techniques and can support different *Quality-of-Services* (QoS) [10].

The links are characterized by the communication protocol (synchronization between sender and receiver), width (number of bit per transmission), presence or absence of error detecting/correction scheme and voltage scaling [11, 12]. In general NoC links are unidirectional.

The internal structure of the router has the heaviest impact on network performance. The router has input ports and output ports where messages enter into and go out from router; each *flit* (FLow control digIT), that represents the information quantum circulating in the network, is stored in internal *buffers* that can be close to input ports and/or output ports; the *routing* module indicates to the *switch* module how flits advance from input stage to output stage; contentions are resolved by specific *arbitering* rules and the *DPM* (Dynamic Power Management) module implements power saving policies by slowing down or speeding up or turning off the whole router or some part of it; the *flow control* indicates how the mentioned router resources are coordinated.

Buffer dimensions, structure (shift register or inserting register [13]) and parallelism degree are parameters that can be chosen. The *switch* structure could be complete or incomplete crossbar between input and output port and/or it can have some additional ports for delayed contention resolution [14, 15]. The *routing* algorithm and *DPM* policy should be implemented in the cheapest and efficient way as possible. Most common *flow control* techniques used in NoCs are virtual channel [16] (VC), Virtual Cut-Through [17] (VCT), Wormhole [18] (WH), and flit-reservation flow control [19].

Compared to bus, NoC has the following advantages:

- bandwidth increases because message transactions take place at the same time in different part of the network;
- arbitering is distributed and it requires simpler and faster hardware;
- regular topologies make NoC scalable and the use of the same components (routers and links) allows a high reuse degree;
- NoC, using GALS (Globally Asynchronous Locally Synchronous) synchronization paradigm, allows communication between module with different clock domains and of the clock skew phenomenon reduction is avoided;
- the network, being a distributed architecture, can be more robust to faults because messages forwarding can be redirected in areas not damaged and/or not congested;
- NoC can adjust power consumption depending on the current communication requirements.

On the other hand, NoC design is a more complex than bus design. Related to bus, new problems and trade-offs arise:

- routing algorithms should verify deadlock and livelock conditions [20–23];
- more complex and fault-tolerant routing schemes improve performances and reliability, but they need more complex, more expensive and slower routers;
- more complex and more efficient power management schemes need additional circuitry;
- routers and interfaces must implement appropriate arbitration schemes and must have suitable hardware in order to manage different QoSs.

Network-on-Chip configuration parameters must be carefully tuned to improve communications, cost and power performances. Early stage tools capable to fast explore solutions space, pruning non-optimal solutions about network and router architecture, will help designers to obtain better time-to-market values.

7.2 The Platform

NOCEXplore (http://sourceforge.net/projects/nocexplorer/) [24] is a SystemC library for NoC performance comparisons and investigations. It has two main aims: to provide a platform for comparison of NoCs chosen in a high and upgradeable design space and to provide to designer a high level of detail of events that happen in the network during simulation. Nevertheless, the CPU time required for simulation and data post-processing are acceptable.

In the NOCEXplore platform models of the networks are configurable by a set of 19 parameters that define the possible configurations of the networks. The traffic description involves three additional parameters. Globally, the configurations space has the 22 dimensions listed below, so the exploration of the solutions is done over a 22 dimensional space. Each dimension could have a physical value, a numeric value or a identification.

1. The network *quality of service* is an identifier and describes global network services and main router architecture.
2. The *network size* is a numeric value and indicates how many modules are connected to the network.
3. *Topology* is a number that identifies how routers are connected by links.

4–7. *Link type*, *link width*, *link delay* and *number of physical link per topological arc* are four parameters that describe the link; the first one identifies link protocol and communication scheme, the second one is the dimension of the flit circulating in the network, the third one represents inter-router flit latency and the last one allows to use more than one link between a couple of routers for a topological arc.

8–9. *flit_per_packet* and *packet_per_message* take into account how many flits correspond to each packet and, in bursty communications, how many packets are in a message. The flits of the same packet move in the same path; different packets, even if belonging to the same message, can be routed in different ways. The distinction between packet and message is necessary when a huge amount of data must be transferred (bigger than the maximum value that the network protocol can provide) from an IP to another in an efficient way.

10. Each router, seen as a synchronous machine, has a local clock generator of a certain *frequency*; each generator has its own starting delay independent from the others.
11. *Routing algorithm* implemented in each router is modeled by three functions:

- the "routing function" is responsible of avoiding deadlock and livelock and it can use topological information to get the distance between current node and destination node crossing the particular output port.
- the "selection function" selects one of the admitted output ports carried out by previous function: the choice can be taken considering the actual status of the router congestion, the actual congestion of the neighbour routers and/or the overall or partial network status based on the elaboration of control messages produced and consumed by routers. The selection is responsible of the degree of adaptability of the algorithm.
- the "header function" modifies, if necessary, some field of the header flit of the packet.

12. *Arbitering scheme* solves contentions where resources are shared in the router. The points of contention are two: different input buffers request the same output buffer; and different output buffers request the same output port or output channel. The arbitering scheme can be based on some user defined fields in the header flit (transaction identifier, priority related to arrival time or packet living time).
13. The *switch structure* refers to the architecture of the component that performs the flit crossover from input stage of the router to the output stage; it affects the number of flit crossovers that can be performed in a router in a single clock cycle.
14. The *DPM policy* is a set of rules determining the router power state for getting the best trade-off between communication performances and energy consumption. Workload conditions are estimated by measuring buffer occupancy and flit rate. The techniques considered are DVS (Dynamic Voltage Scaling) and DFS (Dynamic Frequency Scaling). The designer must provide the power models, the power states and status changing rules.
15. The *flow control* coordinates router internal resources.

16–17. *Input port buffer length* and *input port buffer number* describe the number and the depth of the buffers in input stage: several buffers can be placed in parallel and flit insertion in buffer can take place via insertion or shifting.

18–19. *Output port buffer length* and *output port buffer number* describe the number and the depth of the buffers in output stage.

The traffic is described by three parameters:

20. The *traffic intensity* is the mean value of flit injection operated by the IPs; this value is normalized to the maximum value of one flit per clock cycle, so the traffic offered in the entire network is given by the product of the following three terms: traffic intensity, number of IPs connected to the network and link width.
21. The *traffic scenario* describes the spatial distribution of message flows: the tool offers a small set of traffic primitives where designers can define any mixture message flows given by the CTG (Communication Task Graph) of the applications under exam.
22. The *burstyness* is a percentage value of bursty traffic over total traffic emitted by each source node.

The set of the 22 parameter values is defined as *network configuration*, the nodes attached to the network are traffic generators and they are source and sink at the same time.

The platform has been created for being easily upgradeable: adding a new numeric or physical value, for example a new buffer length, simply needs to insert the new value in the list of this parameter; adding a new behavior, for example a new topology, designers must create a new topology class derived by the topology base class and they must overload one or more virtual methods that describe the topology.

7.3 Investigations

The information extracted by the platform from the simulations of the network stimulated under a defined traffic scenario can be saved in files and/or can be submitted to the postprocessing phase. Four types of analysis can be carried out: global statistical analysis, global and detailed probabilistic analysis, dynamic analysis, power analysis.

7.3.1 Statistical Analysis

The statistical analysis allows a comparison between the performances of different network configurations under identical or different scenario. We can distinguish three types of global performances: the communication performances are based on message delays: minimum, maximum, standard deviation and mean delay of all steady state messages. Moreover throughput is also evaluated. A second nature of overall performances regards collections of events like routing calls, number of routed flits, commutations on links (seen as a parallel bus of wire of length equal to flit length) and flit shift when the shift register structure is adopted in buffers; these events are collected during the simulation of the network.

The third nature of global performances is related to power consumption: based on user defined power models of router and link, an overall estimation of energy consumption is provided taking into account all routers and all links together. In this way, for example, the designer can verify the goodness of dynamic power management policy on routers and links.

As an example some simulation results are reported in the following. Figure 7.1 compares message mean latency of the same network under different routing algorithms. The network consists of 16 nodes in a 4 × 4 mesh topology. The traffic scenario contains four hotspots, where the flows overlap themselves on three links; the other nodes inject uniform traffic of intensity five times lower than the others involved in the hotspots. Two routing algorithms are used: the first one is deterministic and it performs the dimensional ordering routing; the second one is partially adaptive and it performs the west-first algorithm. The graph indicates the mean delay of all the packets received from and sent to all nodes in steady state condition versus the traffic intensity. The overlapping flows yield the saturation threshold to half link bandwidth and network saturates at the intensity of 38% of

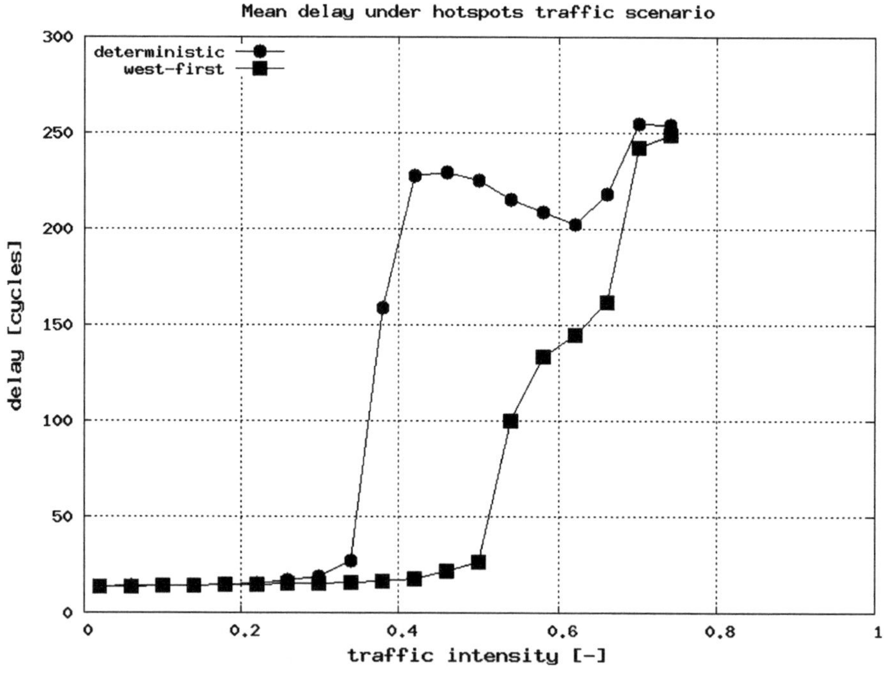

Fig. 7.1 Example of statistical analysis of the communication performances of a 4 × 4 mesh network subject to a overlapped hotspot and uniform traffic scenario under two different routing algorithms: the adaptive one perform an increment of the saturation threshold of 50% about, bringing it from 34 to 50% of traffic intensity

the maximum in the case of deterministic routing; the adaptive algorithm saturates at about 50% of the maximum intensity and with a slower slope.

7.3.2 Probabilistic Analysis

The second type of analysis is detailed and local. A probabilistic analysis of message delays highlights the traffic flows that do not match task constraints. A distribution of message delays is performed taking into account:

- all messages circulating in the network;
- all messages emitted by a specified source node;
- all messages consumed by a specified sink node;
- all messages emitted and consumed by a specified couple of source/sink nodes.

Moreover, the four main statistical indexes (mean value, standard deviation, skewness and kurtosis) of each distribution are calculated. This kind of analysis evidences message flows that satisfy specific communication performance constraints at a specific traffic intensity.

The example of Fig. 7.2 shows the probability density function of the delay of the network with adaptive routing at the begin of its saturation, at traffic intensity of 54%. Despite the more important peak occurs at low delays, the not negligible density peaks at 40, 80, 200, 250, 320 and 500 clock cycles yield the mean value of latency to 125 clock cycles. However, the majority of the packets are delivered in reasonable delays.

7.3.3 Dynamic Analysis

The third type of analysis investigates the temporal evolution of some "events" that take place during the simulation. If the previous analysis indicates which flow does not match constraints at the terminal part of the network, designer could discover which part of the network are involved in the implied flows. These "events" are:

- utilization level of buffers of each router;
- moving average of switch transversal flit;
- moving average of link writings;
- moving average of routing activities.

Moreover, a statistical analysis of these occurrences is provided.

This kind of investigation allows the designer to discover which resources could be oversized or undersized and which traffic conditions influence some temporal congestion and for how long the congestion status is maintained. Each elaboration performs overall, steady state and transient plots of utilization of the buffers.

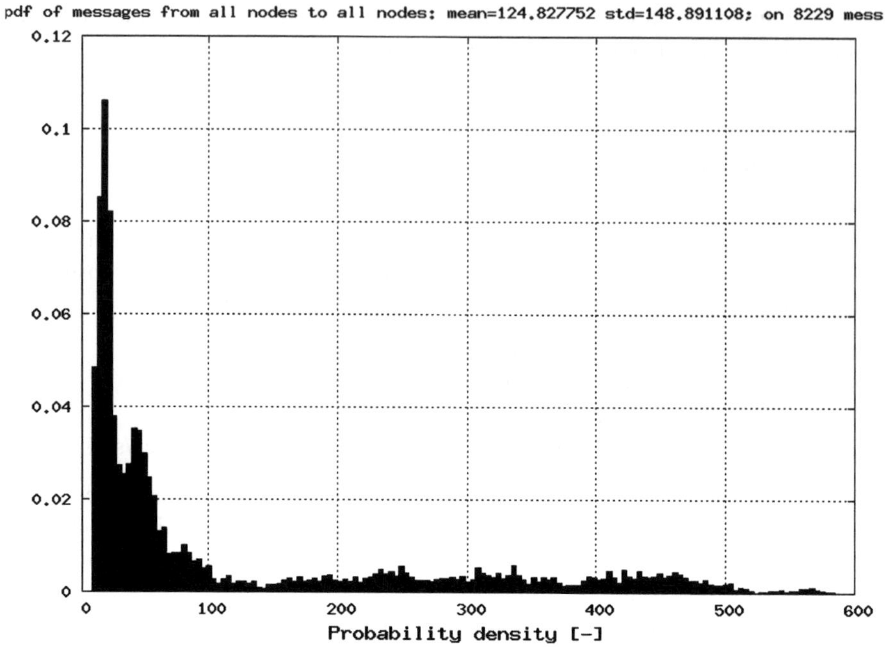

Fig. 7.2 Example of probabilistic analysis of the communication performances of a 4 × 4 mesh network subject to a overlapped hotspot and uniform traffic scenario under west-first partially adaptive routing algorithm

7.3.4 Power Analysis

The power analysis of Network-on-Chip involves energy estimation and dynamic power management and it can be performed in two possible ways depending on the detail required and energetic models accuracy. The coarse analysis considers routers having a power dynamic management module and a power consumption is associated to each power state of the router.

The total energy dissipated by the network is obtained by summing up each energy consumed in each state by each router plus the energy necessary for state changes.

In the parameter 14 described in previous section, the *DPM policy*, the designer has to define both power models and the power state machine implementing the DPM policy.

Figure 7.3 shows a graph where the power state of each router is reported during time: time and router identification number is reported in the x axis and y axis, respectively. The colour indicates the state and colour-bar on the right side indicates the legend with colours and numbers corresponding to the power states.

This analysis highlights repercussions of some router energy and performance states on the neighbour routers and evaluates the goodness of DPM policy

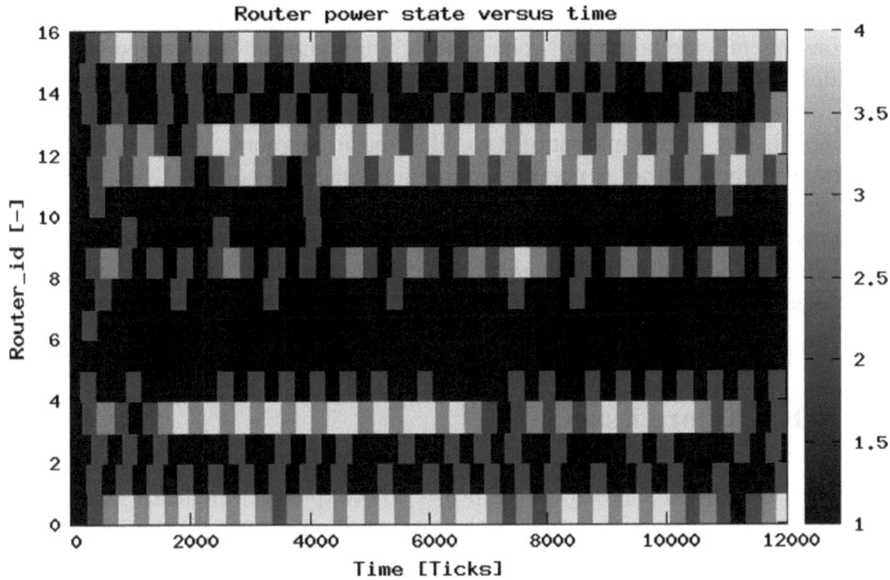

Fig. 7.3 Example of power analysis where energetic states of the router are plotted over the time. Darker colours means more performing and power consuming state while lighter colours less performing and more power saving states. Router power state machine has nine power states and it follows ACPI standard (http://www.acpi.info)

performed on the triple topology/routing/scenario; policy goodness is measured in effective power saving, in sensitiveness to local and temporal congestion and stability of power state.

The second way of evaluation of Network-on-Chip power consumption, as mentioned above, can be done by collecting "activities" related to energy dissipation. These activities are: commutations inside the link based on data value, incoming to and outgoing from router of a flit, routing function calls and flit crossings in the switch. These measures, mixed with technological constants and power models provided by designer, give information about power consumption.

Accuracy of communication and power performances depends on the accuracy of models. Simple models let the designer to get performance trend of architectural choice and more detailed model will improve performance accuracy. At the moment router model needs three clock cycles for a flit crossing without contention and routing decision is done in one clock cycle; inter-router link latency can vary from one to ten nanoseconds depending of the circuitry complexity modeled.

Some words about simulation time. A performance profiling of the platform is planned and the developers are aware that some performance improvement can be done. Simulation time strongly depends on network size and traffic intensity; SystemC simulation time strongly depends by the number of simulation kernel context switchings.

The increment of network size increases the number of modules instantiated and consequently the number of context switchings increases causing performance penalties. Moreover, we have registered a not negligible simulation time increment when traffic intensity increases and when networks start to be congested; this behaviour is due to the amount of data allocated in memory.

Actually, about 2 min is needed on a commercial notebook (64 bit—1.5 GHz CPU with 4 GB of RAM) for simulating and postprocessing a 16 nodes and 16 routers NoC in worst case condition. We consider this computation performance quite good because simulations are cycle accurate and user can access to lots of event details for more detailed investigations.

7.4 NOCEXplore Usage

The platform (http://sourceforge.net/projects/nocexplorer/) includes executable code, a library and some shell scripts to automate analysis by using few commands.

The main executable code allows to perform a set of simulations in a unique shell command specifying the network architecture and the "actions" to be accomplish. Structured and simple text files, called session files, indicates the models of NoC to be analysed. The "actions" that must be done for each simulation are the following:

- create folders for the simulation;
- create the configuration file of the network architecture: it contains the values of the 22 parameters and it completely defines the model;
- run the simulation: some intermediate data can be logged in files for further and manual investigations; this data can be saved in text or binary access;
- run the elaboration: depending on how many options are present in the command line, designer can set the level of detail of the analysis; the first level of detail, only the statistical analysis will be performed; the second level adds probabilistic analysis, while the third level performs a full analysis. Increasing the types of analysis performed, disk usage and elaboration time need increase in a not negligible way;
- group in a text file all the statistical performances;
- group the set of simulation folders in other projects folder according to the arrangement of analysis.

Due to the great amount of data written in a simulation that perform full analysis, we give the following suggestions:

- define carefully the configuration space of exploration and define the sessions of simulation;
- perform simulations of the networks with low level of detail of analysis and without saving intermediate data. Successively, compare overall performances;

- depending on the result of the previous step, create a set of other session files in a second phase of analysis according to the subspace of exploration that should be investigated in deeper detail; if necessary, enable the saving of temporary data in files;
- run the simulation of this analysis;
- repeat iteratively the previous two steps until the investigation is completed.

NOCEXplore provides some diagnosis tools that allow the traceability of all packets and messages and enable the designer to discover potential deadlock.

7.5 Post-processing

In the following the post-processing data generated by the tool will be described.

Intermediate files that can be generated in each simulation are **.log* files and they collect all the events that occur during the simulation.

Since every node connected to the network is source and sink of messages, *ip_*_source.log* and *ip_*_sink.log* files collect events of all the messages generated and consumed by the nodes, respectively. Data are structured in records of one line each; the fields of the record collect the identifier of the message, the time instant of its production or consumption and its destination and length in terms of number of packets. The filename identifies the node.

The write operation on links, operated by the routers, are logged in *link_*.log* file. Each record contains the flit identifier, the time instant of the write event and a value of the energy consumption of the transmission of that flit on the link. The filename identifies the link.

The files named *router_*_stax.log* collects events of arrival and departure of flits in the specified router; the fields of the records are the flit identifier, the time instant of the event, the destination node of the flit, a flag indicating the direction of the event.

The level of utilization of the buffer of each port of the router is stored in the files *router_*_buffer_port.log*. Each record contains the buffer identifier, the time instant of the event and the level of occupancy of the buffer normalized to its capacity.

The *router_*_routing.log* files collect information about routed packet, namely which packets have been routed toward which port of the router in a certain instant. The files *router_*_demux_port*.log*, *router_*_switch.log* and *router_*_mux_port*.log* have the traces of the contentions of the resources in the routers between input port and input buffer, at crossbar and between output buffer and output port, respectively. They store the identifier of the packet that must advance, the resource contended, the result of the contention and a time stamp.

The data stored in the *.log* files are elaborated in the post-processing phase, allowing an analysis of the NoC performances through synthetic parameters.

If only the statistical analysis is performed, only the file *result.net* is generated, which holds global indexes on communication, energetic and activities performances.

The file contains the mean, maximum, minimum and standard deviation of the delays and the throughput of the messages delivered at steady state condition on the NoC.

Based on power consumed by routers in each state, an estimation of power consumption is done and it is normalized to the maximum power consumption if no power management policy is adopted.

Some activities are collected during simulation: the number of packet routings, the number of flit that have crossed a router, the number of shift in buffer if the structure is considered as shift register and an estimate of the total number of commutation in the links.

The probabilistic analysis elaborates the communication performances, that is latency and throughput, for each couple of nodes connected to the network; the analysis generates two groups of file: the former performs some statical indexes for each flow of traffic and the latter shows the distribution of the delay of the traffic flow.

The six files with name *matrix_*_data.elab* are tables and the element on the line "i" and column "j" refers to the couple of nodes identified as "ith" node source and "jth" sink node. The data stored are: mean and standard deviation of latency, throughput, number of messages, packets and flit delivered. Each of these table has an associated colour-plot for an immediate and visual trend of communication performances for each flow of messages.

Data files with the name *prob_source*_sink*.elab* contain the distribution of the latencies of the messages composing the traffic flow considered at steady state condition. Each of the 200 records defines the probability "p" of message having delay "d". Each of these function has an associated plot.

The most detailed analysis generates many postprocessing files containing data able to describe the evolution of some parameter during the simulation. Each one of these data file has an associated plot.

The files *util_router*_inport*.elab* and *util_router*_outport*.elab* contain the temporal evolution of the level of occupancy of flit in a specified router in a specified input port or output port, respectively. The files with name *util_router*_allport.elab* contain the same parameter, but data are referred to flit in all the buffers in the router.

The other parameters monitored refer to the activities related to statistical analysis. The dynamics of the events of the crossings operated by the switch, the routings decided by the routing module and the writings in link are elaborated with a moving average having windows 10 clock cycles wide. The file containing these data are: *switch_ma_*.elab, routing_ma_*.elab* and *link_ma_*.elab,* respectively; the substring "*_ma_*" in the filenames stands for "moving average".

These data are saved in text and graphics mode and in three different time intervals: transient window comprises the first 800 clock cycles; the steady state interval is 800 clock cycles wide and it is located in the middle of the simulation;

the last windows includes all events occurring from the begin to the end of the simulation.

Further elaborations of simulation data stored in the *.log* files can be easily added.

7.6 Conclusions

NOCEXplore, the Network-on-Chip simulator presented in this chapter, allows the designer to compare the communication and power performances of many Network-on-Chip configurations in different traffic conditions. Furthermore, it allows the investigation on possible bottlenecks. Configuration space is easily upgradeable and postprocessing is customizable. A huge set of network topologies with deep level of investigation and comparison is possible and, at the same time, simulations are performed in a reasonable CPU time. Existing tools do not perform at the same time cycle level analysis and huge design space exploration.

On the other hand, power estimation is technology independent and power models must be entirely provided by users.

This platform will be improved by adding following features:

- new topologies, routing algorithms and traffic scenario will be added;
- accuracy of the estimation of power and performance will be given;
- network hierarchical topologies and mixed bus-NoC architectures will be considered;
- new TLM modules will speed-up simulations time;
- a database will be developed for easy and fast managing and performance consulting of previously simulated networks.

References

1. de Micheli G, Benini L (2002) Networks on chip: a new paradigm for systems on chip design. In: DATE '02: Proceedings of the conference on design, automation and test in Europe. IEEE Computer Society, Washington, p 418
2. Kornaros G (2010) Multi-core embedded systems. Taylor & Francis, Boca Raton
3. Bononi L, Concer N (2006) Simulation and analysis of network on chip architectures: ring, spidergon and 2D mesh. In: Proceedings of design, automation and test in Europe (DATE), March
4. Dolter JW, Ramanathan P, Shin KG (1991) Performance analysis of virtual cut-through switching in HARTS: a hexagonal mesh multicomputer. IEEE Trans Multicomput 40(6):669–680
5. Zhao Y-J, Yue Z-H, Wu JP (2008) Research on next-generation scalable routers implemented with H-torus topology. J Comput Sci Technol 23(4):684
6. Guerrier P, Greiner A (2000) A generic architecture for on-chip packet switched interconnections. In: Proceedings of DATE. ACM Press, pp 250–256

7. Kariniemi H, Nurmi J (2003) New adaptive routing algorithm for extended generalized fat trees on-chip. In: Proceedings of the international symposium on system-on-chip, Tampere, Finland, pp 113–188
8. Ohring SR, Ibel M, Das SK, Kumar M (1995) On generalized fat trees. In: Proceedings on 9th international parallel processing symposium
9. Moussa H, Muller O, Baghdadi A, Jezequel M (2007) Butterfly and benes-based on-chip communication networks for multiprocessor turbo decoding. In: Design automation and test in Europe conference
10. Bolotin E, Cidon I, Ginosar R, Kolodny A (2004) QNoC: QoS architecture and design process for network on chip. J Syst Arch Spec Issue Netw Chip 50:105–128
11. Soteriou V, Peh L-S (2004) Design-space exploration for power-aware on/off interconnection networks. In: Proceedings of the 22nd international conference on computer design (ICCD)
12. Shang L, Peh L-S, Jha NK (2002) Power-efficient interconnection networks: dynamic voltage scaling with links. Comput Arch Lett 1(2):1–4
13. Bhat S (2005) Energy models for network on chip components. Ph.D. dissertation, Technische universiteit Eindhoven
14. Laffely A, Liang J, Jain P, Weng N, Burleson W, Tessier R (2001) Adaptive system on a chip (aSoC) for lowpower signal processing. In: Thirty-fifth asilomar conference on signals, systems, and computers, November
15. Liang J, Swaminathan S, Tessier R (2000) aSOC: a scalable, single-chip communications architecture. In: IEEE international conference on parallel architectures and compilation techniques, October, pp 524–529
16. Dally WJ (1990) Virtual-channel flow control. In: Proceedings of the 17th annual international symposium on computer architecture (ISCA), Seattle, Washington, May, pp 60–68
17. Kermani P, Kleinrock L (1979) Virtual cut-through: a new computer communication switching technique. Comput Netw 3:267–286
18. Dally WJ, Seitz CL (1986) The torus routing chip. J Parallel Distrib Comput 1(3):187–196
19. Peh L-S, Dally WJ (2000) Flit-reservation flow control. In: Proceedings of the 6th international symposium on high-performance computer architecture (HPCA), January, pp 73–84
20. Dally W, Seitz C (1987) Deadlock-free message routing in multiprocessor interconnection networks. IEEE Trans Comput C-36(5):547–553
21. Glass C, Ni L (1994) The turn model for adaptive routing. J ACM 5:874–902
22. Duato J (1995) A necessary and sufficient condition for deadlock-free adaptive routing in wormhole networks. IEEE Trans Parallel Distrib Process 6(10):1055–1067
23. Duato J (1996) A necessary and sufficient condition for deadlock-free routing in cut-through and store-and-forward networks. IEEE Trans Parallel Distrib Process 7:841–854
24. Gigli S, Conti M (2009) A SystemC platform for Network-on-Chip performance/power evaluation and comparison. In: Proceedings of the IEEE seventh international workshop on intelligent solutions in embedded systems WISES09, pp 63–69, Ancona, Italy, June 25–26

Chapter 8
Coverage-Driven Verification of HDL IP Cores

Case Study of a Router for Network-on-Chip Communication in Embedded Systems

Sergio Saponara, Francesco Vitullo, Esa Petri, Luca Fanucci, Marcello Coppola and Riccardo Locatelli

8.1 Introduction

The progress of nanometric CMOS technologies and design methodologies has fostered the development of complex digital designs, thanks to the capability of integrating an increasing number of IP cores within a single chip. This trend made one typical issue of the embedded and digital systems design flow to become more and more critical: the exhaustive functional verification of complex systems. In fact, as system complexity grows, the same is for verification tasks which have two main crucial points: (i) generating proper testing scenarios that stress key features of the DUT (Design Under Test) and (ii) determining the amount of different tests needed to reach enough coverage (code and functional) to assert that the DUT is bug free w.r.t. the foreseen utilization scenarios. While code coverage is about ensuring that all part of the RTL netlist (statements, expressions, branches, finite machine states, block instantiations) have been stimulated by the test vectors (automatic measure of achieved code coverage is already supported in recent EDA tools for HDL IP simulations), functional verification requires a major verification engineering effort [1–5]. Indeed, functional verification is about (i) catching functional behaviour of the DUT from the specifications document and (ii) validating that DUT behaviour for all possible working scenarios (input traffic and internal IP state) is consistent with its specification. To be noted that code and

S. Saponara (✉) · F. Vitullo · L. Fanucci
Department of Information Engineering, Universita di Pisa, Pisa, Italy
e-mail: sergio.saponara@iet.unipi.it

E. Petri
Consorzio Pisa Ricerche- Electronic Systems and Microelectronics Division, Pisa, Italy

M. Coppola · R. Locatelli
AST Grenoble Lab, STMicroelectronics, Grenoble, France

M. Conti et al. (eds.), *Solutions on Embedded Systems*,
Lecture Notes in Electrical Engineering, 81, DOI: 10.1007/978-94-007-0638-5_8,

functional verification, on which this work is focused, is the first step of the whole testing flow which entails further steps for IP core emulation on prototyping platforms (typically based on FPGA), back-end gate-level functional/timing verification and finally validation and characterization of the implemented integrated circuit. Failure mode analysis and fault-robustness verification [6, 7] (particularly important for designs conceived for harsh environment applications) are orthogonal methods not included in what discussed in this work. Today the major part of development time and costs for a new IP core are spent on verification rather than on HDL design. To reduce development time and design cost, reusability, configurability and scalability of the functional verification environment have a crucial role, also for subsequent verification steps.

Traditional verification techniques based on direct testbenches (where the test traffic is typically hand-written in HDL as a sequence of input vectors; output vectors are calculated a priori and then matched with the ones monitored from the DUT) or on formal demonstrations are inefficient when dealing with complex designs made up of multiple heterogeneous IP cores [1–5]. Direct testbenches are applied to the DUT by means of simulation; they have a poor level of automation since most testing traffic scenarios are usually hand-written; even using more high-level programming languages (such as C++ or SystemC) to abstract the set of possible significant stimuli, the problem of checking (i.e. catching DUT outputs and establishing whether they are correct or not) is still to be solved. When direct testbenches are used, the output checking tends to be simplified since the user knows what to expect from the DUT; still, this approach is time consuming and cannot be exploited for complex designs. Formal verification techniques, on the other hand, are not based on simulations; instead, the verification engineer tries to extract deterministic laws and relationships internal to the DUT exploiting its HDL description. Then, under the assumption of some hypotheses (typically represented by some set of stimuli) and with the help of an analysis tool, the formal verification approach tries to prove one or more theorems (on those stimuli, the DUT always behaves in the desired manner). This approach does not need simulations (though recently also symbolic simulations have been introduced combining formal techniques with standard simulation [1]) and is general enough to treat also corner cases. However, formal verification proved to be too complex for medium or large sized designs because the set of properties that the verification engineer needs to formally demonstrate is huge; the well known state explosion problem limits model checking, and the cost of theorem proving is prohibitive because of the amount of skilled manual guidance it requires. The limitations of these verification techniques is the lack of reusability and the excessive amount of time, if compared to Time-To-Market needs, which is required to provide a good confidence level that the design is bug free.

To solve the above issues, a coverage-driven methodology for functional verification based on pseudo-random simulations is discussed in Sect. 8.2 and applied to the case study of a Router IP core for Spidergon NoC communication. Section 8.3 briefly describes the NoC approach and presents our novel Spidergon STNoC. Section 8.4 first describes the functionalities of the new Spidergon

STNoC Router IP and its implementation results in 65 nm CMOS technology and than discusses how the verification methodology was customized and applied to it. Section 8.5 comments the achieved results and draws some conclusions.

8.2 Reusable Methodology for the Functional Verification of Platforms

Though many hybrid verification techniques have been explored, trying to combine the strength points of different approaches, an emerging approach for functional verification is represented by the constrained random (or pseudo-random) simulations. It is already partially supported by tools and languages such as SystemVerilog or the aspect-oriented [8–10] programming language ***e*** (recently formalized in the IEEE1647 standard) with Specman by Cadence. The Specman tool is the one used in the case study of this work. The basic idea we exploited for functional verification is to build an ***e***VC (Verification Component in the ***e*** language) starting from the DUT specifications and ending up with a software able to perform the following tasks: generating user-defined traffic patterns to be driven into the DUT; monitoring the DUT outputs and checking them according to the rules programmed in the ***e***VC; parsing collected outputs into a functional coverage scheme to let the user understand if all possible cases have been stressed. The latter is a very important issue and enables a coverage-driven verification: i.e., the user continues developing tests and running simulations until there are no holes left in the defined functional coverage plan. Therefore, to achieve full functional coverage the design of pattern generators, addressed in literature also for the target NoC case study [11, 12], is just one of the steps of a complete verification methodology and on its own is not enough. Figure 8.1 shows the conceptual organization we followed to design an ***e***VC around the DUT. To be noted that,

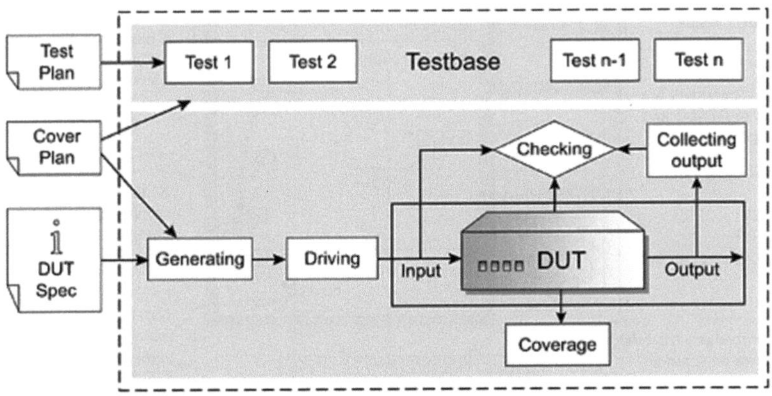

Fig. 8.1 Activities in a functional verification environment

starting from DUT specifications, not only the ***e***VC architecture, but also a test plan and a coverage plan have to be defined.

The philosophy underlying ***e***VCs differs significantly from traditional verification methodologies. Rather than using thousands of directed tests, ***e***VCs employ automatic generation and a coverage driven methodology. Using automated scenarios generation, ***e***VCs can typically achieve higher coverage percentages of the protocol w.r.t. hand-written tests. For instance, in our work 100% code coverage and functional coverage was achieved enhancing the test base with the addition of few directed tests, derived after coverage analysis, which allowed to exercise the remaining corner cases. To this aim, besides ***e***VC development, some HDL and scripts were also developed and HDL probes were defined to increase observability and controllability of the netlist. An ***e***VC for a given IP core or a given protocol can be thought as final product and it does not need to be rewritten from project to project, allowing for significant reusability. By following proper coding guidelines and design best practices, the ***e***VC of an IP core can be easily extended, reused and integrated as part of bigger ***e***VCs for more complex designs when moving from module to system level verification. This kind of approach to functional verification overcomes the limitations of traditional verification techniques and improves time-to-market. Figure 8.2 shows the main eVC blocks that have to be designed to properly stimulate and check the DUT. The *env* block represents an instance of the entire verification environment. The *Config* unit inside it is the user front-end for the configuration of environment's attributes and behavior. For each port of the interface, the ***e***VC typically implements an agent, instantiating it in the environment. Agents can emulate the behavior of a legal

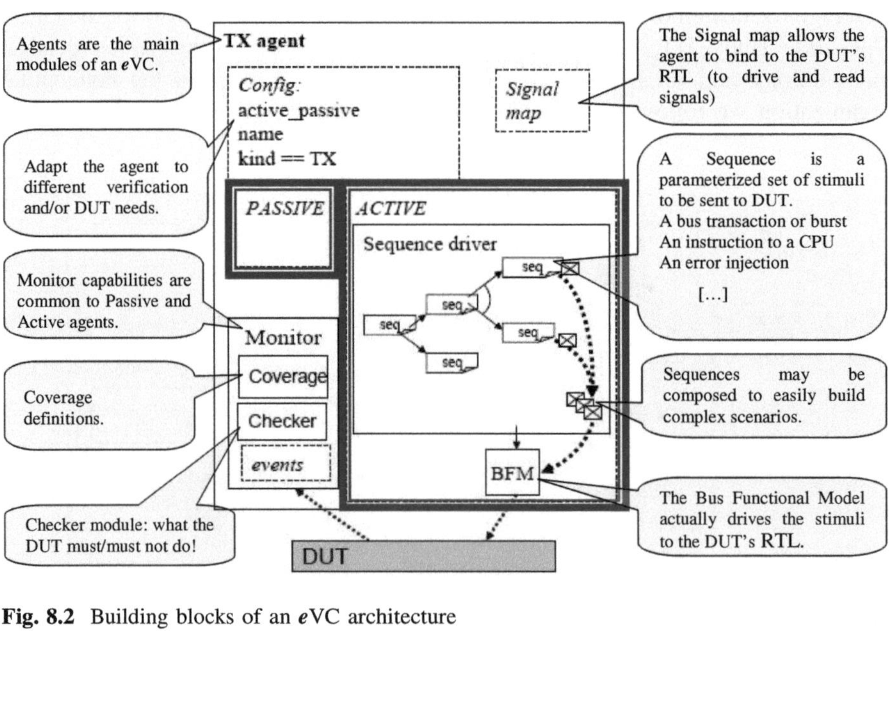

Fig. 8.2 Building blocks of an ***e***VC architecture

device, and they have standard construction and functionality; they are usually bound to a sub-set of the DUT port map and are built on top of building blocks that implement agent specific functionality:

Config: a group of fields that allow configuration of the agent's attributes and behavior.

Signal Map: a unit that contains external ports for each of the HW signals that the agent must access as it interacts with the DUT.

Sequence Driver: a unit instance that serves as a coordinator for running user-defined tests; traffic patterns are implemented as sequences (*seq*).

BFM (Bus Functional Model): a unit instance that interacts with the DUT and both drives and samples the DUT signals.

Monitor: a unit instance that passively monitors (samples) the DUT signals and supplies interpretation of the monitored activity to the other components of the agent. Monitors can emit events when they notice interesting things happening in the DUT or on the DUT interface. They can also check for correct behavior or collect coverage data.

8.3 Spidergon NoC Communication Design

NoC is an emerging design paradigm for building scalable packet-switched communication infrastructures connecting hundreds of IP cores. Its utilization in MPSoCs which strongly depend on the on-chip communication architecture will overcome the scalability limitations of traditional solutions like point to point communication and centralized on-chip busses [13–16]. In fact, as the number of components grows, traditional interconnect systems can degrade the global system performances in terms of area, power or throughput [15]. Beyond that, the floor-planning of long communication wires in presence of so many IP cores is also very problematic because bad wire routing lowers circuit timing performance. NoCs adopt some of the networking ISO-OSI abstraction layers (physical, data-link, network, transport) for decoupling the design of the IP cores from the physical implementation of the interconnect infrastructure, thus speeding up the design flow and increasing scalability and reusability, see Fig. 8.3. A NoC is a distributed network consisting of some main building blocks, the most important of them being the Network Interface (NI) and the Router (R). The way these blocks are interconnected determines the NoC topology (e.g. 2D mesh and Ring have been proposed in the past) which is responsible for packet switching efficiency between the start and the end point of communication and must be designed to avoid traffic congestion. Also the Quality of Service (QoS) applied to the routing policy of packets is a key element.

The NoC infrastructure we designed, called Spidergon STNoC, is based on the vertex-symmetric topology shown in Fig. 8.3b (note that Spidergon topology includes the Ring one). In our design, Routers are in charge of delivering packets towards the destination IP while providing buffering and QoS services by means of

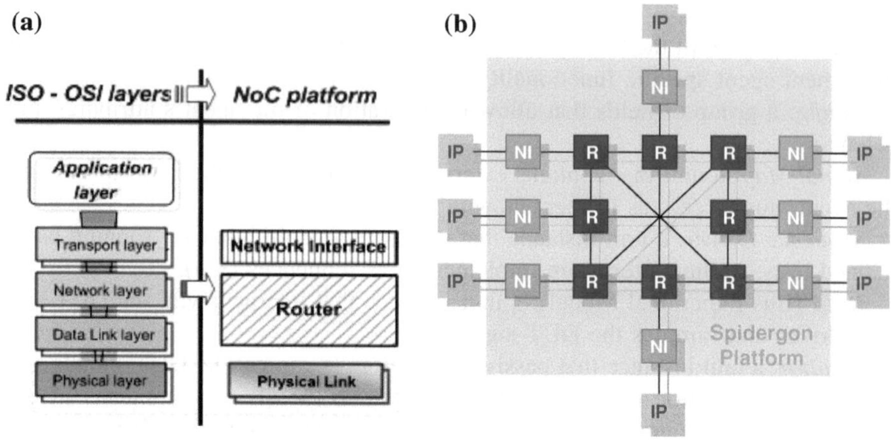

Fig. 8.3 **a** Internet ISO-OSI layers and mapping onto NoC, **b** Spidergon STNoC architecture

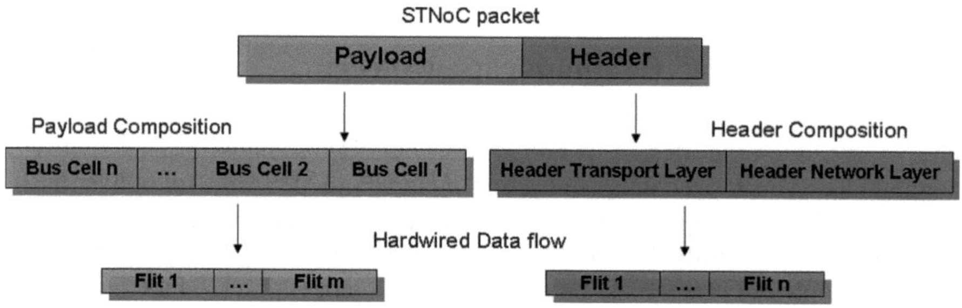

Fig. 8.4 Spidergon STNoC packet format

traffic arbitration policies and link scheduling on packets crossing their network. All issues related to security, error management and clock frequency, bus size and protocol conversions are managed at network boundaries by the NIs. Indeed the NIs are the peripheral building blocks of the NoC, decoupling computation from communication, which provide protocol abstraction by encoding in the packet's header all data to guarantee successful end to end data delivery between cores (transport layer) and all QoS information needed by the router at network layer. Figure 8.4 shows the format of the STNoC packet carrying header and payload data which are physically split in header and payload flits; they are all routed through the same path across the network.

Such architecture avoids many of the problems affecting older communication systems. Since it is a distributed network, data packets switch from a building block to another never being carried by long physical wires which are basically split and distributed along the network. Each NI collects traffic from the core it is connected to, independently from the others, and then converts such traffic into packets sending them to the network of routers, where they move along different paths. This allows parallel communication flows, which was not possible with shared on-chip bus lines. The Spidergon STNoC is also suitable to implement a globally-asynchronous-locally-synchronous (GALS) communication paradigm [17, 18]. First of all, this means that long global clock wiring is not needed; furthermore, each part of the SoC can be fed with a different clock source. It is task of the distributed network to synchronize one clock domain with another with special modules in its building blocks. In this scenario the distributed network decouples the working frequencies of the plugged IP cores/sub-systems. Different kinds of links (synchronous, mesochronous and asynchronous) have been designed for the Spidergon STNoC [17]. All the problems avoided by the NoC are not charge free and the price to pay is the complexity of the network. Indeed, it is important that in the whole computing system the NoC only contributes to a small percentage of overall area and power consumption. Thus the Spidergon STNoC has been conceived as scalable, depending on the surrounding computing infrastructure, in terms of topology, building blocks number and implemented services. The Spidergon STNoC has been also designed to provide compatibility with affirmed bus standards largely adopted in IP cores such as RISC processors, microcontrollers and DSPs (e.g. STBus and AXI NIs have been designed). Such feature ensures a smooth transition from old bus-centric systems to MPSoCs with the novel NoC interconnect.

8.4 Verification Environment for the Spidergon STNoC Router IP

The functional verification of Spidergon STNoC building blocks has been carried out by building a coverage driven simulation environment. In this case study, the DUT is a Spidergon STNoC Router IP core but the same approach has been followed for NIs and links. In this Section, first the functional features of the designed Router and its implementation results in submicron CMOS technology are described and then the design of the verification environment is presented.

The Router architecture has been defined according to a parametric and modular approach using VHDL language. The Spidergon router, see Fig. 8.5, can be connected through two unidirectional links with three other routers into directions Right (R), Left (L) and Across (A), plus the fourth connection to the local Network Interface, used as the network entry/exit point. The physical link consists of two unidirectional data channels, Downstream (DS) and Upstream (US), with the relevant handshake signals to realize a credit-based hop-by-hop flow control (*val* and *credit* in Fig. 8.5). The router adopts wormhole packet-switching, where a packet is subdivided into flits and all of them follow the same path reserved for the header. The routing algorithm is deterministic, so that always the same path is chosen between a source and a destination node, even if multiple paths exist. This choice avoids costly flit reordering at packet reception. The idea is to move along the ring, in the proper direction, to reach nodes which are near the source node,

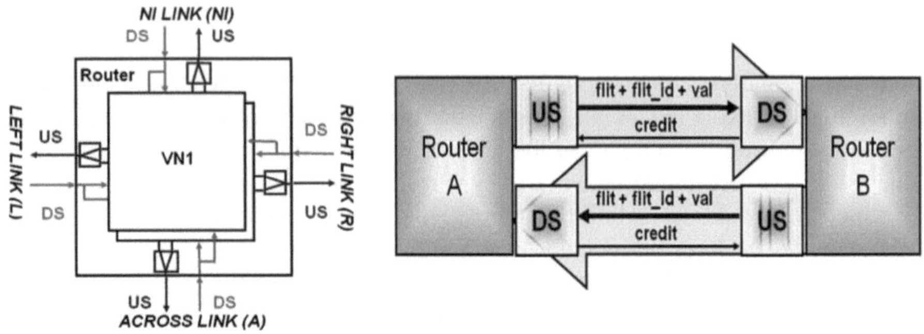

Fig. 8.5 S-STNoC router ports breakdown and Downstream (DS) and Upstream (US) channels with the relevant handshake signals

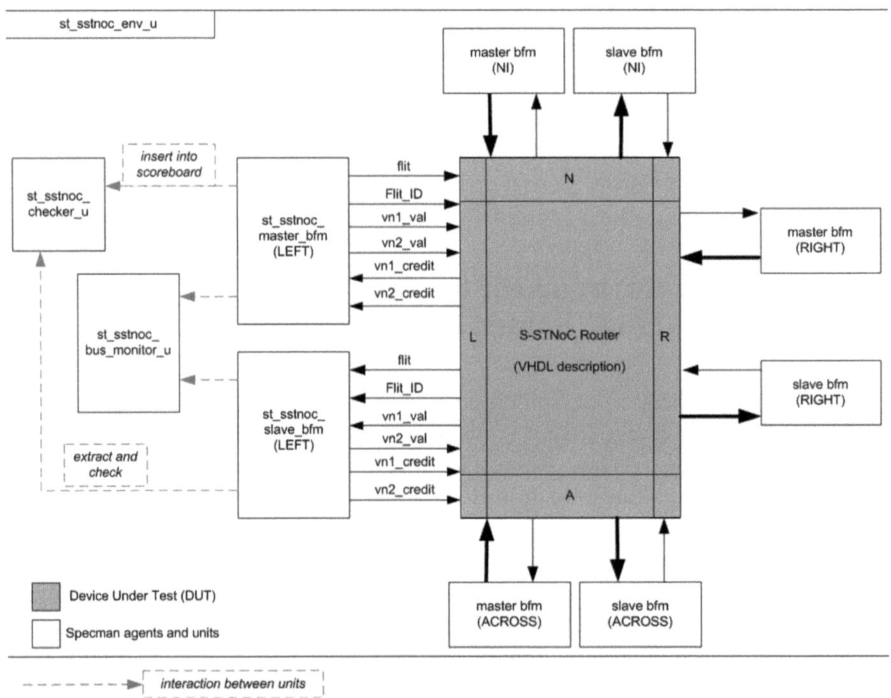

Fig. 8.6 Environment adopted for the Router functional verification

using the Across link as first or last hop to jump to a part of the network that is too far away. The router uses a simple source-based routing: the entire path is encoded in the packet header, so each router has just to extract the forward information, without any need of computation or any look-up table. The routing scheme, along with a proper QoS scheduling policy, is free of starvation issues. The router avoids deadlock also by deploying Virtual Networks (VNs in Figs. 8.5 and 8.6, also called Virtual Channels, VCs). VNs provide logical links over the same shared physical channels, by establishing a number of independently allocated flit buffers in the corresponding transmitter/receiver nodes. Currently the two request and response logical paths are implemented on top of two disjoint VNs for sharing the physical link bandwidth and maximizing wire efficiency. The parametric number of VNs supported by the router can lead to advanced routing schemes or independent QoS traffic classes for real time and low latency flows. The credit-based flow control works on a per flit basis. Flits can be sent in the US direction only if there are enough credits, i.e. the DS interface of the receiving component has enough free locations in its input buffer to store incoming flits. Output Queues on US ports can be instantiated for enhanced performance, avoiding head-of-line blocking. Queues are shared among input flows to limit costly time/space speed up factors and they have the bypass feature to reduce the minimum router crossing latency in case of low traffic conditions. The architecture also supports the possibility of not instantiating the Output Queue for low cost implementations, when performance or traffic types do not require output buffering. It is optionally possible to instantiate a separate Output Queue for each input port directed to that output. This configuration increases global network performance when a lot of traffic is concentrated towards the considered output. The applied QoS mechanism is the Fair Bandwidth Allocation (FBA). It allows for a flexible, scalable and low cost management of the allocation of the available bandwidth. The requested bandwidth value is programmed at injection point (Network Interface) and is not explicitly linked to the path of a data flow through the router like in other NoC architectures. It avoids complexity inside the router by providing all necessary information in the network header and limiting the router behavior to a simple two-step arbitration. When all data flows have the same bandwidth reservation, the arbitration algorithm becomes one of the following: Round Robin (RR), Least Recently Used (LRU) or fixed priority schemes, configurable by the user.

The router has been implemented for different configurations in different (90 nm, 65 nm and 45 nm) STMicroelectronics CMOS standard-cells technologies always achieving optimal trade-off between performance and complexity. As example in 65 nm 1.1 V standard-cells CMOS technology a full Router configuration with all 4 ports enabled (Spidergon topology) and all with 2 VNs, a size of 72 bits on VN1 (request path) and 64 bits on the VN2 (response path), using input buffers (IB) and output queues (OQ) able to store respectively 4 and 5 flits, with LRU arbitration and FBA management, has a circuit complexity of roughly 70 Kgates (including the flip-flop implementation of IB and OQ memory resources). It achieves a clock frequency of 500 MHz, i.e. at least 32 Gbps data transfer for US and DS channels, with a low-leakage library version ensuring a static power consumption less than 100 μW. By using a standard-cells library version optimized for high-speed, with the same IP configuration and CMOS technology node, clock frequencies up to 1 GHz are met, i.e. up to 64 Gbps data transfer per channel, but with an increased static power of 1 mW. Obviously, by changing the Router configuration different results are achieved: as example a basic Router with 3 ports (Ring topology without the Across link), 36-bit size for the flits, 1 VN, no OQs instantiated, has a circuit complexity lower than 9 Kgates and achieves a clock frequency up to 1 GHz, i.e. at least 36 Gbps data rate, with a static power consumption of roughly 200 μW (the static power is 10 μW if targeting 500 MHz frequency).

Proper component operation should be assessed for every Router configuration. However, only a subset of all possible configurations has been actually exploited for assembling platforms to synthesize. For such Router configurations a full regression set of test simulations has been carried out to check correct component operation when stressed with several different traffic scenarios. The DUT of the simulation was the single Router block. Figure 8.6 shows a sample Router DUT surrounded by the corresponding verification environment; in this example, a 4–port Router with 2 VNs and both US/DS directions on each port is considered. Each Router port has its own relevant BFM units which drive and monitor traffic (master agents are connected to DS ports and slave agents are connected to US ports). All BFM units are connected with both the monitor unit and the checker unit. The former is in charge of protocol checking and data coverage, the latter implements a scoreboard for checking correct routing and other traffic properties. At the interface level, the following categories of checks were implemented:

- Routing: (i) each transmitted packet exits one and only one time; (ii) each transmitted packet is output from the correct port (according to header information); (iii) flits within a packet are kept in the correct order and are not interleaved with flits from other packets.
- Credit-based protocol: (i) when a flit is read from the Input Buffer, a credit is sent back by the router; the *valid* signal is high on an output port when a significant flit is transmitted.
- Network Layer Header: FBA bit management (the FBA bits of a packet are correctly updated when it exits the router).

The different BFMs may be configured to generate different kinds of traffic scenarios so to reach the desired functional coverage. For example, configuring properly a test file, it is possible to generate packets that are sourced from 3 ports and all having the same destination port; this is useful for stressing arbiters and output queues as well as for achieving some corner cases coverage points. The developed environment discovered a number of bugs that it was not possible to find with hand-written HDL testbenches.

To achieve full code and functional coverage by exercising some corner cases, for some Router configurations a deeper level of checking has been implemented by means of internal probes (e.g. monitoring internal DUT signals). Indeed, while some DUT functionalities may be easily checked/covered without knowledge of timing such as the data integrity from one port to another, check rules for other DUT functionalities depend on the timing of what is happening on the various ports; as example, the buffers status (empty/full) depends on the rate with which packets are injected into the DUT, besides the destination of those packets; also the correct behavior of an arbitration algorithm depends on the timing with which the different packets accessing the same resource are served by the Router.

Therefore, to achieve 100% functional verification, the basic coverage-driven approach should be enhanced either through the design of a software golden model able to predict timing-dependent properties (very time consuming approach) or alternatively, as we have done in this work, using internal probes to rapidly achieve detailed information about operation of internal router blocks and state machines. An internal probe means monitoring a hardware signal within the router: for example monitoring the inputs of an arbiter block allows to gather extensive coverage information about arbitration scenarios and successful application of a specific arbitration algorithm. The drawback of this approach is that using probes requires a deep knowledge of Router VHDL implementation and it is a hard-to-reuse solution. Figure 8.7 shows the UML diagram of the probes portion of the Router ***e***VC. Thanks to these additional ***e***VC units, some internal arbitrations and QoS mechanisms have been verified such as the LRU arbitration of the

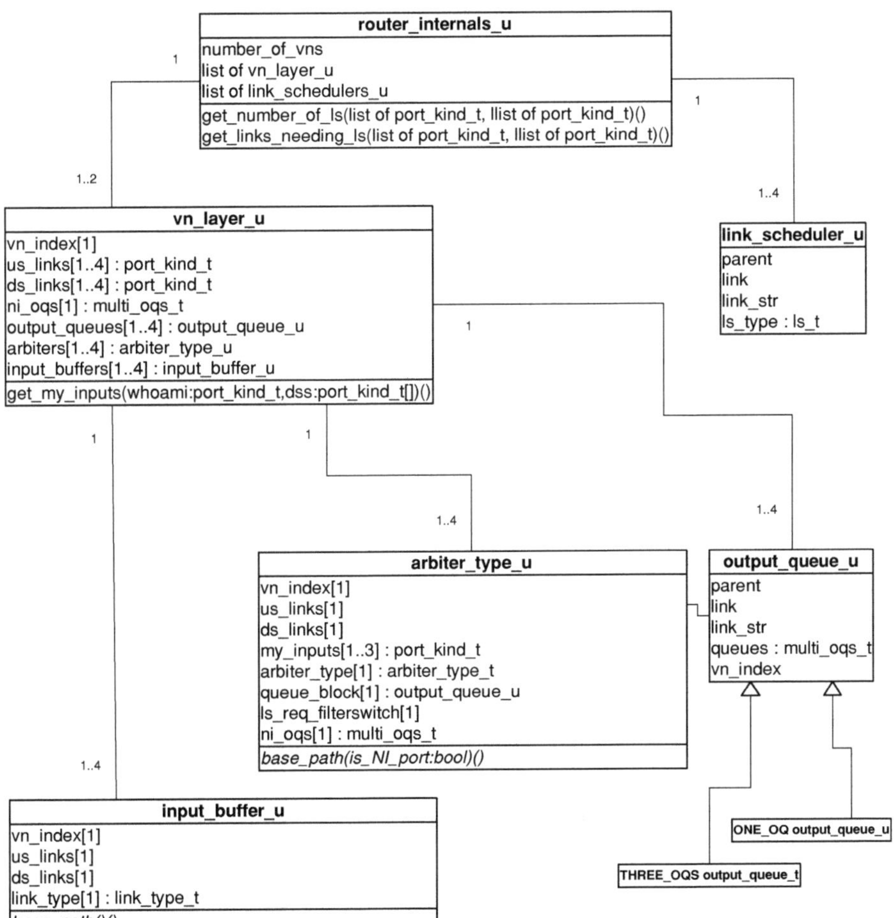

Fig. 8.7 UML diagram of the probes portion of the Router eVC

link scheduler; furthermore, this allowed for the implementation of additional coverage points related to arbiters and internal buffers.

To be noted that many checks must be performed independently for each port, so the *e*VC architecture needs to be highly modular and configurable. It is worth noting that all the checks and coverage points may be selectively enabled to meet various user requirements. For example, if there is no interest in testing queue utilization, simulations can be speeded up by disabling the internal probes portion of the *e*VC (which works with cycle level accuracy), leaving only the transaction level architecture. It is also possible to disable a single check; for example, by disabling the check about routing the Router *e*VC can be used as traffic monitor on a single US or DS bus. To conclude the verification flow, several coverage points have been defined to describe all possible traffic scenarios that can stress the Router; they may be conceptually grouped into the three main categories reported in Table 8.1: Traffic flow, queue utilization, arbitration mechanisms. Some coverage points (such as the ones in queue utilization) are available only as part of the internal probes *e*VC extension, because it is possible to easily retrieve some information only accessing the micro-architecture of some blocks.

8.5 Results and Conclusions

The developed *e*VC has been used to test several router configurations; enough test cases have been implemented to achieve 100% full functional and code coverage for all points defined in the coverage plan. Table 8.1 shows some details about the coverage points in the plan with their corresponding range (e.g. the values that need to be induced in the DUT) and the result achieved. In Table 8.1 we also highlight the coverage points for which probes have been used. After the time spent for developing the *e*VC software, such results were achieved in a relatively short time thanks to the constrained-random traffic generation.

After functional and code verification by simulations, the effectiveness of the verification flow was also demonstrated by several STNoC platforms implemented on FPGA that were successfully emulated with real-life scenarios where multiple ARM11 processors and memory modules were connected through the Spidergon STNoC.

The Router *e*VC has also been used as a component for the functional verification of platforms involving several NIs and Routers; different *e*VCs, all developed according the same paradigm, were put together obtaining a complex platform verification environment.

Finally, it is worth noting that the Spidergon STNoC architecture has been chosen as the inter-tile interconnect of the SHAPES MPSoC European project [17, 19–21] involving ATMEL, University of Pisa and STMicroelectronics. In SHAPES, 8 identical IP tiles building a complex scalable multiprocessor are interconnected by means of a packet-switched Spidergon STNoC network. A typical SHAPES tile contains a VLIW floating-point DSP, a RISC processor based on the ARM926 core,

Table 8.1 Implemented coverage groups and test results

Category	Coverage description	Range	Results
Traffic flow	Routing paths	Routing paths between each pair of ports	100% (8 groups, 1 per port, per VN)
	Packet length (2 levels cross, combination of routing paths, number of flits in a packet)	1–10	100% (8 cross groups, 1 per port, per VN)
	Routing fields in packet header (4 levels cross, combination of DS ports, directions, destination ID[a])	All routing decision scenarios on all DS ports	100% (2 cross groups, 1 per VN)
Queue utilization	Input Buffer utilization (*probes*)	Empty–Full	100% (8 groups, 1 per port, per VN)
	Output Queue utilization (*probes*)	Empty–Full	100% (8 groups, 1 per port, per VN)
	Bypass operation (*probes*)	Queue bypassed/not bypassed	100% (8 groups, 1 per port, per VN)
	Credit availability (*probes*)	0…5	100% (4 groups, 1 per US port)
Arbitration mechanisms	Fair Bandwidth Allocation (FBA) (3 levels cross cover: combination of US port, FBA status of incoming packet, FBA status of outgoing packet)	FBA transitions (0 → 0, 0 → 1, 1 → 0, 1 → 1)	100% (8 cross groups, 1 per port, per VN)
	Concurrent requests scenarios (*probes*)	All possible combinations of concurrent packet requests to the arbiters	100% (8 groups, 1 per VN)

[a] The directions and the destination ID are network layer header fields which are encoded in the head of each packet and are used by the Router to take routing decisions

a Distributed Network Processor (DNP) for extra tile communication and includes the interface to the NoC (NI), on-chip memories and a set of peripherals for off-chip communication. The back-end of the 8-tile SHAPES architecture in 45 nm CMOS technology has been successfully realized.

This work has extended the WISES2009 conference paper [22].

References

1. Bhadra J, Abadir MS, Ray S, Wang L-C (2007) A survey of hybrid techniques for functional verification. IEEE Des Test Comput 24(2):112–122
2. Bartley MG, Galpin D, Blackmore T (2002) A comparison of three verification techniques: directed testing, pseudo-random testing and property checking. In: Proc. 39th Design Automation Conf. (DAC02), ACM Press, New York, 2002, pp 819–823
3. OSCI (2003) SystemC Verification standard specification version 1.0e. http://www.systemc.org, May, 2003
4. Yuan J, Shen J, Abraham J, Aziz A (1997) On combining formal and informal verification. In: Proc. Int'l Conf. Computer-Aided Verific., LNCS 1254, Springer, Heidelberg, 1997, pp 376–387
5. Sumners R, Bhadra J, Abraham J (2000) Automatic validation test generation using extracted control models. In: Proc. IEEE Int'l Conf. VLSI Design, pp 312–320
6. Eghbal A et al (2009) Fault injection-based evaluation of a synchronous NoC router. In: Proc. IEEE IOLTS'09, pp 212–214
7. Mariani R, Boschi G (2007) A systematic approach for Failure Modes and Effects Analysis of System-On-Chips. In: Proc. IEEE IOLTS'07, pp 187–188
8. Berman V (2005) An update on IEEE P1647: the e system verification language. IEEE Des Test Comput 22(5):484–486
9. Murphy G, Schwanninger C (2006) Guest editors' introduction: aspect-oriented programming. IEEE Softw 23(1):20–23
10. Palnitkar S (2003) Design verification with e. Prentice Hall, Upper Saddle River
11. Al-Badi R et al (2009) A parameterized NoC simulator using OMNet++. In: Proc. IEEE ICUMT'09, pp 1–7
12. Wen H-H et al (2009) Design of an on-line configurable traffic generator for NoC. In: Proc. IEEE ASID, pp 556–559
13. Benini L, De Micheli G (2002) Networks on chip: A new SoC paradigm. IEEE Comput 35(1):70–78
14. Muttersbach J, Villiger T, Fichtner W (2000) Practical design of globally-asynchronous locally-synchronous systems. In: IEEE ASYNC pp 52–59
15. Gyu Lee H et al (2007) On-chip communication architecture exploration: a quantitative evaluation of point-to-point, bus, and network-on-chip approaches. ACM Transactions on Design Automation of Electronic Systems 12(3)
16. Grammatikakis MD, Coppola M, Maruccia G, Locatelli R, Pieralisi L (2008) Design of cost-efficient interconnect processing units: Spidergon STNoC. CRC Press, Boca Raton
17. Vitullo FM, L'insalata NE, Petri E, Saponara S, Fanucci L, Casula M, Locatelli R, Coppola M (2008) Low-complexity link microarchitecture for mesochronous communication in networks on chip. IEEE Trans Comput 57:1196–2203
18. Rahman M et al (2009) Efficient 2DMesh Network on Chip (NoC) considering GALS approach. In: Proc. IEEE ICCIT, pp 841–846
19. Paolucci PS, Lo Cicero F, Lonardo A, Perra M, Rossetti D, Sidore C, Vicini P, Coppola M, Raffo L, Mereu G, Palumbo F, Fanucci L, Saponara S, Vitullo F (2007) Introduction to the tiled HW architecture of SHAPES. Proc Int Conf Des Autom Test Eur 1:77–82

20. Paolucci PS, Jerraya A, Leupers R, Thiele L, Vicini P (2006) SHAPES: a tiled scalable software hardware architecture platform for embedded systems. In: Proc. Fourth Int'l Conf. Hardware/Software Codesign and System Synthesis, pp 167–172
21. Saponara S, Martina M, Casula M, Fanucci L, Masera G (2010) Motion estimation and CABAC VLSI co-processors for real-time high-quality H.264/AVC video coding. Microprocess microsyst 34(7–8):316–328
22. Saponara S et al (2009) A reusable coverage-driven verification environment for Network-on-Chip communication in embedded system platforms. In: IEEE WISES 2009, pp 71–77

Chapter 9
A Multiprocessor Platform for Efficient Data Processing in Electronic Musical Instruments

A Case Study

Marco Caldari, Franco Ripa and Massimo Conti

9.1 Introduction

The market for high-end microcontrollers shows a steady growth due to the possibility of increasing performance at ever decreasing costs. Performance and cost should not be considered simply as the computing power and the net price of the device itself, but they should include key factors to be considered in order to take the right choice between a wished component and another.

Two key issues are the architecture of the system and the set of peripherals integrated in the microcontroller: only intelligent peripherals design, a reactive interrupt system and the efficient partition of busses and memory hierarchy can actually provide really high performance. These aspects are specific to high-end devices, typically with 32-bit architectures; such devices are also preferential targets of C compilers and other high-level descriptive languages that allow the optimization of the resources needed to develop an application with the planned performance. An advanced set of peripherals can also reduce the bill of materials, so the final cost of the product, allowing the integration into a single chip of various types of components (e.g. power stages, oscillators, A/D and D/A converters and PWM modules), various serial and parallel communication interfaces (e.g. I2C, SPI, CAN, USB and SD/MMC) and various types of memory with different sizes (e.g. FLASH, E2PROM and SRAM).

It should also be considered that the specifications of a product are never really set up until the first item leaves the warehouse and even then there are a number of wished features to be implemented in subsequent revisions of the project.

M. Caldari (✉) · F. Ripa
KORG Italy S.p.A., Osimo, Italy
e-mail: caldari@korg.it

M. Conti
DIBET, Università Politecnica delle Marche, Ancona, Italy

M. Conti et al. (eds.), *Solutions on Embedded Systems*,
Lecture Notes in Electrical Engineering, 81, DOI: 10.1007/978-94-007-0638-5_9,

To minimize the risk, many organizations are adopting platforms with 32-bit architecture, enabling them to develop future generations of their products while maintaining high possibilities to expand their systems.

Based on these considerations one may think that the market for 32-bit microcontrollers will soon begin to erode the market shares of more traditional devices based on simpler 8- and 16-bit architectures.

In an existing platform, which already includes some microcontrollers, the move to a device with higher performance offers the designer a supplementary computation bandwidth that can be allocated to perform the work so far delegated entirely to the main processor. The additional tasks may be represented by the execution of more complex main application procedures to be assigned to each processor, so it is important to develop an efficient method for dividing the processing load from the application processor to peripheral controllers, when they are not employed for routine operations.

Moreover, in a multiprocessor structure [1–4] the efficiency of the communication channel is really fundamental [5–7], consisting of the particular type of physical interface and protocols that will be used. A kind of multiprocessor structure is already used in KORG products, where the tasks assigned to the peripheral MCUs are mainly simple control applications (e.g. scanning of switches, analogue sliders and LEDs).

The purpose of this work, partially reported in [8], is to identify a multiprocessor architecture for the efficient management of the applications in a musical instrument that can be an electronic keyboard or digital piano. This activity is part of a collaboration between the Department of Biomedical Engineering, Electronics and Telecommunications at the Università Politecnica delle Marche, with headquarters in Ancona, and KORG Italy S.p.A., a company established in Osimo. KORG Italy is a design laboratory for research and experimentation in sound technology aiming to explore new scenarios in music through the application of KORG technology within everyday life.

9.2 Multiprocessor Architecture

The activity is conducted on the basis of an existing platform with a structure consisting of a high-performance application controller (AP) and a set of peripheral microcontrollers (MCUs), with reduced functionality, properly connected through a serial communication channel, as in Fig. 9.1.

Using this type of architecture it was possible to optimize the connections between the various boards a musical instrument is made of and at the same time it was possible to reduce electromagnetic emissions caused by long-distance signals with high harmonic content.

In this architecture each MCU is running a series of specific tasks: some are executed in consequence to commands received on the bus, others are continuously

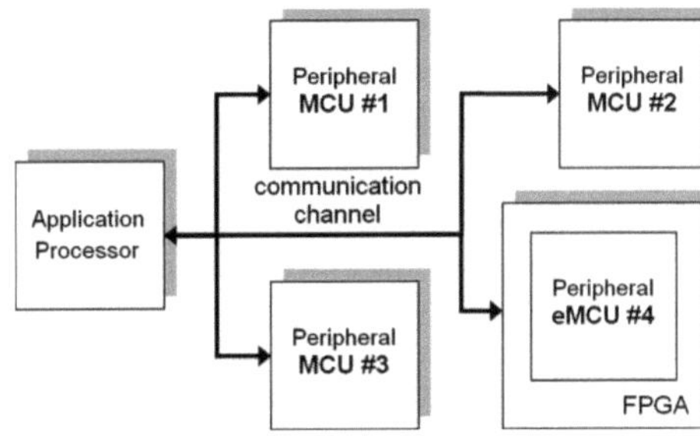

Fig. 9.1 Block diagram of the multiprocessor platform

running (e.g. keys and LEDs scanning), some tasks produce data that, at the right time, are read by the main microcontroller.

It is interesting to evaluate the possibility for the MCUs to perform some additional functions (more complex with respect to the routines that they normally execute) originally intended for the main controller; in this way part of its processing bandwidth can be freed and reallocated to the execution of higher interest tasks. The final goal of this research is to evaluate a number of possible architectures that can efficiently achieve this kind of functionality.

It was decided to model the entire architecture on an FPGA, to have the opportunity to easily test different solutions and analyze in detail each element of the network:

- The type of core used to model the peripheral controllers.
- The type of connection between MCUs and the AP trying to minimize the connections between them.
- Possible strategies for routing of data packets used for communication between different devices.

In particular to model the peripheral processors different options were possible [9–11], but at the end it was decided to use the Xilinx MicroBlaze softcore [12]: Harvard RISC architecture uses 32-bit, it is optimized for Xilinx FPGA family. The board used for the development is a XUPV2P (Xilinx University Program, Virtex-II Pro) that includes a Virtex-II Pro FPGA with two PowerPC405 hard-cores on board, a 100 MHz system clock, some connectors to connect external devices to the FPGA and the possibility of being configured using an USB connection.

9.3 Developing the Communication

Initially the activity was focused on the communication between the different processing elements, i.e., the main processor and the peripheral controllers. The use of a set of point-to-point connections was pointed out instead of a common bus. It was planned to model a simple communication system using VHDL starting from the implementation of an SPI interface. It was thought that the best solution for the connection of the various processors is the one in which the AP can communicate with any peripheral MCU, and each MCU is in touch only with its neighbors and with the main processor, so it must provide different SPI interfaces in order to be able to communicate with multiple MCUs.

The focus has then been shifted on the cleverness of the different processors. These were possible choices:

- The AP uses the MCUs as a simple extension: the AP conveys tasks and collects data.
- The AP gives the MCUs the ability to decide how to split and share the desired workload.
- The AP conveys instructions and commands with the possibility to distribute the load among different MCUs.

In the first situation the MCUs have no decision-making power, their task is to perform the operations required by the master, on data provided by itself. In the second, each MCU can decide, depending on the complexity of the request by the master, if to serve the request or to distribute it partially to one or more MCUs. In this way each of them acts as the AP with respect to those it is requesting some services.

In the third situation, finally, the AP is the only processor to know whether an operation must be conducted on a particular MCU or distributed among several MCUs; according to the type of transaction, the AP will send the appropriate information to the recipients. In cases where the execution has to be distributed, an exchange of information between the peripheral MCUs is required.

The system is built on this last approach, which ensures better performance for very complex operations with respect to the first situation where a command is carried out by a single processor that typically has not a great computation power; also the multitasking nature of the multiprocessor system is almost lost.

The third situation keeps the idea of the MCUs as not clever, calculating machines. This is a limitation of the second situation in which there is a constraint on the complexity of the various processors, which must be treated almost as the AP, being able to take decisions in quite a similar way; furthermore the second situation is the most complex in the planning of the entire system.

The first goal is to model an SPI communication device, based on that proposed by Xilinx, interfaced on the OPB bus of the MicroBlaze processor. The device is composed by three parts basically:

- An IP interface: it allows the connection of devices to the OPB bus in a standard way.
- A set of registers: required by any device that needs to communicate with the processor through the OPB bus. Each register is 32-bit wide and it is addressed using the signals of the IP interface.
- The SPI module: this module is responsible for transmission and reception of data and commands, and is basically created with a shift register, some memory elements and suitable control elements.

It was decided to divide the SPI controller into three finite-state machines: one responsible for the transmission, one for memory management and interrupts and the last one that acts as a master for the other two. Design parameters that should be evaluated are mainly represented by the length of the words to be used for communication and the size of the buffers to be able to support different traffic conditions; such estimations must be conducted according to the type of elaboration that the platform will be requested to perform.

9.4 Exploring the Functionality

One of the most important stages of this work was to evaluate different aspects of MicroBlaze, the softcore chosen to model the peripheral microcontroller, such as the achievable performance, the use of its peripherals and the efficiency of the development tools. To do so the implementation of an application typically performed by an MCU in musical keyboard was planned: the scanning of all the keys the keyboard is made of. To perform this task it was necessary to create a structure to interface the controller with the keyboard, to implement an algorithm which manages the scanning, to process the results and communicate them to the main processor.

An important part of the study is to analyze these different aspects:

- The processing load due to the execution of the scanning routine, it is important in order to estimate what is the free computational bandwidth that can be used to perform additional tasks.
- The use of memory in terms of the size of the executable file that contains the code/data to be loaded on MicroBlaze, this parameter determines the memory resource that the microcontroller should provide.
- Evaluation of resources required to synthesize the hardware.

The hardware system is composed by the MicroBlaze core and some logic used for keyboard scanning. The scanning algorithm was written in C language. The result of the compilation of the hardware and the software part of the project is a single file used to configure the FPGA on the board. The following Fig. 9.2 shows the block diagram of the interfacing between the main board of a KORG keyboard, the FPGA board and a 76-keys keyboard.

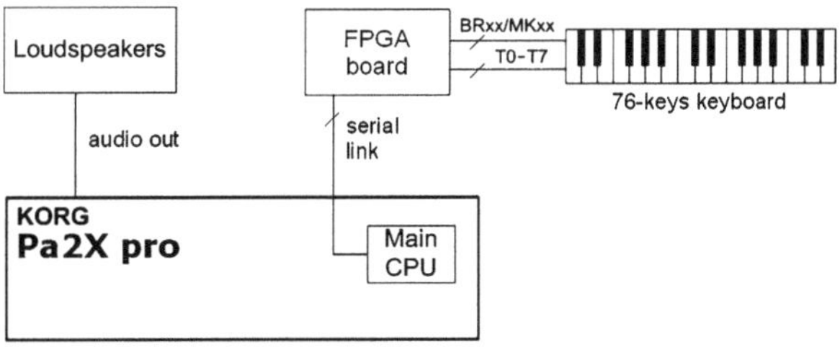

Fig. 9.2 Interfacing diagram: main board KORG → FPGA board → 76-keys keyboard

In general the goal of an algorithm used to scan a music keyboard is made of these consecutive steps:

- Scan the keyboard at regular intervals.
- Detect when a key is pressed or released.
- Measure the time between the transition released → pressed for all keys.
- Manage the stabilization of the reading of the keyboard data lines.
- Send information to the main processor.

9.4.1 Keyboard Scanning Implementation

The hardware part (see Fig. 9.3), synthesized in the FPGA, consists of

- An instance of the MicroBlaze softcore.
- A timer that generates an interrupt at regular intervals.
- A generic I/O device (GPIO) that drives the keys matrix and reads the state of switches (eight at a time).
- A decoder connected to GPIO lines that select one of the scanning lines of the keyboard.
- An UART used to send data to the main-board of the musical keyboard used to generate the sound.

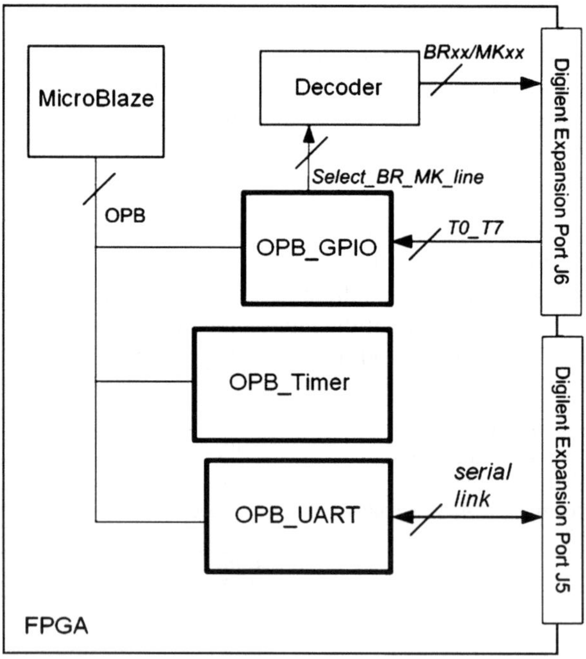

Fig. 9.3 Schematic of the components synthesized in the FPGA

The keyboard uses a set of keys organized in a matrix typology; by convention, the labels *BR* and *MK* will be used to indicate, respectively, the first (break) and the second (make) contacts that are closed while a key is being pressed. The use of two switches allows to estimate the intensity of the switch operation as it appears to be inversely proportional to the time between the closure of the two contacts *BR* and *MK*. If the time of the range is short, this indicates a strong intensity of pressure, while if the time of the range is long it indicates a weak intensity of pressure.

The result of the keyboard scanning is communicated to the main processor through the delivery of some data packets representing two events: *KEY_DOWN* and *KEY_UP*. The *KEY_DOWN* event is fired when a key is pressed and the corresponding data packet reports the key identifier and the corresponding intensity parameter; the *KEY_UP* event is fired when a key is released and the corresponding data packet simply indicates which key has been released. Two different versions of the scanning algorithm will be presented: standard and optimized.

9.4.2 The Standard Scanning Algorithm

The main feature of this first scanning algorithm version is that within each scanning period, all *BR* and *MK* lines enabled sequentially regardless of the current state of the keys on the keyboard. The finite-state machine in Fig. 9.4 visually describes this algorithm.

Initially all the keys are assumed to be in the state *IDLE*. As soon as a *BR* switch is detected as closed for a particular key, the state of that key becomes the state *TOUCHING*. From that moment a counter related to the key is periodically updated until the closure of the *MK* switch is detected for that key. At that point, the counter is no longer updated, a message related to the new *KEY_DOWN* event is dispatched containing the identifier of the pressed key and the value of the counter; the state of the key moves to the state *WAITING1* used to avoid the effects of possible key bounces. Only after a preset time period is elapsed, the state of the key passes to the state *PRESSED* in this state both *MK* and *BR* switches are closed.

If in subsequent scannings the *MK* switch is detected as open the key state advances to the state *RELEASING*. When also the *BR* switch is detected as open a message related to the new *KEY_UP* event is dispatched containing the identifier of the key. Then the state is advanced to the state *WAITING2* to further avoid possible key bounces. After a preset time period is elapsed, the key state can return to the *IDLE* state.

9.4.3 The Optimized Scanning Algorithm

The idea that led to the study of an optimized scanning algorithm derives from the observation that in the standard algorithm the activation of the *BR* and *MK* lines is

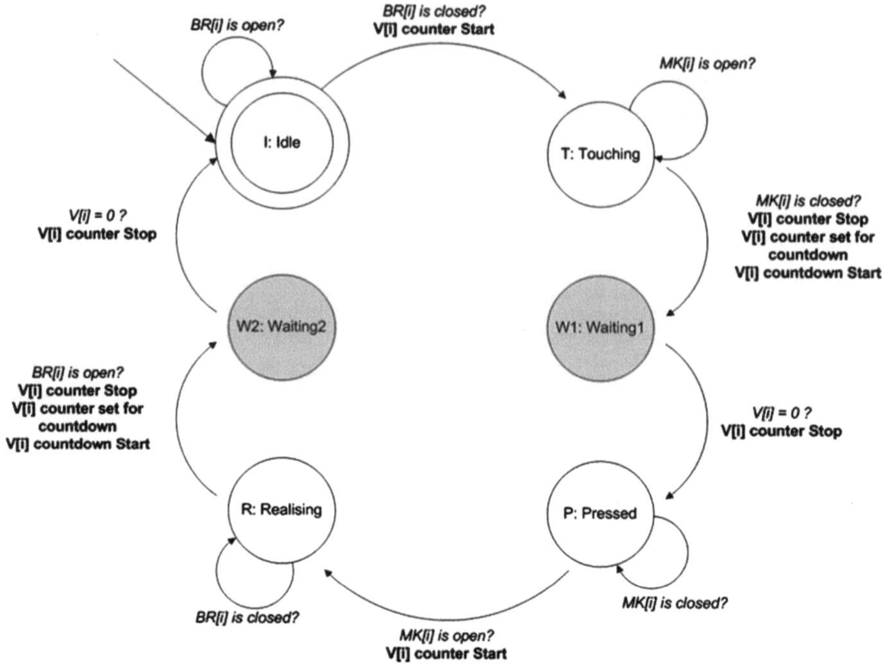

Fig. 9.4 Standard scanning algorithm FSM

performed in any case for each set of eight keys, in order to read the corresponding switches state; this reading is very costly in terms of time. Actually at times it may be sufficient to consider only the reading of the lines associated with *BR* or *MK* depending on the state of a particular switch. As can be seen in Fig. 9.4, in the states *IDLE* and *RELEASING* only the state of *BR* switches is controlled, ignoring *MK* lines, while in the states *TOUCHING* and *PRESSED* only the state of *MK* switches is checked, ignoring *BR* lines.

The goal of the optimized scanning algorithm is to read the position of the switches corresponding to *BR* and/or *MK* in a smart way, on the basis of the current state of the keys avoiding unnecessary readings. Since the reading of the position associated with *BR* or *MK* is performed on a group of eight keys, it is required to read *BR* and *MK* lines even if only one in eight keys needs its state to be updated.

In the best case, for each group of eight keys, all the keys belonging to that group are in a state that requires the reading of the same *BR* or *MK* line. This allows to gain 50% of the time required by the standard algorithm. In the worst case, for each group of eight keys, there are at least two keys belonging to that group which are in two particular states that require the reading of *BR* and *MK*. In this case the gain compared to the standard algorithm, with respect to the time spent for reading, is zero. In general, the gain introduced by the optimized scanning algorithm over the standard one is statistically very high.

It is possible to assign a particular meaning to the bits representing the states of the FSM in Fig. 9.5:

- Bit0 identifies which switch line is needed to be checked: bit0 = 0: check the *BR* line and ignore *MK* line; bit0 = 1: ignore the *BR* line and check the *MK* line.
- Bit1 identifies the particular state of the key: bit1 = 0: the key is being pressed; bit1 = 1: the key is being released.

9.4.4 Some Variations to the Scanning Algorithms

In addition to the standard and optimized scanning algorithms some features were implemented trying to maximize the efficiency of the algorithm itself. These features may be applied to each of the two kind of algorithm and represent a particular method of reading a switch line (*BR* or *MK*), they have been called '*once reading method*' and '*voting reading method*'.

Using the 'once reading method' a particular line is enabled and after a few microseconds the state of the corresponding switches is read. Using the 'voting reading method' a particular line is enabled and two successive readings of the state of the corresponding switches are taken at a time interval of hundreds of nanoseconds; if these two first readings match, then the result of the reading is considered as good otherwise, if there is any difference in at least one bit, a third reading is taken to define the bits that have been marked as uncertain.

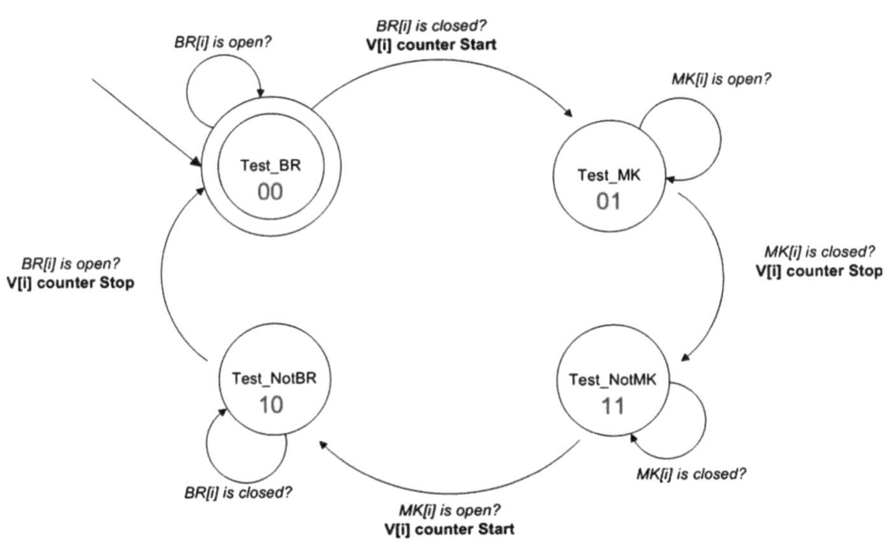

Fig. 9.5 Optimized scanning algorithm FSM

For both the variations a buffer can be introduced to store the outgoing data packets related to *KEY DOWN* and *KEY_UP* events waiting to be sent to the UART for transmission.

9.4.5 Simulations and Results

This paragraph collects the results of simulations carried out on the project of the keyboard scanning implementation. The measurements were related to

- The duration of a full scan of the keyboard.
- The length of the *ELF* (Executable Linkable Format) file generated by the C code compilation.

During the analysis it was possible to change the value of some parameters:

- The type of algorithm being used: standard or optimized.
- The scan reading method: once or voting.
- Data buffer insertion: enabled or disabled.

The different combinations of parameters of the scanning algorithm are shown in the following Table 9.1.

The setup used to perform the different measures is reported in the following Fig. 9.6.

With respect to both the total time required for the scanning routine and the size of the *ELF* file the compilation option '-oS' resulted to be the best, even if only slightly, provided that the debugging symbols are not included, as shown in Fig. 9.7. This parameter has been chosen for the execution of the performance tests that follow. A comparison of the eight algorithm combinations as been performed for different frequencies of the MicroBlaze clock: 33, 50, 66 and 100 MHz.

From the data obtained, shown in Fig. 9.8, it is possible to say that the performance using a clock of 50 MHz is not acceptable because, in the perspective of implementing the keyboard scanning task in a multiprocessor system using a scanning period of 200 μs, the time the CPU is occupied is so high that it impossible to do anything else between a scan and the next. The performance

Table 9.1 Different scanning algorithm combinations

Combination	Algorithm	Scan reading method	Buffer
Comb1	Standard	Once	Disabled
Comb2	Standard	Once	Enabled
Comb3	Standard	Voting	Disabled
Comb4	Standard	Voting	Enabled
Comb5	Optimized	Once	Disabled
Comb6	Optimized	Once	Enabled
Comb7	Optimized	Voting	Disabled
Comb8	Optimized	Voting	Enabled

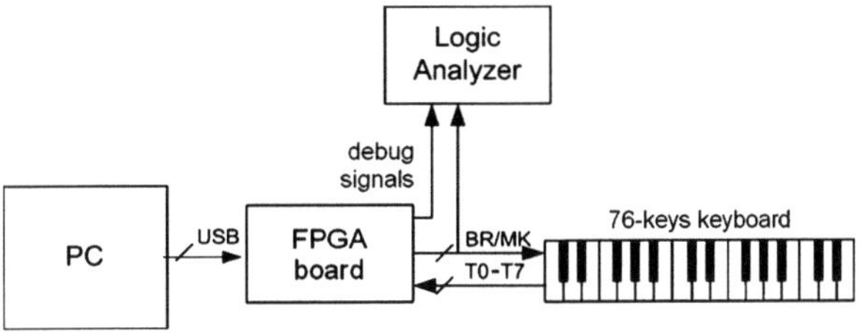

Fig. 9.6 Block diagram of the measurement setup

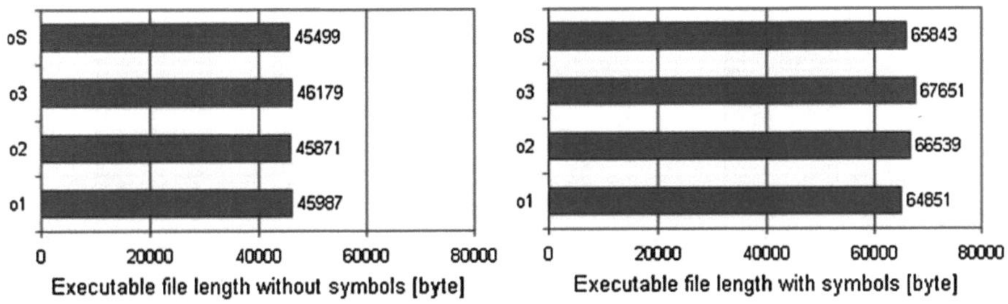

Fig. 9.7 Executable.elf file length variation with respect to different compilation options

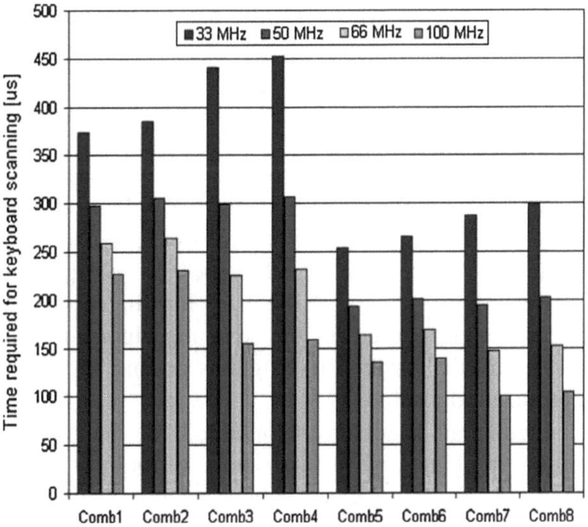

Fig. 9.8 Scanning time at different microblaze clock frequencies

obtained with a 100 MHz system clock are surely the best among the considered cases, however, it is important to remember that higher frequencies may create more problems with respect to electromagnetic compatibility issues.

A further analysis was performed to derive the percentage of occupation of the MicroBlaze computing power during the execution of the keyboard scanning task repeated every 800 μs.

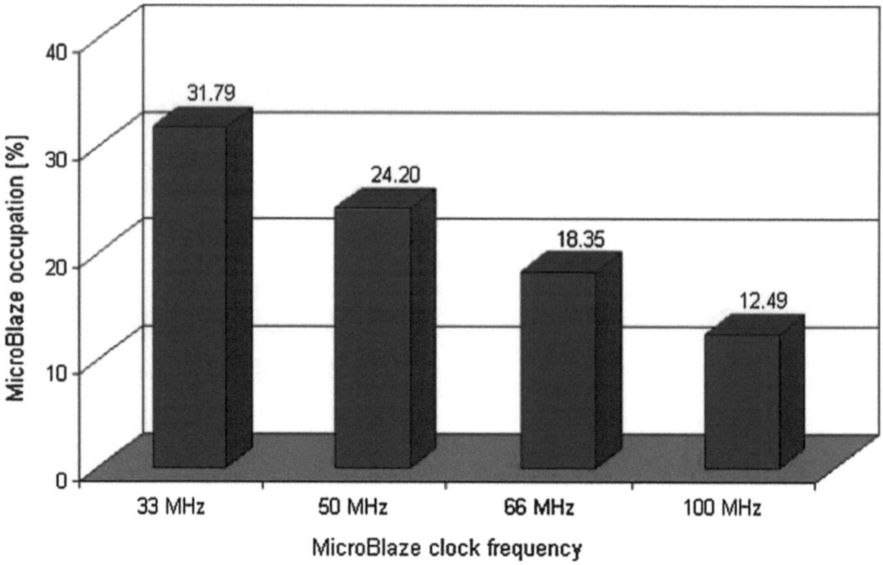

Fig. 9.9 Microblaze percentage occupation during a scan period of 800 μs

This measure, whose results are presented in Fig. 9.9, is very important to understand the availability of the microcontroller to perform additional functions, possibly provided by the main processor.

In conclusion two types of scanning algorithms have been taken into account, the standard and the optimized one. The second was built on the optimization of the initial standard. In the analysis of performance several compilation options were tested and the more efficient '-oS without debug symbols' allows a reduction of about 20% of the execution time of the scanning routine scan and a reduction of about 30% on code size, with respect to the option '-o1 with debug symbols'.

Regarding the size of the *ELF* file the results indicate that among the eight variations considered for the scanning algorithm, there are no remarkable differences and the binary file length is around 45 Kbytes for both the standard and optimized algorithms.

Significant results were obtained for the timing parameter. Using the algorithms implemented in this work, to obtain a scanning routine duration of less than 200 μs, MicroBlaze should be provided with a clock frequency of more than 33 MHz; in fact using a 50 MHz clock the minimum scanning routine duration corresponds to around 193.6 μs. In case this application is going to be executed with a scan period equal to 200 μs too little time remains to serve other tasks possibly assigned by the main processor; using a clock frequency of 66 or 100 MHz performance obviously increase significantly. These results show that it is possible to continue the research on a multiprocessor structure in which each processing element may execute different tasks assigned by the main controller.

9.5 Conclusion

In this paper a preliminary work was presented regarding a multiprocessor architecture for the efficient processing of applications for electronic musical instruments. This activity is still incomplete, but nonetheless some projects were completed in order to understand some key elements of the architecture as the strategy of communication between processors and their maximum achievable performance. The results obtained so far are quite interesting and the search will continue to integrate all elements of the architecture and to test it with real traffic conditions generated by different kind of applications.

References

1. Xing J, Zhao W, Hu H (2008) An FPGA-based experiment platform for multi-core system. In: The 9th international conference for young computer scientists, 18–21 November 2008, pp 2567–2571
2. Gao RX, Fan Z (2006) Architectural design of a sensory node controller for optimized energy utilization in sensor networks. IEEE Trans Instrum Meas 55(2):415–428
3. Maslennikov O, Shevtshenko J, Sergyienko A (2002) Configurable microcontroller array. In: Proceedings of the international conference on parallel computing in electrical engineering, pp 47–49
4. Marton L (2008) Distributed controller architecture for advanced robot control. In: International symposium on industrial electronics, 30 June 2008–2 July 2008, pp 1412–1417
5. Kim S-H, Seo S-H, Kim J-H, Moon T-M, Son C-W, Hwang S-H, Jeon JW (2008) A gateway system for an automotive system: LIN, CAN, and FlexRay. In: 6th IEEE international conference on industrial informatics, 13–16 July 2008, pp 967–972
6. Wobschall D, Prasad HS (2002) Esbus—a sensor bus based on the SPI serial interface. In: Proceedings of IEEE sensors vol 2. pp 1516–1519
7. Cucej Z, Gleich D, Kaiser M, Planinsi P (2004) Industrial networks. In: Proceedings of the 46th international symposium electronics in marine, 16–18 June 2004, pp 59–66
8. Paggi G, Ortolani M, Gigli S, Conti M, Caldari M, Ripa F (2009) Development of a multiprocessor architecture for efficient processing allocation in electronic musical instruments. In: Proceedings of the IEEE 7th international workshop on intelligent solutions in embedded systems WISES09 Ancona, Italy, June 2009, pp 79–86
9. Berekovic M, Heistermann D, Pirsch P (1998) A core generator for fully synthesizable and highly parameterizable RISC-cores for system-on-chip designs. In: IEEE Workshop on signal processing systems, 8–10 October 1998, pp 561–568
10. Hempel G, Hochberger C (2007) A resource optimized processor core for FPGA based SoCs. In: 10th euromicro conference on digital system design architectures, methods and tools, 29–31 August 2007, pp 51–58
11. Gschwind M, Salapura V, Maurer D (2001) FPGA prototyping of a RISC processor core for embedded applications. IEEE Transac Very Large Scale Integr Syst 9(2):241–250
12. MicroBlaze core, see http://www.xilinx.com

9.5 Conclusion

In this paper a preliminary work was presented regarding a multiprocessor architecture for the efficient processing of applications for electronic musical instruments. [illegible] completed in [illegible] of the [illegible] the [illegible] evaluated [illegible] with real traffic conditions generated by a different kind of applications.

References

[illegible]

Chapter 10
A Distributed Hardware Algorithm for Scheduling Dependent Tasks on Multicore Architectures

Lorenzo Di Gregorio

10.1 Introduction

Recently the industry has been moving toward multithreaded and multicore architectures in the hope of exploiting parallelism rather than pushing on crude performance: there is growing evidence across the semiconductor industry that the number of cores per chip doubles at least every three years.

Many embedded multicore architectures resemble the block diagram of Fig. 10.1: one *control* processor executes management software and delegates the processing of data streams by dispatching tasks to a *data plane* of specialized cores. It has been just the need for introducing processor cores into the data-planes of programmable chip architectures, which has driven several developments of the last decade in the field of embedded microprocessors. These data-plane processor cores may be equipped with memories, accelerators or coprocessors which can be shared within restricted local pools, hence the control flow of an application must actually migrate across the processors and may itself dispatch further tasks to the data-plane. Such architectures are typical of network processors and graphic processors, but get widely employed whenever the applications provide enough parallelism and the computation demands exceed the capabilities of standard processors.

For programming such parallel architectures, the threads model of computation seemed a straightforward and relatively small step from the conventional models, but it has faced surprising difficulties in establishing itself within the mainstream programming practices. According to Lee [1], the large amount of concurrency allowed within the thread model is actually excessive: he points out that synchronization primitives such as semaphores or barriers have turned out to be alien

L. Di Gregorio (✉)
Lantiq Deutschland GmbH, Neubiberg, Germany
e-mail: Lorenzo.DiGregorio@lantiq.com

M. Conti et al. (eds.), *Solutions on Embedded Systems*,
Lecture Notes in Electrical Engineering, 81, DOI: 10.1007/978-94-007-0638-5_10,

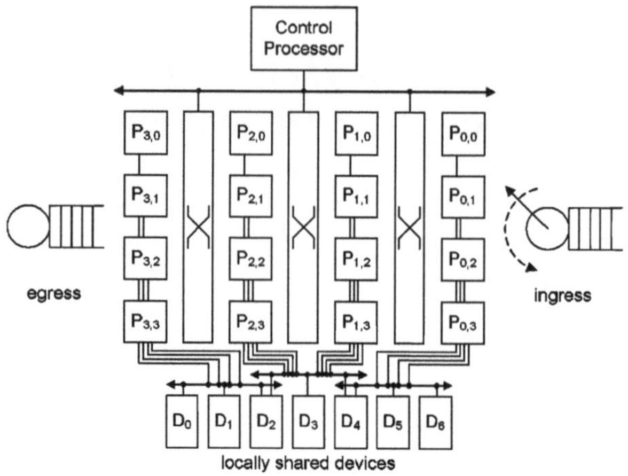

Fig. 10.1 Example of one embedded multicore architecture

and deceptive to programmers, while techniques for the automatic extraction of concurrency are still far from achieving maturity.

As a consequence of the lacking focus and magnitude in demand, multithreading has typically received only indispensable hardware support and most of its problems must be handled by software layers. Rather than insisting on thread parallelism, developers have been recently focusing on *task parallelism*, also known as function parallelism. For example, the recent OpenMP 3.0 specification introduces a task model [2].

We stress the point that the bookkeeping and sequencing activities required to schedule tasks should be a hardware duty and to this purpose we propose a novel hardware-based scheduling management infrastructure with a twofold purpose:

- retaining backward compatibility with legacy firmware and traditional programming models,
- offering the possibility of a gradual transition to a different programming model, by delegating part of the event management to the hardware layer.

We employ a distributed network of small hardware devices associated to the processing cores of the data plane. Every processing core interacts with its associated hardware unit by means of three operations: “declare”, “provide” and “require”. These operations are directly related to the conventional concepts of function call, value write and value read and can be as simple as memory-mapped accesses to the associated hardware units. Other structures for increased efficiency are equally possible, e.g. these units can be connected to the exception mechanism or to the context switch services of more sophisticated processor cores.

In contrast to conventional *synchronization* techniques, what this setup actually provides is a *sequencing* capability in hardware: this is key to hiding from the software all the event passing and enforcing sequences over code chunks. Yet, this sequencing support can still provide conventional mutex and barrier synchronization.

10.2 Examples

In this section we show some programming use cases. We want to schedule functions over a generic network of processing cores and in order to operate, we require that these functions get annotated to identify which shared resources are accessed by them.

On issuing a function to one core of the network, we first state which resources *might* need to be read within the body of the function and which resources *shall* be released by the function. This purpose is served by the operation DECLARE $((p_1, \ldots, p_n),(r_1, \ldots, r_m))$, which states that the function being entered *might* request access to the resources associated to the variables $r_1, \ldots, r_m$ and at any time before terminating it *shall* release the resources associated to the variables $p_1, \ldots, p_n$. It is legal to release a resource which has not been requested.

It is the operation REQUIRE$(r_i, \ldots, r_j)$ which actually requests access to the resources associated to the variables $r_i, \ldots, r_j$ and stalls the execution until all these resources are released. A non-blocking variant can be implemented as well. The operation PROVIDE$(p_i, \ldots, p_j)$ releases the resources associated to the variables $p_i, \ldots, p_j$.

This programming model is known as *task model* and every properly annotated function is considered a task. Tasks can call other functions or further tasks.

The algorithm in Fig. 10.2 shows an example of an endless loop on a task in charge of accessing two devices. If the iterations are distributed to parallel processing entities, it is well known that Dijkstra's classical *dining philosophers'* problem could lead to a deadlock or a livelock in the system if the two devices are interacting.

To prevent this situation, our algorithm ensures in hardware that whenever any task accesses x_0, that same task is guaranteed at some point in time in the future to get access to x_1 with the same task access order which has been applied to x_0. In this example we *declare* that in the body of the task we *might* require x_0 and x_1 and we *shall* provide x_0 and x_1. Subsequently, we do actually *require* access to device 0 (REQUIRE(x_0)) and release it (PROVIDE(x_0)) after having used it. When we will

Fig. 10.2 Declaration, requirements and provisions

```
loop
    (x0, x1) ⇐ task(x0, x1)
end loop
(x0, x1) = task(x0, x1) {
    DECLARE((x0, x1),(x0, x1))
    REQUIRE(x0)
    use device 0
    PROVIDE(x0)
    ...
    REQUIRE(x1)
    use device 1
    PROVIDE(x1)
    ...
}
```

Fig. 10.3 An example of in-order processing around a lock

```
Require: packet arrival indication
  DECLARE((x3, x4),(x3, x4))
  REQUIRE(x3) // packet available on input channel
  repeat // search for a free queue entry to load the packet
    DECLARE((x5),( x5))
    REQUIRE(x5)
    if queue entry is locked then
      PROVIDE(x5)
      select another queue entry
    end if
  until unlocked queue entry found
  lock queue entry
  PROVIDE(x5)
  transfer from input channel to queue entry
  PROVIDE(x3)
  ...
  process packet
  ...
  REQUIRE(x4)
  forward packet to output channel
  PROVIDE(x4)
  unlock queue entry
```

get to the point of requiring to access device 1 (REQUIRE(x_1)), we can be sure that this access gets granted with the same task order which has been applied to device 0.

A more complex example is provided by the algorithm in Fig. 10.3: a packet handler is triggered on packet arrivals and starts concurrently on different processors. It must fetch the packet from an input channel, store it into a queue, process it and forward it to an output channel. To avoid reassembly on the communication peers, the departure order of the packets must be the same as their arrival order. Still, the enqueuing and dequeuing must be carried out of order to exploit the memory bandwidth.

We associate x_3 to one input channel, x_4 to one output channel and x_5 to the memory queue. x_3 and x_4 are handled as x_0 and x_1 in the algorithm of Fig. 10.2. The **repeat–until** loop looks for a free entry to store the packet in a queue and forms an exclusive lock. Nevertheless, parallel instances of this algorithm will not deadlock while locking both the input channel and the queue because they present mutual exclusion with respect to x_5 but also present sequencing with respect to x_3. Furthermore, the packet order will not be changed, despite of different processing times, because the sequencing on x_3 also applies to x_4.

10.3 Related Work

Our hardware algorithm schedules sequences of tasks of the form $(p_{i,1}, \ldots, p_{i,n}) = f_i\ (r_{i,1}, \ldots, r_{i,m})$ with dependencies expressed by equalities $r_{j,x} = p_{i,y}, j > i$.

Our approach merely requires that the f_i get annotated with respect to the inputs and outputs which are relevant for the synchronization.

The idea of annotating functions and scheduling them on the basis of data dependencies, instead of scheduling threads on the basis of synchronization barriers, has been proposed by Bellens et al. in [3] for extracting parallelism in the compilation of software for the Cell BE architecture. Since our work focuses on a hardware implementation rather than on compilation, we employ an elaboration for multicore systems of the classical hardware algorithm by *Tomasulo* [4, 5]. In terms of the original algorithm's formulation, we regard every f_i as a large microcoded instruction whose inputs are $r_{i,1}, \ldots, r_{i,m}$ and whose outputs are $p_{i,1}, \ldots, p_{i,n}$.

Interestingly, Duran et al. propose in [6] to extend the tasking model of OpenMP 3.0 to dependent tasks and to detect the dependencies at runtime. For this purpose, Perez et al. present in [7] a bundle of compiler and runtime library, called SMPSs, which employs dependency renaming as provided by Tomasulo's algorithm. They indicate that for a good performance with their software solution, a granularity of considerably more than 105 cycles execution time is required. While we are aware from own experience that a much smaller granularity is well performing on protocol stack workloads, Stensland et al. provide in [8] a strong indication that, in order to reduce the overhead of the inter-core communication, this much smaller granularity should be the one of choice also for scheduling media applications on multicore architectures.

Within the software domain, the handling of both nested tasks and dependent tasks is still a partially open issue: Cilk [9] (now commercially evolved in Cilk ++), a task-based programming environment for recursive decomposition, supports nested tasks with task dependencies, but requires barriers to return values across the task recursion levels. OpenMP 3.0 [10] supports nested tasks but no task dependencies, while SMPSs [7] supports task dependencies but replaces nested tasks with conventional function calls. The hardware algorithm that we propose supports the scheduling of nested tasks along with task dependencies and does not strictly require barriers, although it requires including the remaining of a task after a spawning within a subtask in order to correctly receive values from the spawned task.

Our work is also very loosely related to the analysis of Salverda and Zilles [11] about instruction scheduling to multiple cores from typical general purpose workloads: indeed one could regard our work as core fusion at the granularity of small or medium size functions rather than at the instruction level granularity as in [11].

10.4 Algorithm

A task is described by its functionality $(p_{i,1}, \ldots, p_{i,n}) = f_i\ (r_{i,1}, \ldots, r_{i,m})$, the times $t_1^r, \ldots, t_m^r$ associated to the reads of its inputs $r_{i,1}, \ldots, r_{i,m}$ and the times $t_1^p, \ldots, t_n^p$ associated to writes of its outputs $p_{i,1}, \ldots, p_{i,n}$. The form $\bar{p} = f(\bar{r})$ represents a function and we term its input and output variables "*events*" in accordance with much literature on concurrent computation: we highlight that in this paper we disregard the values of the events and are only concerned with their access times. In the context of this work, we define as *declaration* of a function its issue to the multicore data plane, as *requirement* the reading of r_y at t_y^r and as *provision* the writing of p_x at t_x^p. With this notation we can pose the scheduling problem with the implicit requirement that tasks may not deadlock or livelock.

Definition (scheduling problem): For every pair of tasks f_i and f_j in an ordered sequence $f_1, \ldots, f_k$ with $i < j$, for every provision p_x of f_i and requirement r_y of f_j such that $p_x = r_y$, a valid schedule must hold f_j on its requirement of r_y until $t_y^r \geq t_x^p$. □

The scheme in Fig. 10.4 represents the components of the hardware layer for supporting the scheduling. The DECLARE, REQUIRE and PROVIDE operations are issued by the cores and control the *M* units. The *M* units interact with each other and decide whether to stall their associated cores until *all outstanding requirements* have been *provided.* The components shown in Fig. 10.4 are:

- a plurality of (*virtual*) *processing cores*, which can be a hardware accelerator, a processor or a virtual processor, intended as thread-reduced version of the underlying physical processor.
- a *multicore unit M* per processing core, which contains the proper hardware implementation of the scheduling algorithm, based on a very small content-addressed memory, called *requirement table R*, contained in *M* and addressed by event.
- an *event file E*, which is a central store area for *events* to get passed across the network of *M* units.
- a *sequencing* bus (Q-bus), which is a generic serial bus for serially issuing *declarations* over the network of *M* units.

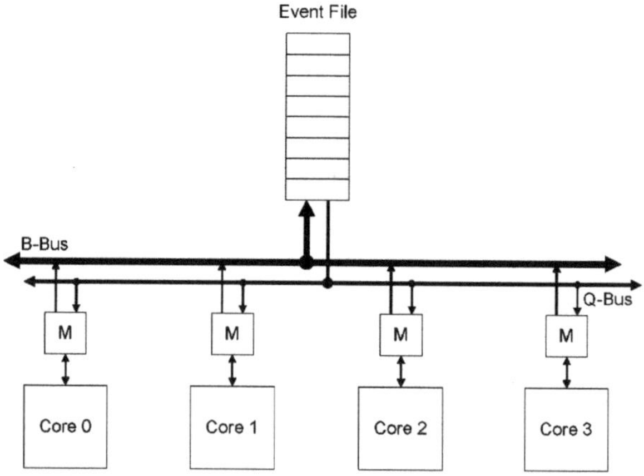

Fig. 10.4 Hardware units employed for supporting the scheduling

- a *broadcasting* bus (B-bus), which is a generic parallel bus for broadcasting several *provisions* in any order over the network.

Two further abstract agents are necessary to get the system running:

- a *communication backbone* between the cores, which is any communication structure for allowing the cores to pass data and control to each other.
- a *task dispatcher*, which is any software or hardware structure to dispatch tasks to the cores, e.g. one program running on a control processor.

It is pretty straightforward to compare this structure with Tomasulo's one and observe that the event file plays the role of the register file, the *M* units the role of the reservation stations and the B-bus the role of the common data bus. The main differences originate from the introduction of the declaration phase, the supporting Q-bus and from the generalization provided by the *colored* events, which we are going to present.

Events are associated to the resources to be employed: in order to employ a given shared resource, a core must require the associated event and provide it on releasing the resource. An event *e* is identified by a number and bears the following information:

- *e*.provided: indication that the event has been provided
- *e*.src: last declared provider of the event, i.e. last task which has accessed the Q-bus with event *e* among its *provisions*.

A colored event needs no ".src" field and bears the following information instead:

- *e*.capacity: much like the top of a counting semaphore, it is the maximum capacity of a shared resource to accept concurrent accesses.
- *e*.color: a qualifier which represents how many times this event has been declared for provision.

Colored events are associated to resources of corresponding capacity and are called "colored" because we picture that on using them, they change color and they are considered *provided* when they get back to their initial color. In our implementation we do not explicitly associate one capacity per event. Instead we regard all events belonging to a given range as bearing one capacity and put the ".color" field in place of the ".src" field within the event file.

Colored events are a different concept than standard events, because they provide sequencing without dependency renaming. In principle one could employ also multiple standard events to regulate the access to shared resources, but this approach clashes with the need to declare all events in the declaration phase of the task. This could be circumvented again by issuing multiple declarations in a similar way as it has been done in the example of the algorithm in Fig. 10.3 for a lock, but the whole handling would destroy the simplicity of the task-based scheduling which we achieve by the colored events.

In the next three sections we provide an abstract description of how the fundamental operations are implemented in the M units. Obviously, the physical implementation requires additional logic for bus access etc.

10.4.1 DECLARE Operation

The operation DECLARE((p_1, …, p_n), (r_1, …, r_m)) initiates a task by stating that the subsequent code *might* require events r_1, …, r_m and *shall* provide events p_1, …, p_n.

On a DECLARE, M assigns to the task a unique identifier i, e.g. by reading it from a counter and adding a unique prefix, and gains access on the Q-bus to perform the *atomic* bus transfer described in the algorithm of Fig. 10.5.

This operation locks the Q-bus for one burst $m + n$ transaction with the event file E: this guarantees that all DECLARE operations are seen serially by E.

DECLARE represents one entry point of a task and corresponds to the call of a function $\bar{p} = f(\bar{r})$. In a function, all code paths reachable from the call entry point must belong to the function until they provide valid outputs. In the same sense, all code paths reachable from a DECLARE and not providing all p_1, …, p_n must belong to the task. In OpenMP terminology this corresponds to a *task region* [10, p. 8] whose boundary is determined by the end of the *structured block* of the task generating construct. This observation provides an exact definition of what a task is and it is entirely possible for a task to call and also contain other tasks: it just need to contain additional DECLARE operations. It is also possible for a task to terminate (i.e. provide p_1, …, p_n) while called sub-tasks are still being executed.

10.4.2 REQUIRE Operation

The operation REQUIRE(r_a, …, r_z) holds the task until the events r_a, …, r_z get provided.

Fig. 10.5 DECLARE operation

```
R(r1),..., R(rm) ← E(r1),..., E(rm)
for all p ∈ {p1,...,pn} do
    if p is no colored event then
        E(p).provided ← False
        E(p).src ← i // i is the current task identifier
    else // p is a colored event
        E(p).color ← E(p).color+1
        if E(p).capacity ≤ E(p).color then
            E(p).provided ← False
        end if
    end if
end for
```

On a REQUIRE, M shall hold the core until the condition shown in the algorithm of Fig. 10.6 is met.

REQUIRE consults the local R table and not the global event file E. A task may REQUIRE only events which have been loaded into the R table by DECLARE, but it is not necessary to do so, e.g. the following code is legal:

```
DECLARE((r1, r2),(r1, r2))
REQUIRE(r1)
use device 1
if condition is true then
    REQUIRE(r2)
    use device 2
end if
PROVIDE(r1,r2)
```

Fig. 10.6 REQUIRE operation

```
while ∧_{e∈{r_a,...,r_z}} R(e).provided = False do
    hold core
end while
release core
```

10.4.3 PROVIDE Operation

The operation PROVIDE(p_a, ..., p_z) broadcasts the notifications that the events p_a, ..., p_z are being provided by the task i over the B-bus to all other M units. Furthermore it updates the event file E.

On a PROVIDE, all M units snooping on the B-bus update their R tables on receiving an event p according to the algorithm of Fig. 10.7.

Fig. 10.7 PROVIDE operation

```
if p.src ≠ i then
    if R(p) exists and p is no colored event then
        if R(p).src = p.src then
            R(p).provided ← True
        end if
    else if R(p) exists and
            p is a colored event and
            R(p).color > R(p).capacity then
        R(p).color ← R(p).color − 1
        if R(p).capacity > R(p).color then
            R(p).provided ← True
        end if
    end if
end if
```

Fig. 10.8 Provision to the event file

```
if p is no colored event then
    if E(p).src = p.src then
        E(p).provided ← True
    end if
else // p is a colored event
    E(p).color ← E(p).color − 1
    if E(p).capacity > E(p).color then
        E(p).provided ← True
    end if
end if
```

10.4.4 Event File

The event file E processes the provision of event p in a similar way as the M units do, as shown in the algorithm of Fig. 10.8.

10.4.5 Migration

In order to exploit the performance acceleration of local coprocessors and increase the reaction times of a task, we need to let the control flow migrate to different processing cores. Since a multithreaded processor is a resource of a task, we employ the distributed algorithm proposed in this paper to schedule migrations over the processing cores.

Assuming that some cores C_m, …, C_n are associated to the *colored* events x_m, …, x_n, the following code implements a migration from the core C_i to the core C_j by first obtaining access to a processor's context and then transferring the context-specific contents of the M unit. This transfer could have also been accomplished by a dedicated bus structure rather than in software.

```
DECLARE((x_1,...,x_{m-1},x_m,...,x_n),(...,x_m,...,x_n))
...
if migration to x_j then // note that j ∈ {m,...,n}
    REQUIRE(x_j)
    C_j.capacity ← C_i.capacity
    for all x ∈ {x_1,...,x_n} do
        C_j.M.R(x) ← C_i.M.R(x)
    end for
    PROVIDE(x_i)
else // we are sure that we will not migrate to x_j
    PROVIDE(x_j)
end if
```

To simplify matters, in this code we have omitted three features which we describe here in text:

1. in line 7, the transfer is affected by a race condition because $C_i.M.R(x)$ might get provided after having been read from $C_i.M$ but before being written into $C_j.M$. This race condition can be avoided by any of several well known techniques.
2. in line 9, on leaving the core C_i, the executing task must issue PROVIDE(x_i) only if the task had migrated onto C_i previously. A new task, which gets *initiated* on C_i, has not *migrated* onto it and does not need to release it with a PROVIDE(), in fact the corresponding variable x_i would not be in the task's DECLARE().
3. on terminating, if the task has migrated at all, it issue a PROVIDE() to release the last core it has migrated onto.

10.5 Experimental Setup

We have modeled the proposed algorithm for a generic multicore system of multithreaded processor cores as shown in Fig. 10.9. A distributor agent dispatches tasks to a subset of "entry" processors and these tasks are then free to migrate through the remaining "data plane" cores.

The figures of interest are:

- *makespan*: the time required to complete all the tasks divided by the total number of scheduled tasks.
- *sojourn time*: the time elapsed between the start and termination of a task.
- *execution time*: the time necessary for executing a task, including the peripheral access times but excluding the scheduling delays caused by thread preemption.
- *CPU time*: the time in which the task keeps the CPU busy.

These figures have been measured for two topologies which we have modeled: *parallel pipelines of processors* and *pipelines of parallel processors*. Our goal has

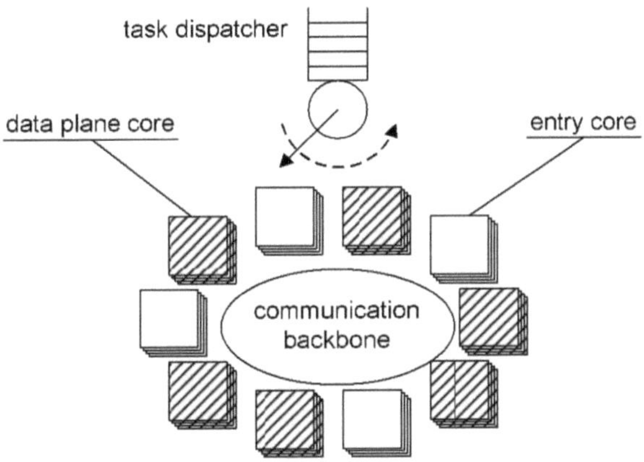

Fig. 10.9 Scheme of the generic multicore system employed in simulation

Table 10.1 Workload characteristics

Instruction	%	Characteristics
Execution	88	Takes up to 2,000 instructions
Access	7	Random latency up to 8 cycles
Synchronization	3	REQUIRE up to 16 events out of 64
		PROVIDE up to 32 events out of 64
Migration	1	Random migration points
Figures for the tasks in isolation		
Average execution time		1,026 cycles
Average CPU time		826 cycles
Average utilization		76%

been to investigate how the scheduling of tasks over these processor clusters can be improved. The basic topology consists of four lanes with eight stages each. Every processing core bears four contexts in its basic configuration. In the case of parallel pipelines, tasks are not allowed to move from one lane to the next. In the case of pipelines of parallel processors, they may do so.

With respect to Fig. 10.1, in parallel pipelines of processor, every processor $P_{i,j}$ can communicate only with $P_{i+1,j}$. In a pipeline of parallel processors, every processor $P_{i,j}$ can communicate with any processor $P_{i+1,k}, \forall\, k \in \{1, 2, 3, 4\}$.

We have randomized most characteristics to address generality. Both migration points and the destination of the migration are random. The context switch policy is also completely randomized and reflects the *generalized processor sharing* discipline common in many applications which process streaming data. It has the effect of equalizing the sojourn times of the tasks within the system: if tasks T_1 and T_2 are started at the same time, instead of executing task T_1 as first until time Δ_1 and subsequently task T_2 until time Δ_2, the execution of both tasks is distributed over the time max(Δ_1, Δ_2), consequently the average sojourn time will be max(Δ_1, Δ_2) instead of ($\Delta_1 + \Delta_2$)/2.

The figures for the tasks in isolation, reported in Table 10.1, correspond to the case in which 32 simultaneous tasks are executed on 32 parallel processors and show that 24% of the idle time in this workload is caused by dependencies between the tasks. Figure 10.10 shows how the sojourn time of 32 simultaneous tasks decreases and the processor idle time increases when moving from 32 contexts on a single processor to 32 single processors. It demonstrates that the idle time in the workload can be eliminated by multithreading.

10.6 Results

Our main results are summarized in Table 10.2. The scheduling performance achieved by the colored events is considerably higher than the one achieved by the standard events, i.e. pure dependency-based scheduling. Quadruplicating the width

Fig. 10.10 Workload sensibility to multithreading

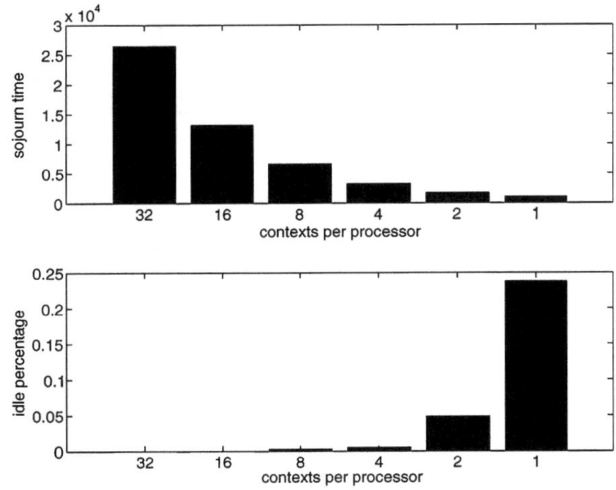

Table 10.2 Effect of task wormhole

Topology	Makespan	Sojourn	Utilization (%)
Parallel pipelines (colored)	35.87	3,273.43	73
Parallel pipelines (standard)	90.05	1,697.96	29
Pipeline of parallels (colored)	40.94	3,689.08	63
Pipeline of parallels (standard)	187.32	2,477.33	14
Pipeline of parallels (double size)	91.70	2,742.70	14
Pipeline of parallels (quad size)	55.79	3,250.93	12

of the pipeline, and hence the number of processors, still does not cope completely with the task congestion.

The parallel pipelines deliver a better performance than their equivalent pipelines of parallel processors because there is less traffic. In the case of pipelines of parallel processors, tasks may need to wait longer because their destinations can be occupied by tasks from other lanes. This penalty is not compensated by the fact that some lanes increase their availability due to the tasks which leave them.

The reason why the colored events perform better is that they allow wormhole routing of tasks while retaining deadlock freedom. The problem is shown in Fig. 10.11: task A may overtake task B and fill up the free context in the stage below B. If A depends on B and B shall provide its dependency only after having moved to the subsequent stage, a deadlock happens because B cannot move to the next stage occupied by A and A cannot leave it without B having provided the dependency first.

Without carrying out a finer functional partition to solve the problem "manually", the overtaking of tasks must be disabled to avoid deadlocks.

Instead, the colored events sequence only dependent tasks over the available contexts; therefore they provide a less strict policy for a deadlock-free routing than just disabling the overtaking.

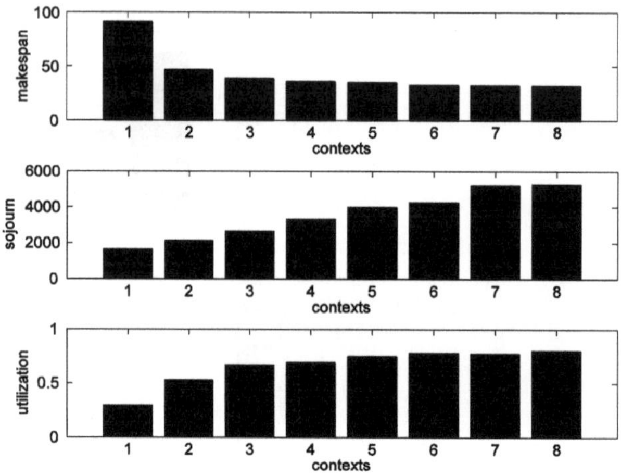

Fig. 10.11 Deadlock in wormhole routing of tasks

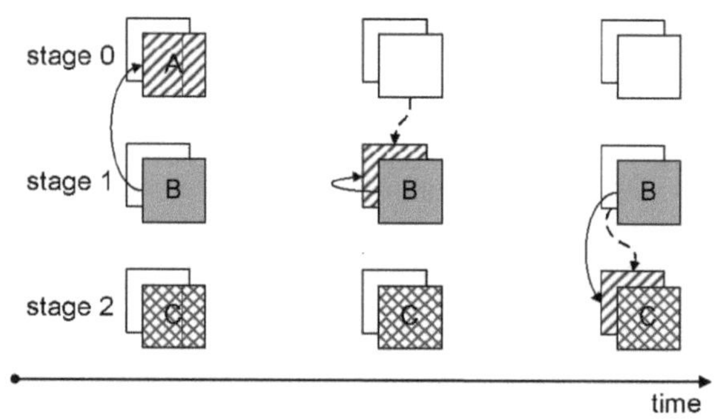

Fig. 10.12 Effect of increasing the number of contexts on a cluster of four parallel pipelines of eight processors each

In Fig. 10.12 we report the effect of increasing the number of contexts in a cluster of parallel pipelines. The makespan can be largely reduced by moving from one context to two, but it does not improve much by adding more than three contexts: further increases in the sojourn time of the tasks do not eliminate further idle time.

Subsequently, we have analyzed the effect of increasing the depth of several *parallel pipelines* and *pipelines of parallel* processors. Figure 10.13 represents the outcome of the measurements for a set of 64 processors bearing four contexts each. The processors have been initially organized in 32 parallel groups of two stages each and subsequently in 16, 8, 4 and 2 groups of respectively 2, 4, 8, 16 and 32 stages each. From the data in Fig. 10.13, we can estimate an increase of the makespan by about 5% for every halving of the number of parallel groups and doubling of the groups depth.

The additional flexibility of a pipeline of parallel processors costs from 13.5% (narrowest configuration: 2 groups of 32 stages each) to 25% (widest configuration: 32 groups of two stages each) in terms makespan for a random workload. The sojourn time increases about 1% slower than the makespan because of the lower utilization achieved in the last stages of the narrower configurations.

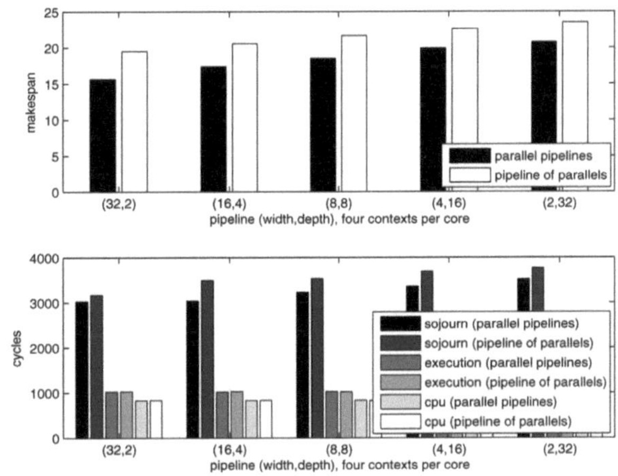

Fig. 10.13 Effect of task congestion in a pipeline of parallel processors

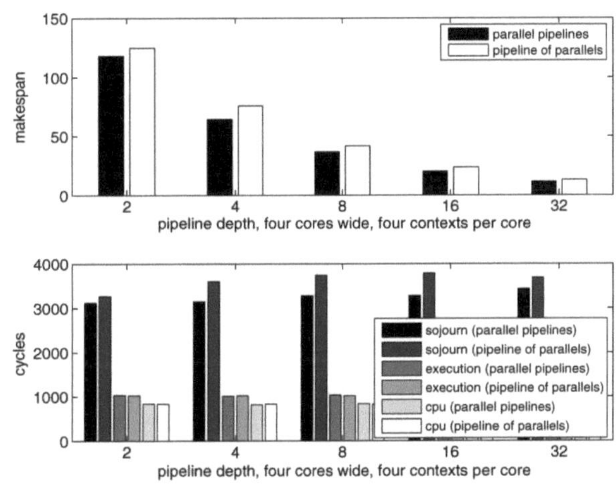

Fig. 10.14 Delay caused by deeper parallel pipelines

The results of Fig. 10.14 show the performance increase achieved by adding stages of four processors each to a four processors wide configuration. Every doubling of the pipeline depth leads to a performance increase of about 80%, with the pipeline of parallel processors delivering between 13.5 and 17.5% less performance than its equivalent parallel pipelines of processors.

10.7 Conclusions

We have presented a novel algorithm for scheduling tasks on multicore architectures. Its most striking feature is the hardware support for avoiding deadlocks and livelocks. In comparison to the fundamental algorithm by Tomasulo in [4, 5], we have introduced a separated *declaration* stage on a dedicated serial bus (*Q-bus*) and multiple *requirement* and *provision* stages. This generalization allows us to employ the algorithm for detecting and renaming data dependencies across multiple concurrent tasks, rather than across single instructions.

The approach of employing dependency renaming for scheduling tasks has been proposed in software by Perez et al. in [7], but it requires tasks of coarse granularity (10^5 cycles or more) to deliver a good performance. Instead, our hardware approach can efficiently schedule tasks of much finer granularity (down to a few tens of cycles), which are much more performing on embedded applications like the ones examined by Stensland et al. in [8].

Within our generalization, we have introduced the *colored events* for dealing with hardware resources supporting multiple concurrent accesses.

We have applied the colored events in the scheduling of tasks over pipelines of processors and we have shown that we can allow a deadlock-free *wormhole* scheduling of tasks across multithreaded processor networks. We have presented numerical evidence of how this scheduling can deliver more performance than a large increase in the number of processors.

This algorithm has been validated by intensive simulation. We have also carried out some hardware implementations, but they are not final and shall be a subject for future work.

This approach provides a *partial sequencing* of tasks with regard to selected resources, but it does not clash with other existing scheduling techniques, e.g. for increasing performance. As the number of processing cores per chip keeps increasing, traditional synchronization techniques will not cope with the scaling and we believe that this approach provides a more advanced and distributed sequencing technique, enabling a smooth transition from existing legacy code.

Acknowledgments This work has been partially supported by the German Federal Ministry of Education and Research (BMBF) under the project RapidMPSoC, grant number BMBF-01M3085B.

References

1. Lee EA (2006) The problem with threads. Computer 39(5):33–42
2. Ayguadé E, Copty N, Duran A, Hoeflinger J, Lin Y, Massaioli F, Teruel X, Unnikrishnan P, Zhang G (2009) The design of OpenMP tasks. IEEE Trans Parallel Distributed Syst 20(3):404–418
3. Bellens P, Perez JM, Badia RM, Labarta J (2006) CellSs: a programming model for the Cell BE architecture. In: SC '06: Proceedings of the 2006 ACM/IEEE conference on supercomputing. ACM, New York
4. Tomasulo RM (1967) An efficient algorithm for exploiting multiple arithmetic units. IBM J Res Dev 11(1):25–33
5. Tomasulo RM, Anderson DW, Powers DM (1969) Execution unit with a common operand and resulting bussing system. United States Patent, August, number US3462744
6. Duran A, Pérez JM, Ayguadé E, Badia RM, Labarta J (2008) Extending the OpenMP tasking model to allow dependent tasks. In: International workshop on OpenMP '08, pp 111–122

7. Perez J, Badia R, Labarta J (2008) A dependency-aware task-based programming environment for multi-core architectures. In: IEEE international conference on cluster computing, October 2008, pp 142–151
8. Stensland HK, Griwodz C, Halvorsen P (2008) Evaluation of multicore scheduling mechanisms for heterogeneous processing architectures. In: NOSSDAV '08: Proceedings of the 18th international workshop on network and operating systems support for digital audio and video. ACM, New York, pp 33–38
9. Frigo M, Leiserson CE, Randall KH (1998) The implementation of the Cilk-5 multithreaded language. In: Proceedings of the ACM SIGPLAN '98 conference on programming language design and implementation, Montreal, Quebec, Canada, June, 1998, pp 212–223 (proceedings published ACM SIGPLAN Notices, vol 33(5), May 2008)
10. OpenMP Architecture Review Board (2008) OpenMP application program interface-version 3.0. Avaliable online: http://www.openmp.org/mp-documents/spec30.pdf
11. Salverda P, Zilles C (2008) Fundamental performance constraints in horizontal fusion of in-order cores. In: 14th international symposium on high performance computer architecture (HPCA), pp 252–263

Part III
Automotive

Chapter 11
Automotive Embedded Systems

The Migration Challenges to a Time Triggered Paradigm

Eric Armengaud, Allan Tengg, Mario Driussi, Michael Karner, Christian Steger and Reinhold Weiß

11.1 Introduction

Cars have become highly distributed systems implementing up to 70 Electronics Control Units (ECUs) and exchanging up to 2,500 different messages (e.g. speed sensor) [1]. The network plays a central role as communication enabler between the ECUs and further in maintaining the system in a safe state. Former technologies (e.g. CAN) used to rely on event-triggered communication systems. The introduction of the time-triggered paradigm [2] with technologies such as FlexRay [3] or TTP/C [4] provides different advantages with respect to system design, integration and validation. One of its important attribute is to shift the complexity to an earlier design phase and force the system designer to completely define the system behavior and thus a priori solve the potential conflicts. However, the paradigm change has different fundamental impacts with regards to the communication concepts and to the integration within the software components.

It is the aim of the research project TEODACS[1] (Test, Evaluation and Optimization of Dependable Automotive Communication Systems) to gather expertise for the deployment and validation of FlexRay based distributed systems. The approach is based on the development of a co-simulation framework (FlexRayXpert.Sim) tightly interfaced to a realistic prototype (FlexRayXpert.Lab).

E. Armengaud (✉) · A. Tengg · M. Driussi
Virtual Vehicle Competence Center, Graz, Austria
e-mail: eric.armengaud@v2c2.at

M. Karner · C. Steger · R. Weiß
Graz University of Technology, Graz, Austria

[1] http://www.teodacs.com

M. Conti et al. (eds.), *Solutions on Embedded Systems*,
Lecture Notes in Electrical Engineering, 81, DOI: 10.1007/978-94-007-0638-5_11,

The simulation framework, on one side, provides observability of the internal components and thus enables the efficient analysis of the system. On the other side, the prototype presents a real behavior both in the time and value domain. The aim of this document is to review the philosophy behind the time-triggered architecture and point out the deployment and integration challenges related to FlexRay. The focus is set to the migration challenge as well and the different options for integrating this new paradigm within an existing software environment. Further, we present the development flow used within TEODACS for the efficient design and configuration of our two platforms.

This document is organized as follows: Sect. 11.2 reviews the concepts of the time-triggered architecture. Section 11.3 focuses on the FlexRay protocol and the integration challenges due to the paradigm shift. Then, Sect. 11.4 presents the software development flow enhancement for the development of the two platforms. Finally, Sect. 11.5 concludes this work.

11.2 The Time-Triggered Architecture

11.2.1 The Time-Triggered Computation Model

An event-triggered architecture is characterized by the fact that all system activities are initiated by an event [5] and consequently *reacts* to its environment (an operation is started as soon as the event is received, regardless the current processing status). On the contrary, in the time-triggered architecture, every action is derived solely from the progression of real-time and thus *follows* the progression of its environment (an operation is started at a pre-defined starting point and processes the information that has occurred since the last computation; conflicts about processing resources are avoided per construction).

The time-triggered computational model is based on the representation of the controlled system as a Real-Time entity (RT entity), which represents the system as a subset of significant state variables [6]. This RT entity can be observed at a particular point in time. The observation is then a Real-Time image (RT image)—a current picture of a RT entity that is an accurate representation of the RT entity, both in the value and the time domains [6]. The communication follows a periodic scheme where the system status (e.g. current motor speed) is updated. This differs from event-triggered communication where event messages (e.g. motor speed increased by 500 rpm) are transmitted using an exactly-once semantic [7]. For the time-triggered paradigm, each node is provided with a RT image of the controlled system, which is processed locally according to a static, a priori defined schedule. This deterministic progression of the system enables task multiplexing in the time domain in order to avoid conflicts and race conditions.

The static schedule is based on the concept of the *sparse time base* [8]. Parallel to synchronous microchips that rely on a discrete global clock to update its registers and trigger a new computation step, the clock in the time-triggered architecture is made discrete and distributed within the system. The availability of a globally agreed time base provides *simultaneity property*: the nodes act at about the same time (defined by the precision of the clock synchronization) on the same observation. This aims at keeping temporal coordination between the nodes and avoiding unsynchronized behavior of the system. Second, it provides *temporal ordering*, so that the nodes react on different observations in the temporal order of their occurrence, thus avoiding state divergence. The time-triggered architecture provides the four following properties:

P1 *Independent node development* at the architecture level. This attribute is based on the precise specification of the node both in the time and value domain, and on an abstract model of the node services. This information enables independent development by different teams and supports system integration [9].

P2 *Stability of prior services* at the node level, which means that the validated services of a node—both in the value domain and in the time domain—is not refuted by the integration of the node into an encompassing system-of-systems [9]. This requirement aims at enabling sequential node development and reducing the integration efforts (a module needs to be validated only once and not after every development stage).

P3 *Constructive integration* of the communication system, to ensure that the integration of the $n + 1$ node will not disturb the operation of the n nodes already integrated [2]. This requirement has implication for the management of the network resources and assures that the timing constraints are satisfied even at the critical instant (i.e. when all nodes request the network at the same instant). It supports system development by avoiding sporadic failures during the integration of additional nodes.

P4 *Replica determinism* is required for service replication in order to tolerate faults and improve the system robustness. A fault tolerant unit consists of a set of replicated nodes that are intended to produce the same results at approximately the same time [5]. Their role is twofold [10]: they make the system resilient to transmission errors (since the computation result is transmitted more than once) and to measurement and computation errors (occurring before transmission). For that, they require (1) agreement on inputs, (2) agreement on computation time (replica coordination) and (3) deterministic algorithms [2, 11]. Items (1) and (2) are provided by the time-triggered architecture.

11.2.2 Time-Triggered Versus Event-Triggered Architecture

A lot of comparisons between event- and time-triggered architectures have already been published (e.g., [12–16] for some recent ones) without having clearly identified the "best" solution. In fact, the two architectures focus on different properties. Event-triggered architectures provide *flexibility* and try to improve the *overall performance* while the focus is set to *timeliness* and *worst-case execution time* for the time-triggered architecture [13, 16]. The time-triggered computation model presents different advantages for the efficient design and development of safety-critical systems:

- the system complexity increases more than linear with the system size (number of elements and intensity of the interaction) [17]. Structuring—the description of a system at an abstract level—is required to cope with the complexity. Horizontal structuring, or layering, is related to the representation of the system at different abstraction levels and can be used both in event- and time-triggered systems. Vertical layering, or partitioning, splits a system into a number of nearly independent subsystems with their own resources and well-specified interfaces, both in the temporal and value domain [6]. This concept requires the system to be composable [9, 12] and to adhere to the four properties discussed in Sect. 11.2.1. Event-triggered architectures are not composable, since the temporal behavior of the communication system depends on the application software [9] and on the bus load. The addition of nodes might affect the system and the stability of prior services (properties P2 and P3).
- robustness to their environment: Time-triggered architectures are not driven by interrupts outside their sphere of control, but instead decide autonomously when to observe their environment. Consequently, and contrary to event-triggered architectures, there is no possibility for a malicious device to upset a time-triggered system [17].
- fault containment: the time-triggered architecture provides an interface free of temporal control signals, (*temporal firewall*), thus providing error containment regions within the system. This attribute increases the overall system's dependability since errors are contained and do not lead to a system failure.
- static schedule: all timing and data dependencies are resolved during system design, thus simplifying the inter-task synchronization and avoiding race conditions within the system. Moreover, the static schedule strongly supports timing analysis [18].
- efficient fault detection: the periodic and a priori defined task execution can be used for fast fault detection and the message transmission can be used as "heartbeat'' to detect failed node [15].
- deterministic communication with guaranteed worse-case transmission time and low jitters (required for high-performance control loop).

One important advantage of event-triggered architectures is that fewer assumptions are required to build a system [14]. Adding a node into a system does not require any change in the other nodes, but can invalidate the temporal behavior of the system [9]. This makes the event-triggered architecture more flexible and avoids a restrictive design process as required for the time-triggered architecture [12]. Moreover, event-triggered systems make better use of the bandwidth due to better average transmission time (the messages are transmitted as soon as the communication medium is available). This leads also to a better average system reactivity. To conclude, event-triggered systems are well suited for sporadic transmission, alarm, low-power sleep modes and best effort soft real-time systems. Time-triggered systems, on the other hand, trade the flexibility for more predictability, determinism and guaranteed latencies.

11.2.3 Time-Triggered Communication

Time-Triggered communication protocols such as TTP/C or FlexRay implement a Time Division Multiple Access (TDMA) scheme based on a priori defined time windows ("communication slots"), which are uniquely assigned to the nodes for message transmission within a periodic communication cycle, see Fig. 11.1. The messages are broadcasted above the communication medium and consequently are available for each node of the cluster. A fundamental principle is that the transmission depends only on the time progression and is not triggered by any external (not-deterministic) event.

Communication between two systems is commonly based on a master–slave control scheme. Data exchange can be initiated by the sender (*push* style) or by the receiver (*pull* style). In both cases the requester generates the control flow, and thus can start a transmission at any time. While this scheme is very comfortable for the master, the slave has to stay available at any time, which may result in high resource costs and difficult scheduling. Time-triggered communication protocols are using a combination of push and pull communication model [19]. During a first step, the transmitter implements a push style and transfers its data to a local memory. Then, the communication service autonomously transmits the message according to the pre-defined schedule. On the receiver side, the message is stored into a local memory and stays available for the consumer (pull style).

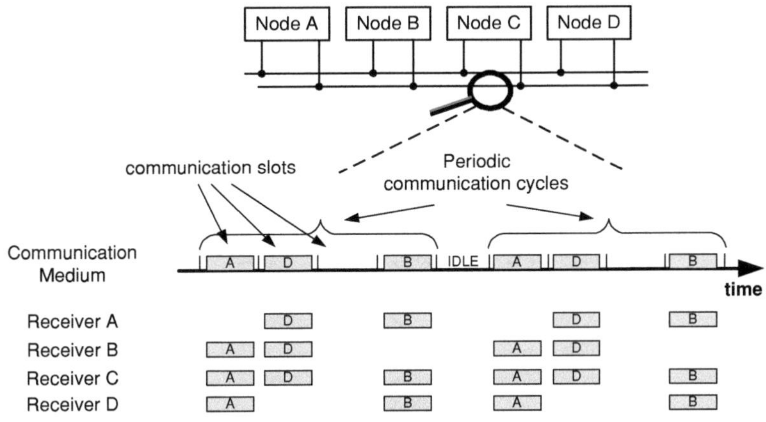

Fig. 11.1 TDMA scheme

This combination is ideal both for the sender and receiver since they can transmit and access the data whenever they want and do not need to be watchful for transmission request. This communication scheme is building a temporal firewall [20], a fully specified interface for the exchange of data. Additionally to the resource saving, this interface is free of end-to-end control signals and thus avoids the possibility of control-error propagation.

11.3 FlexRay and the Integration Challenges

11.3.1 The FlexRay Host Interface

FlexRay defines both a static segment for deterministic communication as well as a dynamic segment for prioritized communication. The FlexRay Controller Host Interface (CHI—see Fig. 11.2) provides protocol and message data interface with the host software for efficient configuration, control and data exchange.

The protocol data interface, on one side, is dedicated to the configuration and control of the FlexRay controller. Contrary to event-triggered communication systems, the timing behavior of time-triggered protocols is a priori defined and must be initialized. The configuration parameters fall into two categories: *global cluster parameters*, which define the communication behavior agreed within the entire network. For FlexRay V2.1A, it consists of 39 parameters (relevant plus related). Further, *local node parameters* define a specific node behavior such as key slot number or last transmission time during the dynamic segment. FlexRay V2.1A defines 29 local parameters that need to be individually configured for each node.

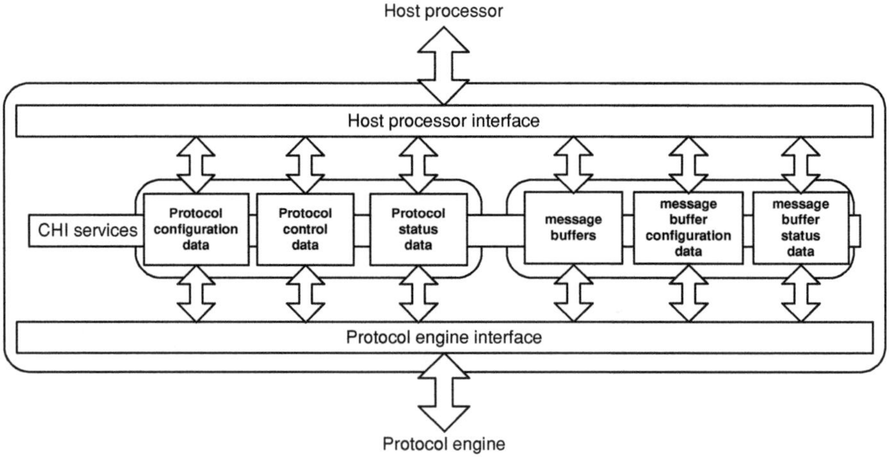

Fig. 11.2 FlexRay controller host interface [3]

The control and status interface is kept quite simple since functionalities such as wake-up, start-up, synchronization, triggering for sending/receiving message are performed autonomously by the FlexRay controller. The control interface consists of (1) mode change request for the FlexRay Protocol Operation Control (e.g. between config mode, run mode, halt request), (2) medium test triggering, and finally (3) the optional control of external clock synchronization (e.g. for synchronizing different FlexRay clusters together). Status information regroups the status of the Protocol Operation Control as well as the FlexRay global time. This last information is relevant when the node's software has to be synchronized with the FlexRay network.

The message data interface is organized as a list of buffers that are configured prior operation. Regarding transmission, the configuration consists of channel, slot ID, header CRC information as well as communication cycle. Notice that FlexRay supports both slot and cycle multiplexing. A frame sending point is defined both by a slot identifier and by a cycle counter. This tuple increases the flexibility with regards to bandwidth management. Concerning the reception, buffers can be assigned to single frames (same as transmission) or organized as FIFO.

11.3.2 Integration Within the Node's Software—The Challenges

An important feature of this protocol is to support both time- and event-triggered communication as well as to present a flexible interface for the efficient integration within time-triggered and event-triggered software architectures. On one side, FlexRay provides different interrupts (transmission, reception, status change) for the integration within an event-triggered operating system. Moreover, the FIFO based buffer configuration is well suited to event-triggered communication, where each signal represents a value change ("event") and should not be missed. On the other side, the FlexRay controller provides global time information for the synchronization between node's operation and communication, as well as buffer–frame mapping for the efficient data access (in a time-triggered architecture the timing relations are known in advance and each node precisely knows when a message is expected/needs to be sent).

The flexible interface makes thus different integration variants possible, with different advantages and disadvantages, see Fig. 11.3 for an overview. Principally we differentiate between event-triggered and time-triggered paradigms, both for the communication architecture and for the node's operating system. Historically, systems used to be purely event-triggered. While providing flexibility and efficient use of the bandwidth, these systems are extremely difficult to validate (see Sect. 11.2.2) which presents negative effects on their robustness. However, the migration from a purely event-triggered to a purely time-triggered system usually requires a complete system re-design that is difficult to perform at once.

		Operating system	
		Event-triggered (interrupts driven)	Time triggered (schedule)
Communication system	Event-triggered (e.g. CAN)	➔ *Priority based communication* + Flexibility, average response time - Complex timing analysis - None of properties P1 to P4	➔ *Static communication scheme supported by the application* + Easy timing analysis + Stability of prior services (property P2) - Application overhead (e.g. for synchronization)
	Time-triggered (e.g. FlexRay)	➔ *static comm. scheme with interrupt based data interface* + Constructive integration (property P3) - No synchronization between the tasks within the system (properties P1, P4 not verified) - No stability of prior services (property P2 not verified)	➔ *Asynchronous systems* + Properties P2, P3 - No synchronization between the tasks within the system (properties P1, P4 not verified) ➔ *Synchronous systems* + All advantages of a time-triggered architecture (properties P1, P2, P3, P4) - Flexibility

Fig. 11.3 Integration variants

Two intermediate solutions exist to improve the system's robustness (see Fig. 11.3): (1) the integration of time-triggered communication into an event-triggered system and (2) the migration to a time-triggered operating system with an underlaying event-triggered communication protocol. Solution (1) improves the communication robustness (property P3): The time-triggered protocol ensures that the introduction of a new functionality (new set of nodes) will not influence the existing communication. Solution (2) represents the intention of introducing a time-triggered architecture without dedicated communication system. The different services (e.g. synchronization) have to be implemented in software, thus introducing an overhead for the system. Recent works relate the problematic of making coexisting time-triggered with event-triggered schemes within a system. In [21], the transmission latency for a system comprising a TTP/C and a CAN network is analyzed. In [7], virtual networks providing both event-triggered and time-triggered schemes are implemented on top of a TTP/C network.

The last migration step goes toward the integration of a synchronized time-triggered communication protocol and operating system as presented in Sect. 11.2.1. The resulting system can be efficiently analyzed and different approaches (e.g. redundancy) can be deployed to improve the system robustness. The main drawback of this approach is the rigidity: the node's timing behavior is related to the communication timing (schedule) and must be correspondingly adapted from an implementation to the other. More especially, the timing information must stay consistent during the entire design flow. Hence, timing consistency is required at node level between application, basic software and communication architecture (cross layer), as well as at system level between the ECUs (cross partition). The deployment of time-triggered architectures therefore strongly requires the formalization (modeling) of the system description (including timing behavior) and the according tool chains to support efficient implementation.

11.4 TEODACS Development Process

11.4.1 TEODACS Development Platforms FlexRayXpert.Lab and FlexRayXpert.Sim

The approach chosen within the TEODACS project is based on the parallel development of a co-simulation environment as well as a hardware prototype which model the entire distributed system, see Fig. 11.4. The tightly interface between the two platforms sums up the advantages of both environments. Despite their large differences, both platforms require a similar system and architecture design, configuration and validation flow.

The main aims of the prototype environment FlexRayXpert.Lab are to provide a realistic network reflecting the current car architecture and to understand the typical design, integration and validation challenges a car supplier is confronted to. Our prototype implements different topologies (active star, bus topology with different cable length) and regroups different suppliers: FlexRay transceivers from NXP (TJA1080) and austriamicrosystems (AS8221 and AS8224) as well as standalone FlexRay controller from Fujitsu (MB88121B) and Infineon (CIC310), and integrated solutions from Freescale (S12XF512 MCU with embedded FlexRay controller).

The FlexRayXpert.Sim co-simulation platform combines models of the different network components in order to simulate the entire communication architecture. The co-simulation framework [22] creates the possibility to implement selectable levels of accuracy according to the requested needs, thus largely reducing the processing resources and making the analysis of such complex systems possible. In our case we are using the System Architect Designer tool from

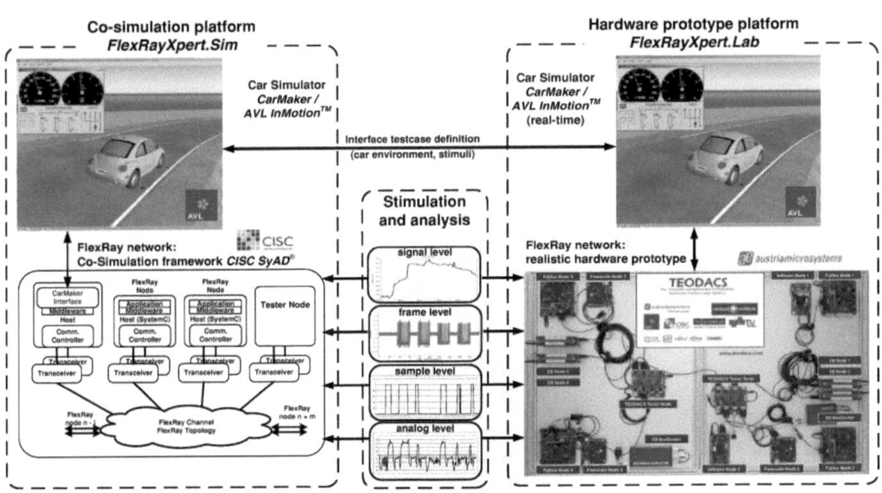

Fig. 11.4 TEODACS development platforms: FlexRayXpert.Lab and FlexRayXpert.Sim

CISC [23]. The simulation of the entire architecture supports the analysis of the interactions between the single components and thus the design exploration of the assembled system. The two platforms are stimulated by the CarMaker/AVL InMotion simulator [24], which simulates the dynamics of a car driving on a road and thus provides a realistic workload for the network.

11.4.2 The TEODACS Development Flow

The development flow chosen within TEODACS is illustrated in Fig. 11.5 and relies on (1) the identification of the main development steps as well as their corresponding meta-models, (2) the identification of the common elements between these development steps and (3) the development of dedicated transformators to efficiently map the information from a development step to the next one. The first step consists of *functional design* and represents the design of the system functionalities independently from the software or hardware architecture. We use for that the CapeMaster tool [25] that enables a hierarchic description of the system and of the dependencies between the functions. This information can be used for automated configuration generation (e.g. configuration of communication protocols), consistency checks (e.g. bus load checks) or for fault diagnosis (e.g. which functions are directly or indirectly affected from a missing or corrupted information).

The second steps represents the *design of the communication architecture* according to the functionalities described previously. Different data models such as Fibex [26] and AUTOSAR [27] have been developed to cope with the growing

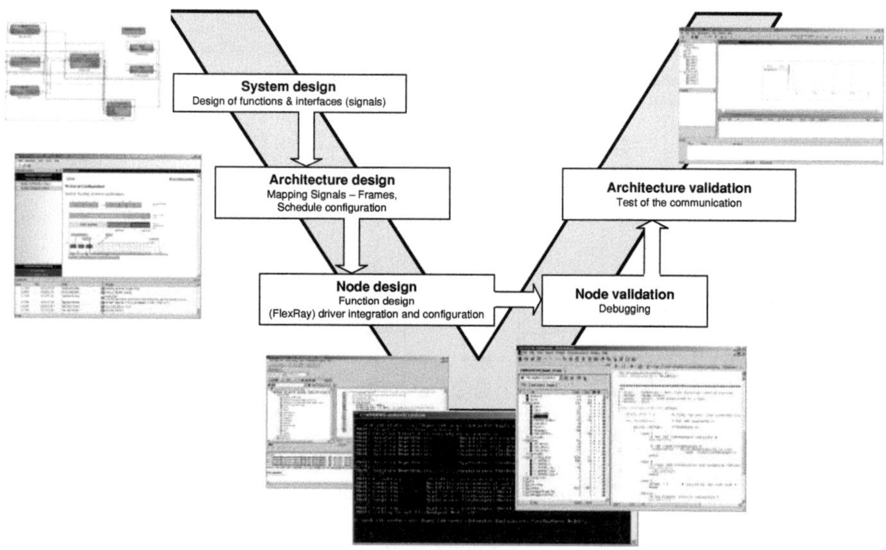

Fig. 11.5 TEODACS development flow

complexity of the communication configuration. They describe the topology (which clusters, ECUs and communication channels exists), the communication (mapping frame–signals), as well as the relation between the different parts (e.g. mapping frame–channel–ECU). In our case we are using the DesignerPro tool from Elektrobit.[2] It enables the efficient configuration of the FlexRay schedule as well as of the communication matrix. It takes as input the hardware topology (which nodes are available), the mapping between ECU and functions as well as the list of signals (application data such as engine speed) to be transmitted. Then, the user is guided during the definition of a FlexRay schedule as well as for the definition of frames and mapping between signals and frames. The output of this tool is a Fibex file.

The following steps represents *node design and validation*. In the case of TEODACS, we rely on existing starter kits from different suppliers. This diversity enables the integration of different chips and require the parallel use of different Integrated Development Environments (IDEs) as well as different FlexRay drivers. This diversity represents a challenge at two levels at least. First, the different IDEs require additional training in order to get fit for the different compilers and environments. Second, the different FlexRay drivers present different configuration interfaces and need to be adapted consequently. A contribution of the TEODACS project is a tool for the efficient export of the Fibex configuration to the different platforms as well as the enhancement of a FlexRay driver. These points are discussed in the following section.

The last step described here is the *validation of the communication architecture*. It represents the validation of the inter-ECU communication using the FlexRay network. For that, the network behavior is monitored at different abstraction levels in parallel and each attribute (e.g. schedule, bus traffic) is tested against the standards and against the configuration. In particular, FlexRay monitoring is performed using the tresos Inspector tool from Elektrobit. The tests from the resulting traces include both schedule information (e.g. cycle length, slot length) and comparison with the expected configuration (Fibex). Missing or unexpected frames can be thus easily detected. The proposed analysis of the communication completeness is useful for system correction and validation. Note that this approach can be easily extended for the analysis of the application if further information about the system functionalities is available (e.g. formal models of the components). Further information about the proposed test approach is available in [28].

11.4.3 FIBEX Database Format

The most important entities defined by a *FIBEX database* are described briefly in the following; for more details please consult the FIBEX specification [26].

[2] http://www.elektrobit.com

- *CLUSTER* A cluster describes the ensemble of processing nodes (ECUs), which are linked by a communication medium sharing the same communication protocol, (i.e. CAN, FlexRay, MOST, ...).
- *CHANNEL* Some communication networks provide more than one communication channel (i.e. FlexRay channel 'A' and 'B').
- *ECU* An electronic control unit represents a real-world processing unit, composed of one or more processors, memory and I/O ports, that execute parts of a distributed application. Each ECU has at least one communication controller.
- *CONTROLLER* A communication controller is a dedicated hardware device that allows ECUs to send and receive messages on the communication medium.
- *CONNECTOR* Connectors are symbolic elements, describing the bus-interface of the ECUs and specify the send and receive behaviour of a node.
- *SIGNAL* Signals are input or output parameters of a function, represented by a contiguous sequence of bits.
- *FRAME* A frame is the smallest piece of information that is exchanged over a communication channel. Typically it is composed of several PDUs (protocol data units) which are basically a collection of several SIGNALs.
- *FRAME-TRIGGERING* represents the condition for which a frame is transmitted on the network. In case of time-triggered protocols, a frame triggering is usually a communication slot (eventually combined with a communication cycle number).

These different entities represent the different system views at different abstraction levels as well as the mapping between these views. Hence, hardware information (e.g. channels, controllers) are mapped with software information (e.g. signals). Moreover, low level views (frames) are mapped to high level views (signals).

11.4.4 Automated Configuration of the FlexRay Communication Stack

The efficient export of the configuration (e.g. from Fibex) to the different platforms represents a challenge. Hence, the correct and complete data structure for a given target platform has to be collected and efficiently exported to a given development environment. Within the TEODACS project, we have developed such an exporter tool in order to automatize the configuration steps, both for the FlexRayXpert.Lab and for the FlexRayXpert.Sim platforms [29]. This tool principally reads the Fibex file and exports the relevant data for a given target platform. In the context of embedded systems, resources like memory and processing power are quite limited. Therefore it is not feasible to load the entire Fibex database onto an ECU and browse for the needed information every time a message has to be received/sent. Hence, a data structure is needed which is derived (offline) from a given Fibex database and which complies with the following requirements:

- Representation by (automatically generated) C source code
- Short and constant time to determine for a given *slot-id* and *cycle-counter* the corresponding signals
- Suitable to prepare incoming and outgoing message buffers
- Low memory consumption
- Readable for humans

A data structure that combines all these requirements is presented in Fig. 11.6. The structures *Cluster FlexRay Parameter* and *ECU FlexRay Parameter* are needed to initialize the FlexRay communication controller in the startup phase. The array *Slot Usage Read/Write A/B* determines for every *slot-id* whether it is used or not. A value of *NULL* indicates that this slot is not used on this ECU and, thus, no message buffer needs to be registered for it. Otherwise a buffer must be registered for the appropriate direction and channel. If a slot is cycle-multiplexed, the repetition counter value tells how many frames share this slot; the pointer to the cycle usage array (*CycXYn*) specifies the frames which should be sent/received in each cycle. It may happen that some slots have to be skipped for some cycles—this is indicated by a *Frame* = *NULL* value. Once the correct frame is discovered, the incoming message can be split up into its signals; the outgoing messages can be compiled from its signals respectively.

It is easy to see that it takes constant time to find for a given *slot-id* and *cycle counter* the corresponding frame by using this data structure. The memory requirements of these structures are very low. Since all values and pointers can be resolved at compile time, the linker is able to store them in the code segment, thus conserving RAM. Because the variable names generated by our tools are chosen carefully, it is quite easy for a human developer to understand the underlying communication matrix even in the C-source-code.

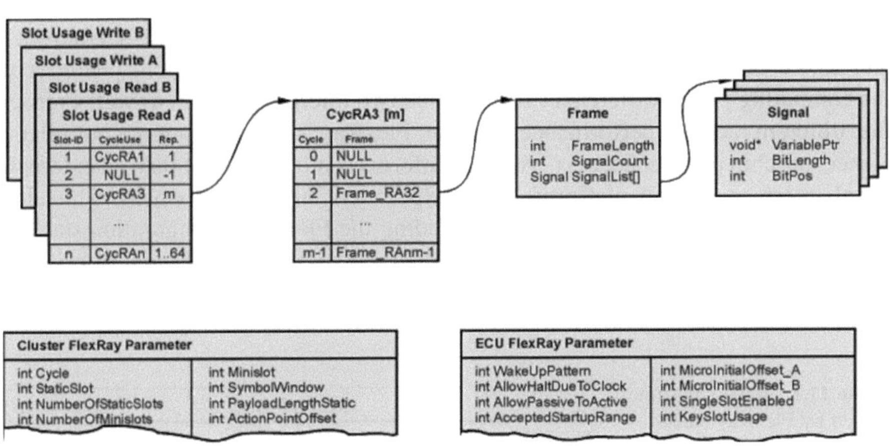

Fig. 11.6 Structure of configuration data

A further development within the TEODACS project concerns the FlexRay drivers. The motivation for that is to (1) efficiently import the configuration generated, and (2) abstract the underlying communication architecture. The advantage of this approach is that the node's application then directly access the variables of interest without requiring information about the communication architecture (mapping frame–signal), and thus does not require any know-how (or have any dependency) concerning the communication technology. This concept is similar to the Virtual Functional Bus from AUTOSAR and can be used for both event-triggered and time-triggered schemes. Only the triggering (when the frames are transmitted, when the signals are updated) changes.

11.4.5 Implementation of the FlexRay Communication Stack

The automated configuration, as presented in the previous chapter, has been applied to two different FlexRay evaluation boards:

- High-end automotive microcontroller: Infineon *Tricore TC1797* 32-bit controller running at 180 MHz with FlexRay CC (CIC310) on chip—starter kit from Infineon
- Low-cost generic microcontroller: Atmel ATMEGA128 8-bit controller running at 16 MHz with Flexray CC (CIC310) connected via SPI—own development

Based on a simple application both implementations have been validated and evaluated. For this simple application a FlexRay cycle time of 5 ms has been used. The static segment was divided into 80 slots à 16 byte messages; the remaining cycle time was configured to host dynamic slots. For this specific FlexRay configuration the overall reception- and transmission-time of a single message has been measured (see Table 11.1).

Our example application occupies four static slots for transmission and five static slots for reception. A cpu-load of 3.4% is caused on the TC1797 controller and 28.4% on the ATMEL respectively. The performance differences result from the computing power differences between the two microcontrollers as well as from the different access performances to the FlexRay communication controllers. Hence, on-chip parallel access (TriCore platform) is more efficient than an external, serial interface (Atmel platform). This experiment highlights two main results. First, the proposed development flow including the FlexRay configuration stack is suitable for different kind of microcontrollers. Second, the Atmel platform better suits the development of intelligent sensors with low requirement on bandwidth

Table 11.1 Processing time caused by FlexRay stack

	Receive (μs)	Transmit (μs)
TC1797	20.8	17.3
ATMEGA 128	179.8	130.2

(only a few messages need to be exchanged). Dedicated microcontrollers (such as TriCore) are required for central ECUs in order to efficiently support the full power of FlexRay.

11.5 Conclusion

The time-triggered paradigm enhances the design flow with the early integration of the timing behavior. This additional system view, on one side, provides interesting properties such as independent node development, stability of prior services or constructive integration at communication level and thus supports the development of (timely) predictable systems. On the other side, this additional timing information has to be efficiently managed during the entire design process. It is outmost important to keep the timing information consistent across the component boundaries. To that aim, seamless modeling approaches as well as the assistance of dedicated tool chains are required to support the development process. We have presented the design flow used within the TEODACS approach, and experimentally evaluated its capacities for a high-end and for a low-cost platform.

Acknowledgments The authors wish to thank the "COMET K2 Forschungsförderungs-Programm" of the Austrian Federal Ministry for Transport, Innovation and Technology (BMVIT), the Austrian Federal Ministry of Economics and Labour (BMWA), Österreichische Forschungsförderungsgesellschaft mbH (FFG), Das Land Steiermark and Steirische Wirtschaftsförderung (SFG) for their financial support.Additionally we would like to thank the supporting companies and project partners austriamicrosystems, AVL List and CISC Semiconductor as well as Graz University of Technology and the University of Applied Sciences FH Joanneum.

References

1. Hansen P (2005) New S-Class Mercedes: pioneering electronics. The Hansen report on automotive electronics, vol 18, no 8, pp 1–2, Oct 2005
2. Kopetz H, Bauer G (2003) The time-triggered architecture. In: Proceedings of the IEEE, vol 91, no 1, pp 112–126, Jan 2003
3. FlexRay Communications System—Protocol Specification V2.1 Rev A (2005). http://www.flexray.com, Flexray Consortium, December 2005
4. Time-triggered protocol TTP/C high level specification, Document Protocol Version 1.1, http://www.tttech.com/technology/specification.htm, 2005
5. Kopetz H (1997) Real-time systems: design principles for distributed embedded applications. Kluwer Academic Publishers, Norwell
6. Kopetz H (1998) The time-triggered model of computation. In: Proceedings of the 19th IEEE Real-Time Systems Symposium, Dec 1998, pp 168–177
7. Obermaisser R (2008) Temporal partitioning of communication resources in an integrated architecture. IEEE Transactions on Dependable and Secure Computing, vol 5, no 2, pp 99–114
8. Kopetz H (1992) Sparse time versus dense time in distributed real-time systems. In: Proceedings of the 12th International Conference on Distributed Computing Systems, June 1992, pp 460–467

9. Kopetz H, Obermaisser R (2002) Temporal composability. Comput Control Eng J 13(4):156–162
10. Wilwert C, Navet N, Song Y-Q, Simonot-Lion F (2004) Design of automotive X-by-wire systems. In: Zurawski R (ed) The industrial communication technology handbook. CRC Press, Boca Raton
11. Bauer G (2001) Transparent fault tolerance in a time-triggered architecture. Ph.D. Dissertation, Technische Universität Wien, Institut für Technische Informatik, Treitlstr. 3/3/182-1, 1040 Vienna, Austria
12. Alber A (2004) Comparison of event-triggered and time-triggered concepts with regard to distributed control systems. In: Embedded World, pp 235–252
13. Almeida L, Pedreiras P, Fonseca JAG (2002) The FTT-CAN Protocol: why and how. IEEE Trans Indus Electron (TIE) 49(6):1189–1201
14. Gwaltney D, Briscoe J (2006) Comparison of communication architectures for spacecraft modular avionics systems. NASA/TM-2006-214431
15. Navet N, Simonot-Lion F (2005) Fault tolerant services for safe in-car embedded systems. The embedded systems handbook. CRC Press/Taylor & Francis, Boca Raton
16. Scarlett JJ, Brennan RW (2006) Re-evaluating event-triggered and time-triggered systems. In: 11th IEEE International Conference on Emerging Technologies and Factory Automation (ETFA'06), Sept 2006, pp 655–661
17. Kopetz H, Braun M, Ebner C, Kruger A, Millinger D, Nossal R, Schedl A (1995) The design of large real-time systems: the time-triggered approach. In: Proceedings of the 16th IEEE Real-Time Systems Symposium, Dec 1995, pp 182–187
18. Ebner C (1998) Efficiency evaluation of a time-triggered architecture for vehicle body-electronics. In: Proceedings of the 10th Euromicro Workshop on Real-Time Systems, June 1998, pp 62–67
19. Elmenreich W, Bauer G, Kopetz H (2003) The time-triggered paradigm. In: Proceedings of the Workshop on Time-Triggered and Real-Time Communication, Manno, Switzerland, Dec 2003
20. Kopetz H, Nossal R (1997) Temporal firewalls in large distributed real-time systems. In: Proceedings of the 6th IEEE Computer Society Workshop on Future Trends of Distributed Computing Systems, Oct 1997, pp 310–315
21. Zug S, Schulze M, Kaiser J (2008) Latency analysis for the cooperation of event and time-triggered networks. In: IEEE International Workshop on Factory Communication Systems, 2008 (WFCS 2008), May 2008, pp 3–9
22. Kajtazovic S, Steger C, Pistauer M (2005) A HDL-independent modeling methodology for heterogeneous system designs. In: Behavioral Modeling and Simulation Workshop, 2005. BMAS 2005. Proceedings of the 2005 IEEE International, pp 88–93
23. System Architect Designer (SyAD®), http://www.cisc.at/SyAD, CISC Semiconductor Design + Consulting GmbH, Lakeside B07, 9020 Klagenfurt, Austria, www.cisc.at, March 2007
24. Schyr C, Schaden T, Schantl R (2008) New frontloading potentials through coupling of HiL-simulation and engine test bed. In: FISITA 2008 World Automotive Congress, September 2008, pp F2008-12-317
25. Watzenig D, Pölzlbauer F, Kaiser J (2007) Fault tracking and failure effect analysis in complex automotive control systems based on a generic modeling approach. In: SAE World Congress 2007, Apr 2007, pp 2007-10-31
26. ASAM MCD-2 NET (2008) Fibex v3.0—data model for ECU network systems. http://www.asam.net, ASAM
27. Fennel H, Bunzel S, Heinecke H, Bielefeld J, Fuerst S, Schnelle K-P, Grote W, Maldener N, Weber T, Wohlgemuth F, Ruh J, Lundh L, Sanden T, Heitkaem-per P, Rimkus R, Leour J, Gilberg A, Virnich U, Voget S, Nishikawa K, Kajio K, Lange K, Scharnhorst T, Kunkel B (2006) Achievements and exploitation of the AUTOSAR Development Partnership. In: Convergence 2006, October 2006, p 10

28. Armengaud E, Tengg A, Karner M, Steger C, Weiss R, Kohl M (2009) Moving beyond the component boundaries for efficient test and diagnosis of automotive communication architectures: In: 14th IEEE International Conference on Emerging Technologies and Factory Automation (ETFA 2009), Sept 2009, 8 pp
29. Armengaud E, Tengg A, Driussi M, Karner M, Steger C, Weiß R (2009) Automotive software architecture: migration challenges from an event-triggered to a time-triggered communication scheme. 7th Workshop on Intelligent solutions in Embedded Systems, pp 95 –103

Chapter 12
An Embedded Datalogger with a Fast Acquisition Rate for In-vehicle Testing and Monitoring

Automotive Testing

Gioacchino Fertitta, Antonio Di Stefano, Giuseppe Fiscelli and Costantino G. Giaconia

12.1 Introduction

The growing complexity of automotive systems and the widespread use of electronics in almost any key vehicle component, ranging from safety critical ones, such as airbag, engine or breaking control to passenger comfort, makes in-vehicle testing a more and more complex task to deploy, and a very time consuming activity to carry out. This is an important issue since the fulfilment of new standards and regulations push manufacturers to increasingly allocate bigger amount of R&D time and budget in tests. In order to address these issues a key factor is the choice of a suitable instrumentation capable of performing these tests. Compared to laboratory measures, in-vehicle and ground tests present a number of challenges and requirements not always addressed by commercial-off-the-shelf test products. These include the need of a great versatility (i.e. adaptability to different measure configurations and interfacing needs as well as ease of use), compact physical dimensions, low power consumption and a rugged design in order to withstand to strong vibrations and harsh environments. Currently available commercial products fall into two main categories: PC-based data acquisition systems [1] and stand-alone dataloggers [2–4]. The former is the most common and flexible solution but it is characterized by higher volume occupation, high power consumption and moderate tolerance to vibrations (mainly due to the presence of a laptop PC). The latter is represented by autonomous units capable of recording signals coming from sensors or vehicle data buses. These products are usually rugged and more compact, but their dimensions do not always allow an easy collocation in every vehicle position. Besides this, both types of instruments

G. Fertitta · A. D. Stefano (✉) · G. Fiscelli · C. G. Giaconia
Dipartimento di Ingegneria Elettrica, Elettronica e delle Telecomunicazioni,
Universita degli Studi di Palermo, Viale delle Scienze, ed. 9, 90128 Palermo, Italy
e-mail: distefano@dieet.unipa.it

M. Conti et al. (eds.), *Solutions on Embedded Systems*,
Lecture Notes in Electrical Engineering, 81, DOI: 10.1007/978-94-007-0638-5_12,

usually provide either few acquisition channels either moderate sampling rate (in the order of few tens of KHz per channel). These may represent major limiting factors when performing high demanding or complex tests.

The proposed datalogger, as previously described in [5], is designed to address all this issues, making so possible deploying either ordinary tests, either more complex and sophisticated measures, not performed by traditional instruments. The most innovative features of the proposed datalogger are the simultaneous availability of 10 analog acquisition channels, 4 high speed digital I/O, fast acquisition rate up to 100 kHz per channel (1 MHz for digital I/Os), a rugged structure, very small size (Fig. 12.1) and its interfacing capabilities, making possible to deploy tests with novel approaches. As better described in the following paragraphs, this great flexibility and high performance is a direct result of a particular hardware and software design and partitioning.

12.2 Application Requirements

Before describing the datalogger hardware and firmware architecture, it is worth to consider typical requirements of the target application. The main purpose of the datalogger is to acquire signals coming from a number of on-board sensors, during bench or on-road tests. The number of required analog inputs usually spans from a few to some tens, and some digital inputs are also required (either for logging binary values, either for counting purpose). Sampled data are recorded on a high capacity storage medium, usually based on flash memories since magnetic hard disks can be easily damaged by in-vehicle vibrations. Recording sessions starts at power up, by an user command or by an external trigger. A problem often found when performing a test session is the wiring set-up. Especially when tens of channels have to be used, setting up the wirings can be a difficult and time-consuming task, since different vehicle subsystems can be located quite far from the

Fig. 12.1 Datalogger prototype

logger position (usually set inside the passenger compartment). Moreover long wires are easily subject to vibration issues and strong electromagnetic interferences. Usually sampling frequencies in the range of 1 to some tens of Ksps per channel are used, but the most demanding or complex measures, such as noise and vibration tests, may requires higher bandwidth, up to 50 kHz. In order to address these requirements the developed datalogger employs a radically different approach, allowing both to satisfy these specifications and to obtain better performance. The datalogger was designed as a small modular and scalable instrument: the device is capable of logging 10 analog channels and 4 digital ones, and to locally store sampled data in a flash memory. Thanks to its small physical dimensions, and to a rugged enclosure, the datalogger can be set wherever in the vehicle, even near the engine or powertrain, in the body or chassis. If more channels have to be monitored, more than one logger can be used to sample the signals in parallel. This approach gives two advantages: the loggers can be set near components to monitor, without the need of setting up long wirings, and the acquisition rate is not decreased when the numbers of channels grows, since it is done in parallel by all the loggers. The loggers can be remotely synchronized, controlled and monitored through the vehicle CAN bus [6], to which they are connected.

The datalogger offers an high acquisition throughput compared to commercial products, sampling each of the 10 analog channels up to 100 Ksps with an ADC resolution of 12 bit. Analog inputs are configurable as single-ended, differential and pseudo-differential, in order to be easily connected to different sensors. The 4 digital inputs are sampled at 1 Msps so allowing to implement fast triggers and counters too. Acquired data are stored in a local CompactFlash card that is also used to store the logger configuration that is loaded during the instrument start-up, thus avoiding the need of a local PC in the harsh test environment.

In order to make this approach convenient compared to conventional ones [3, 4], a very efficient and accurate hardware and firmware design had to be carried out, so to considerably reduce the implementation costs.

12.3 Hardware Architecture

The most challenging constraint in designing the datalogger hardware was the quite high data throughput that had to be handled. Considering a maximum aggregate sampling rate of about 1 Msps (100 Ksps per each of the 10 channels), and a 12 bit resolution of converted analog data plus 4 bit of digital data, an overall continuous data stream of 2 MB/s has to be expected. Such a throughput may represent a significant issue for a small real-time embedded system: firstly the CPU has to control and continuously handle the data acquisition process, so a very little and constrained processing time is left for other tasks such as interfacing and data storage; secondly other tasks, namely storage, become very critical since they have to be performed faster than the data acquisition task, otherwise there will be an increasing data accumulation, leading to a final data loss.

In the sampling phase the CPU has to scan all the selected channels while skipping the unselected ones, properly setting them so to handle single-end and differential channels, and transferring the converted data. Since the aggregate maximum sampling frequency is 1 MHz, this process has to be completed in less than 1 μs, so to reserve an adequate spare processor time for other tasks. The proposed system employs a 32 bit ARM7TDMI [7] based microcontroller from Analog Devices endowed with a 12 bit A/D converter operating up to 1 Msps, and 12 multiplexed input channels [8]. All the peripherals (A/D converter, GPIO, timers, etc.) are mapped into the ARM memory space and are connected to the same bus. This solution guarantees an adequate data throughput and processing capability. On the other hand however, even if the CPU is capable of operating at 41.7 MHz, delivering about 41 MIPS, it was not possible to meet all the constraints due to the interrupt latencies and the large number of clock cycles required to access all the memory mapped registers. As an example, when executing code from RAM, about 20 ARM instructions only can be executed in 1 μs, and the interrupt latency can be as long as 24 clock cycles. For this reason a multiprocessor architecture was adopted, as showed in Fig. 12.2. The second microcontroller was a low cost 8 bit ATMega16 AVR micro from Atmel [9], working at 16 MHz. The AVR microcontroller was chosen for its computational efficiency (about 1 MIPS/MHz), for the very low latency in instruction execution, jump, interrupt and I/O operations. This choice allowed to completely fulfil all the requirements at a very little cost. A special care in designing task partitioning was however required, as explained hereafter. As shown in Fig. 12.2 the ARM microcontroller was used to handle the data acquisition process, while the AVR

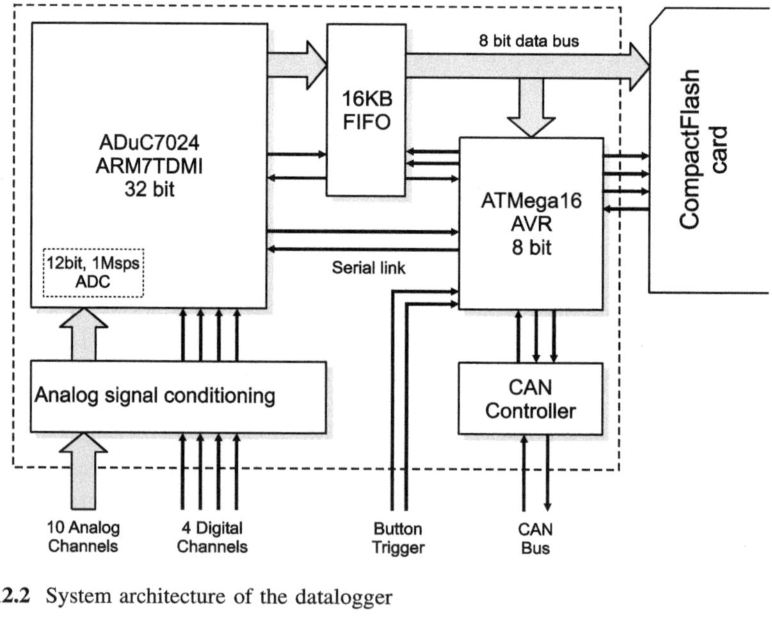

Fig. 12.2 System architecture of the datalogger

was used as a master, controlling and coordinating the system operations, handling the interfacing tasks and data storage with the flash card. The two CPUs need to communicate only during boot phase and at the end of each recording section, so only a bi-directional asynchronous serial link was used to this end. High speed data transfer is instead obtained through a double port FIFO memory.

As already mentioned an essential requirement for proper system operations is that the storage throughput is greater than the data acquisition one. This second problem is closely related not only to the processing time available for this task, but also to the storage medium used. At first a 2 MB/s throughput may seem quite reasonable for any of the currently available storage media (see Table 12.1), but actually this is not the case. Flash media devices specifications in fact are given only for peak, burst or average data transfers [10–12], if real data write throughput is considered, a substantial worst performance is obtained.

This is due to high flash cards latencies in executing commands, in particular write commands. These delays are usually unpredictable, depending on the specific card brand and integrated flash controller, on the previous memory content and finally on the particular address location and data block size. Even if average delays are quite tolerable (being in the order of 100 μs), very long delays, as long as 100 ms, can be occasionally obtained from certain cards. This poses a very important constraint on the choice of the storage media. In order to tolerate these unpredictable delays an asynchronous 16 KB FIFO buffer was used. The FIFO buffer allowed the two CPU to operate in a completely asynchronous and independent fashion. The acquisition task timings are in fact governed by the programmed acquisition rate, the storage tasks by external events or signals (communication bus, storage card response, pushbutton etc.). FIFO status signals (FIFO full and FIFO empty) provides an automatic handshake for coordinating the two part of the system. The choice on the specific flash media to use was done considering the maximum throughput, the maximum capacity, the availability of open specifications and the availability of industrial grade components. This analysis led to the choice of CompactFlash cards [13]. The card was interfaced using the 8 bit IDE mode, considering that the overall throughput is limited by the above mentioned delays rather than the bus capacity.

At system boot the AVR checks for the flash card, reads the previously stored configuration, decodes it and configures the system, also sending the configuration data containing the selected channels and sample rate to the ARM. The acquisition starts according to the programmed trigger: manual click on the pushbutton,

Table 12.1 Flash media comparison

Media	Max throughput (MB/s)	Max capacity (GB)	Open specifications
SD/SDHC	20	32	Partially
MEMStick	20–30	8	No
XD	9	2	No
CompactFlash	60	64	Yes
USB	40	32	Yes

external trigger or reception of a specific CAN command. Once the acquisition is started, the ARM keeps scanning and reading the analog and digital channel values and writing data to the FIFO, while the AVR handles data write to the CompactFlash memory, check for external events (pushbutton click, command on the CAN bus), and sends a low frequency replica of selected channel data to the CAN bus for external logging or monitoring purposes.

The use of the CAN bus is particularly suited for this application, since it employs a broadcast scheme to transmit packets. By exploiting this feature it is possible to send commands to all the loggers on the bus at the same time, also obtaining an easy synchronization and control method. When a stop condition is verified (stop command, memory full, etc.), the AVR CPU issues a halt to the ARM microcontroller, so stopping the data acquisition.

12.4 Firmware Description

Due to the precise and stringent timing requirements, and the hard real-time nature of the application, the firmware design for the two microcontrollers required a particular care and the use of some unconventional programming techniques. Both firmware were written in ANSI C language, so easing and accelerating the development and debug process, but the most time critical routines were written in assembly language. The GCC tool chain was used for both firmware developments. A more detailed description of the firmware structure and operation is provided in the following paragraphs.

12.4.1 AVR Firmware

The AVR firmware tasks and structure are shown in Fig. 12.3. The firmware was designed trying to exploit the AVR core features. In particular, the absence of an instruction pipeline avoid jump penalties, so the use of conditional operations is convenient; interrupt latencies are very small, so interrupt can be used without significantly perturbing the main tasks; registers and port operation timings are deterministic and very fast (requiring only 1 or 2 clock cycles), allowing fast internal data movement and efficient external bit banging operations, useful for interface handling, especially when coupled with read-modify-writing or bit instructions. At system boot, after the preliminary peripheral configuration, the CompactFlash card is initialised and checked. If it is correctly inserted and ready, the stored system configuration is read, decoded and actuated by sending acquisition parameters to the ARM. Then the data write loop is entered, accomplishing the task of reading data from the FIFO and writing them to the CompactFlash. In order to satisfy the main constraint of a continuous writing throughput greater than the acquisition one, this routine was optimised and coded in assembly language and the fastest strategy for driving the CompactFlash was used. Speeding up memory write operations is not an easy and obvious task, in fact there are two main delays imposed by the CompactFlash: one at the reception of the command block and the other while executing the command once all arguments (data) have been transferred.

CompactFlash employs ATA block command set [14] and this requires to send an 8 byte block for each write command, hence in order to minimise these delays either command number has to be minimised, or argument dimension (data) have to be maximized for each command.

Since the standard command set allows different possibilities to perform write operations and cards of different brands may implement a different command subset or may handle in a different way standard commands, a number of commands and command combinations were tested. In particular the following tests were

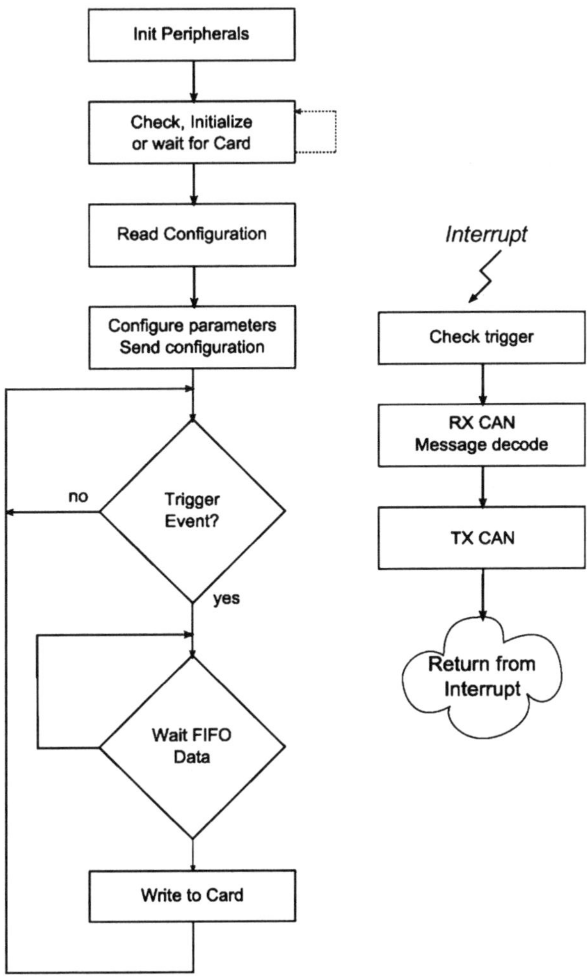

Fig. 12.3 Flow chart of the AVR firmware

performed: "Write Sector(s)" command with one sector only, "Write Sector(s)" command with up to 256 sectors, "Write Multiple Sectors", "Write Sector(s) w/o Erase" with and without pre-formatting. Different card models or brands gave slightly different results, but for most of the tested cards best performances were obtained by using the "Write Sector(s)" command with 256 sectors. This allowed to write blocks of 256 × 512 bytes for each command sent. The assembly coded routine achieved a raw data throughput of more than 4 MB/s, this allowed to tolerate card delays and latencies, and to prevent the FIFO to overflow. Data were sequentially stored in the card, as they were read from the FIFO. Session information such as channels configuration, acquisition rate, start sector, length, etc. were also updated at the end of the recording session in a specific area used as a directory. More recording sessions then are possible in the same CompactFlash card. In the implemented version no explicit file system was used, either for efficiency reasons, either for the limits of the most common file systems in handling very large files (in FAT32 for example maximum file length is 4 GB) [15]. The data write was handled by the firmware main loop while external events such as triggers, CAN data reception and decoding has been handled instead by a small interrupt service routine. To this purpose a relatively slow timer, with a period of about 1 ms, controlled the firmware operation by setting some global flags. The interrupt provided reception and transmission of CAN frames, starting and stopping the acquisition if programmed conditions are met (triggers, CAN messages or pushbutton).

12.4.2 ARM Firmware

The ARM7TDMI has radically different characteristics compared to the AVR core; in particular it features an instruction pipeline, long interrupt delays and quite long access times for memory or peripheral. Moreover the execution of code from internal flash memory is slower than the execution from RAM. These facts make

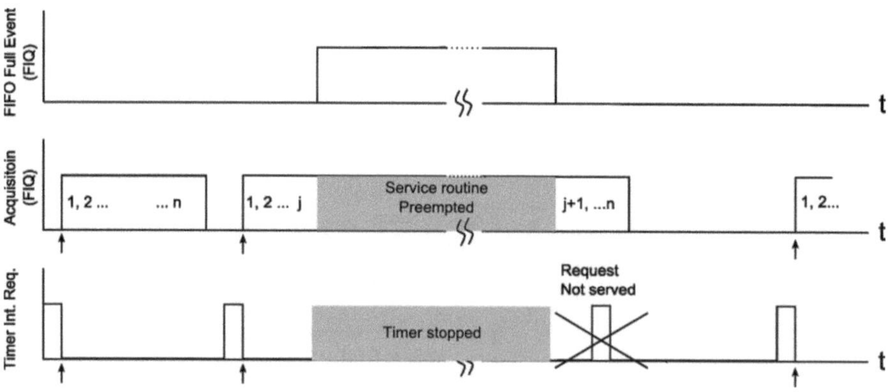

Fig. 12.4 Operation timings in case of FIFO full event

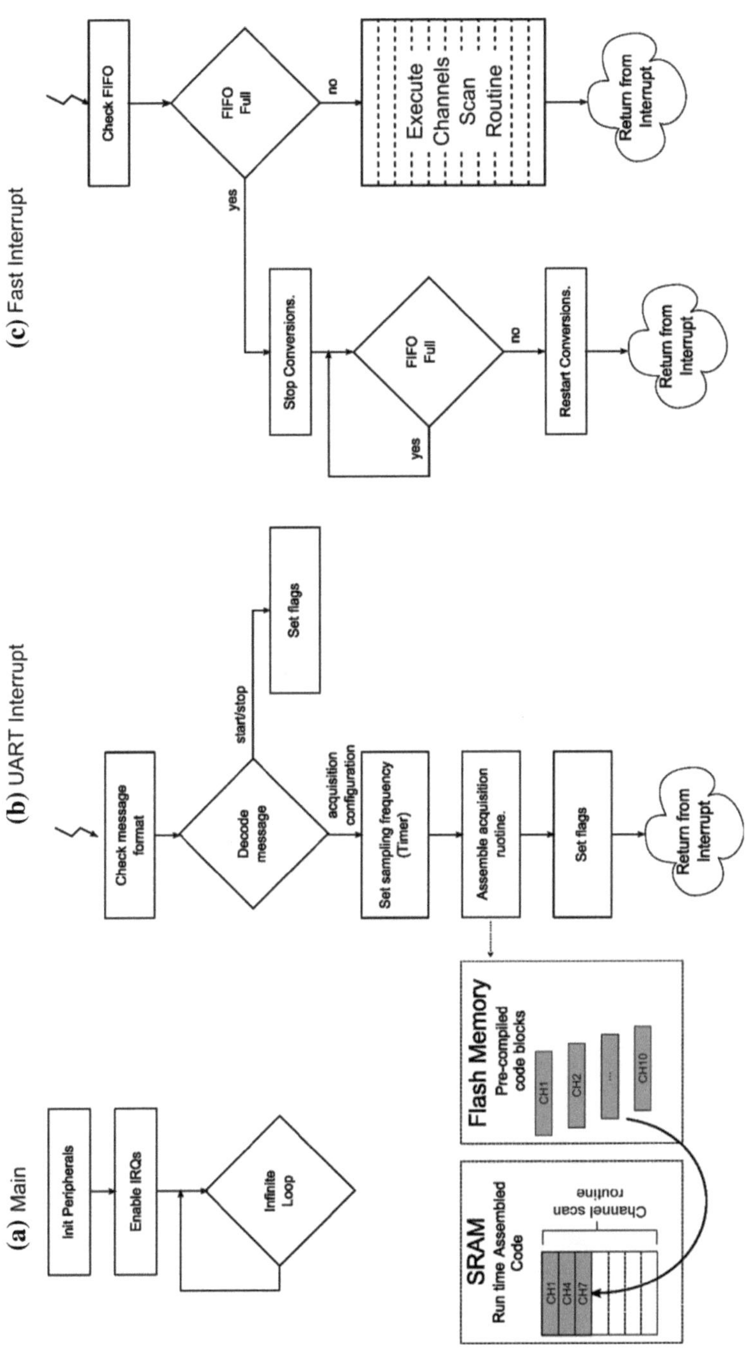

Fig. 12.5 Flowchart of ARM firmware

the use of branches and conditional statements, as well as interrupts very inefficient. The datalogger acquisition parameters are highly configurable, in fact each channel can be selected or not, configured as single-ended, pseudo-differential or differential, and in these last cases it can be coupled with another channel; hence an high number of conditional statements would be needed in each acquisition scan. All these conditions would introduce a considerable delay and would make impossible to meet the acquisition rate constraints of 1 Msps. Since the configuration is decided at start-up only and never changes after, the optimal solution in terms of computational efficiency would be to store in the flash a specific code segment, without any branch, for each possible channels configuration. This solution however would produce a very large code. For this reason a quite unconventional programming technique was used: the main acquisition routine (scan loop) code is generated on the fly at start-up according to the acquisition configuration received from the AVR. This is implemented by storing in the microcontroller code memory (flash) a set of code fragments implementing all the required elementary operations. These fragments are selected and assembled in the internal RAM according the received configuration so to form the main scan routine (see Fig. 12.5b). In this case the routine only contains the exact sequence of operations needed for the specified configuration, without any jump due to conditional instructions. This code is executed from RAM and used to serve the ARM fast interrupt (FIQ, Fig. 12.5c), triggered by a timer programmed according to the desired acquisition rate. The routine basically sets the next analog channel, starts A/D conversions, reads the converted data, formats and writes them to the FIFO. If the FIFO is found to be full (this condition is unlikely, but may happen if the CompactFlash card used is particularly slow) the acquisition is temporarily suspended as shown in Fig. 12.4.

The other interrupt service routine is used to handle the UART messages from the AVR. Even if this routine may perturb the acquisition timings due to its high latency, it is called when the AVR sends messages to the ARM, that is at boot time or on start or stop conditions only.

12.5 Implementations and Tests

Thanks to its simple hardware design the datalogger was realized in a small form factor: the final implementation measures only 53 × 34 × 80 mm and was housed in a rugged extruded aluminum enclosure (Fig. 12.1). In order to increase the resistance against vibrations only low profile surface mount components were used and the board was fastened to the enclosure rail with elastic silicon glue in order to obtain a damping effect. The rear enclosure cap, used to access the CompactFlash was endowed with a polyurethane foam coating in order to keep the CompactFlash inserted while acting as a damper against vibrations.

The chosen enclosure features an IP65 protection rate (resistant to water and dust) and two flanged ends in order to easily fix the datalogger to the vehicle. All the electrical connections (analog and digital channels, CAN bus and power) are accessible through a Deutch automotive grade 22 pins circular connector. Power supply can range from about 10 to 48 V, so it can be directly derived from the vehicle battery (this voltage is internally converted to 5 V by a DC/DC converter and well filtered so to prevent or attenuate undesired transients phenomena and noise).

Power consumption is less than 2 W, even during high speed recording. The datalogger is configured by using a small custom software utility on a PC that allows either to set the desired acquisition parameters, either to download acquired data. Since recording session can easily produce huge files (up to one GB per 10 min), tools for selecting and downloading only partial intervals and for downsampling all the acquired data results very useful for analysis purposes. The datalogger was extensively tested in all the configuration and working conditions and it provided the expected results, even at the maximum acquisition rate. It has to be noted however that card commercially rated up to 66×, roughly corresponding to about 10 MB/s maximum throughput, occasionally produced some data loss at the highest acquisition rates, indicating that the FIFO overflowed due to card very long delays. Faster cards (e.g. SanDisk Extreme III, rated 80×) did not experimentally show this problem. The datalogger was also tested in a test cell and on track by Ducati Corse; it showed very good performance and good immunity to electromagnetic noise and disturbance to the power supply, even when placed close to the engine.

12.6 Conclusions

It was described a fast datalogger specifically designed for in-vehicle testing in automotive or motor sector. The designed datalogger took a radically different approach compared to existing commercial products since it was realized as a very small, autonomous and low-power device allowing the implementation of a distributed logging scheme, when more than one device are used. Loggers can be remotely synchronized and controlled via CAN bus and they experimentally featured very fast recording speed, reaching up to 100 Ksps per channel, and high immunity to electromagnetic noise and vibrations.

These results have been obtained by using a multiprocessor architecture and an optimized coding of the firmware. This allowed both to satisfy all the very demanding requirements while employing low cost components. The combined datalogger performance are, to the best of author's knowledge, among the most advanced available for this kind of instruments.

References

1. Kuranz E (2002) In-vehicle data acquisistion systems. PXI Test and Technology 13(Spring)
2. Lang K (2005) Rugged measurement technology with integrated video playback. VFI Magazine, Feb 2005

3. HEIM Systems GmbH (2007) DATaRec 4—DIC6B/L technical specifications. Apr 2007
4. Elektrobit (2009) Versatile and ruggedized FlexRay, CAN and LIN interface EB 6100/EB 6110/EB 6120. Feb 2009
5. Fertitta G, Di Stefano A, Fiscelli G, Giaconia CG (2009) An embedded datalogger with a fast acquisition rate for in-vehicle testing and monitoring. Intelligent solutions in Embedded Systems, 2009 Seventh Workshop on, pp 105–110, 25–26 June 2009
6. ISO/DIS 11898-1/4 Standard (2003) Road vehicles—Controller area network (CAN). International Standard Organization
7. Arm Ltd. reference manual (2007) ARM7TDMI Technical Reference Manual. Rev 4, 2007, Jun 2007
8. Analog Device Inc. datasheet (2007) Precision analog microcontroller, 12-bit analog I/O, ARM7TDMI MCU
9. Atmel Inc. datasheet (2007) 8-bit AVR microcontroller with 16 K bytes in-system programmable. Aug 2007
10. Transcend Inc. website: www.transcendusa.com
11. Apacer Inc. website: www.apacer.com
12. Sony Memory Stick website: www.memorystick.com
13. CompactFlash Association specifications (2004) CF+ and CompactFlash specification revision 3.0. Dec 2004
14. T13/ANSI 317-1998 standard (1998) Information technology—AT attachment with packet interface extension (ATA/ATAPI-4)
15. Microsoft Co. hardware white paper (2000) Microsoft extensible firmware initiative FAT32 file system specification. Dec 2000

Chapter 13
Secure Gateway Interoperability

Álvaro Reina, Jesús Sáez, Natividad Martínez Madrid and Ralf Seepold

13.1 Introduction

In the area of the Intelligent Transportation Systems (ITS), one challenge for embedded systems is to provide on-board units tuned to automobile requirements, in general to mobile environments and to emerging safety applications. Nowadays, commercial vehicles are equipped with many electromechanical systems often consisting of a sensor network connected by a Controller-Area Network (CAN) bus and one or more Electronic Control Units (ECU). ECUs have such reduced capacity of processing that it cannot be considered to support complex applications. Current tendencies use more advanced on-board units like car PCs with higher capacities of processing and alternative communication interfaces. Different from the CAN-bus and the ECUs, a car PC supports several network protocols and communication technologies mainly used for intelligent environments. For instance, a mobile Bluetooth device could connect to the car PC while the car PC has a parallel connection to a server in the Internet through a UMTS link. Further than supporting small stand-alone applications, in such a scenario, the car PC supports a full featured car-gateway (CGW) which provides connectivity between heterogeneous connected devices and it supports advanced distributed

Á. Reina · J. Sáez
Universidad Carlos III de Madrid, Avda. de la Universidad 30,
Leganes 28911, Spain

N. Martínez Madrid
Reutlingen University, Alteburgstraße 150, Reutlingen 72762,
Germany

R. Seepold (✉)
University of Applied Sciences Konstanz, Brauneggerstr. 55,
Konstanz 78462, Germany
e-mail: ralf.seepold@htwg-konstanz.de

M. Conti et al. (eds.), *Solutions on Embedded Systems*,
Lecture Notes in Electrical Engineering, 81, DOI: 10.1007/978-94-007-0638-5_13,

applications. State-of-the-art communication protocols provide connectivity up to the transport level for most devices that could be considered as part of a vehicular network (IEEE 802.11p [1] for V2V and V2I, Bluetooth and WiFi for PAN devices like mobile phones or PDAs, CAN and IEEE 1451 for sensor networks, etc.). However, how to share services between such different devices from the Vehicular Area Network (VAN) is an open issue.

Thus, this work aims to put forward the advantages of putting together network services and middleware platforms in vehicular networks. Specifically, this paper deals with the interoperability between OSGi [2] (a technology for fast easy deployment of a middleware platform in a gateway) and UPnP (a standard to automatically discover and share services between heterogeneous devices) always focusing on intrinsic problems derived from an open-access mobile environment, e.g. ubiquity, reduced resources and security. Using these challenges as a guideline for designing solutions adapted to a vehicular environment and integrating them into a Network Service Access (NSA) layer as part of a car-gateway OSGi middleware is the main challenge of this work. An NSA system provides a solution for matching a local service model to a distributed one. The local service model corresponds to the OSGi middleware and the distributed one corresponds to the UPnP network services. As it will be demonstrated, any OSGi services concepts can be perfectly matched to the UPnP concepts. Thus, the NSA offers transparent interoperability between OSGi and UPnP. As a consequence, any shared service registered with an OSGi middleware is in fact an UPnP service and vice versa, whenever it was compliant with the NSA requirements. The NSA architecture integrates high performance add-on modules providing quality of service (QoS) and security. In the future, it is planned that QoS will be supported also in the UPnP QoS architecture [3].

13.2 State of the Art

Many protocols deal with services discovery and sharing. Among them, the Universal Plug and Play Protocol (UPnP) defines an architecture that aims to enable auto-discovery and announcement so that other devices in the network can transparently use their capabilities. Different from other service protocols like Devices Profile for Web Services (DPWS), UPnP is not based on a centralized client–server architecture. Instead of this, it is based on the Broker pattern [4] for distributed systems. UPnP defines two entity types, the device and the control point (CP). An UPnP device is an abstract representation of any kind of network services supplier, normally a physical device like a printer, a gateway, a security camera and so on. The CP can be shown as the remote control of a device. The CP automatically searches and registers devices and their services. So, the CP is like a service registry that any user or application may request for consuming a devices' services. UPnP is very suitable for environments like vehicular networks [5].

Some approaches address to integrate network services protocols into a gateway middleware for enhancing a gateway's connectivity capabilities. For instance, the proposals in [6, 7] are focused on the integration of UPnP with the OSGi middleware of advanced home gateways. Other proposals as [8–11] move the gateway, middleware and services concepts from the home networks to the vehicular environment. Also [12] proposes a complete gateway platform description. While the references in this paper are mostly focused on describing an overview of a complete system, the work presented here provides a description in depth of the NSA component of a CGW OSGi middleware which provides transparent interoperability between the CGW bundles and the vehicular UPnP network.

The NSA is not just another bridge engine between OSGi and UPnP as the UPnP Base Driver [13, 14] for the OSGi Felix distribution. While the Base Driver only deals with bundles that fit into the standard OSGi structure of an UPnP service, the NSA converts any bundle's public service into an UPnP service whatever its structure it. The NSA uses the OSGi standard data structure for the UPnP services as a transitional state between the OSGi framework and the UPnP network. Thus, any other implementation of the UPnP stack that would use this OSGi standard interface (org.osgi.service.upnp) [15] will be compliant with the NSA.

13.3 Network Service Access System Overview

The Network Service Access layer provides the CGW middleware with the capability of publishing and consuming network services as operating with local services inside the middleware. This can be done through three complementary interoperability models.

- *Direct synchronous*: The middleware modules invoke the network services through the NSA interface. The middleware needs to be aware of the specific underlying network services technology and the services structure.
- *Direct asynchronous*: The middleware subscribes the network services events through the NSA interface. The middleware is aware of the network services events.
- *Transparent synchronous*: The middleware accesses the network services as they were services of the platform. Likewise, the middleware services are also accessible from remote devices as they were network services. The middleware services are not aware of the network services architecture.

While the direct synchronous approach is equivalent to an ad-hoc Application Programming Interface (API) adapted to the underlying network services technology, the direct asynchronous and the transparent synchronous access models abstract from the details of the specific network services technology. This means that for each remote network service, the NSA generates and dynamically installs a module (agent) fully compliant with the gateway middleware services model. Its functionality consists of translating the invocations within the middleware into the

Table 13.1 UPnP architecture and OSGi framework

UPnP	OSGi	Description
Device	Bundle	The services containers
Service	Service	One service of the platform. An interface
Actions	Methods	The service interface methods
State variable	Data types	The data types of the action arguments
	Class attributes	The variables describing the state of the service
	Bundle properties	Attributes containing service metadata
Events	Event admin	Asynchronous communication channel
Control point	Service registry	A registry of services

corresponding network requests. The NSA registers into the platform the same interface that the corresponding remote network service publishes through the network and thus, it enables a transparent synchronous access model. On the other hand, a centralized event manager in the NSA catches events from the network and notifies them to the middleware subscribers. This mechanism enables a direct asynchronous interoperability model. Different from the agents approach, the subscription mechanism depends on the particularities of the network services technology avoiding transparency.

Besides providing access to a network of services through the previously defined interoperability models, the NSA transparently improves reliability and quality in network communications and protects them against malicious user attacks and resources starvation.

The NSA is based on some rules translating OSGi services into UPnP and vice versa. These rules match the concepts of the ontology of both architectures. Table 13.1 shows such equivalences.

13.4 The NSA Architecture

The core components of the NSA architecture (see Fig. 13.1) are the UPnP Stack and the Control Point (CP).

The UPnP Stack implements the application layer protocols: SOAP, GENA, SSDP, HTTPU and HTTPMU. Besides the basic UPnP functions, the UPnP Stack module provides an object representation of the generic UPnP elements (CP and devices). Thus, the NSA can build a local representation for each of the active network devices. Once a device announces its attachment to the network, the CP gets the service description (SSDP) from the URL the device has published. The CP parses the SSDP and translates it into its own object model. Finally, the CP stores the local representation of the device in a local inventory.

The CP interface towards the NSA components provides the NSA components with functions for accessing the object representation of the UPnP devices. Specifically, the NSA components access the representation of the services and use them for invoking network services and subscribing their asynchronous events. This can

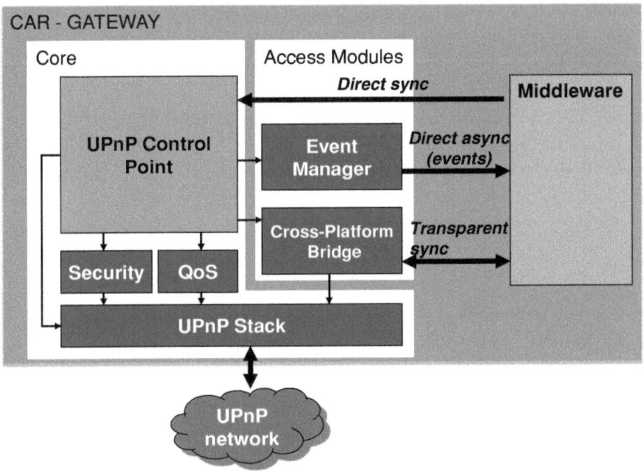

Fig. 13.1 The NSA architecture: the NSA core and the Access Modules

be used also for getting descriptive metadata of the service like the identifier, the name, the device type and other values contained in the service descriptor. Any of the actions invoked for the representing object of an UPnP service are translated by the UPnP stack into an appropriate HTTP message. Thus, additional services of the NSA can interoperate with a remote UPnP services like a local one.

The UPnP specification does not support security or QoS by itself. Thus, the NSA architecture is completed by Security and QoS modules. The security module provides privacy to the UPnP communications. The QoS module supports the CP analyzing the resource consumption before accepting connection requests.

The NSA core enables only the direct synchronous mode. The NSA core is complemented with the Access Modules, namely the Event Manager and the Bridge which enable the asynchronous and the transparent synchronous mode.

13.4.1 The Control Point Behaviour

The control point searches and registers UPnP devices attached to the network. It also subscribes to events announced by the device via a service descriptor. The NSA enriches these two basic functionalities with QoS (via the QoS Manager Service) and security (via the Security Manger). Security is supplied for the UPnP transactions (messages exchanged between the CP and any UPnP devices). Otherwise the QoS is supplied even for the derived connections. Figure 13.2 shows how the control point bundle invokes an UPnP service after enabling QoS and security for this invocation.

The CP receives a service request and tries to forward the petition to the service owner device. In case the invoker supplies a traffic descriptor where the resources

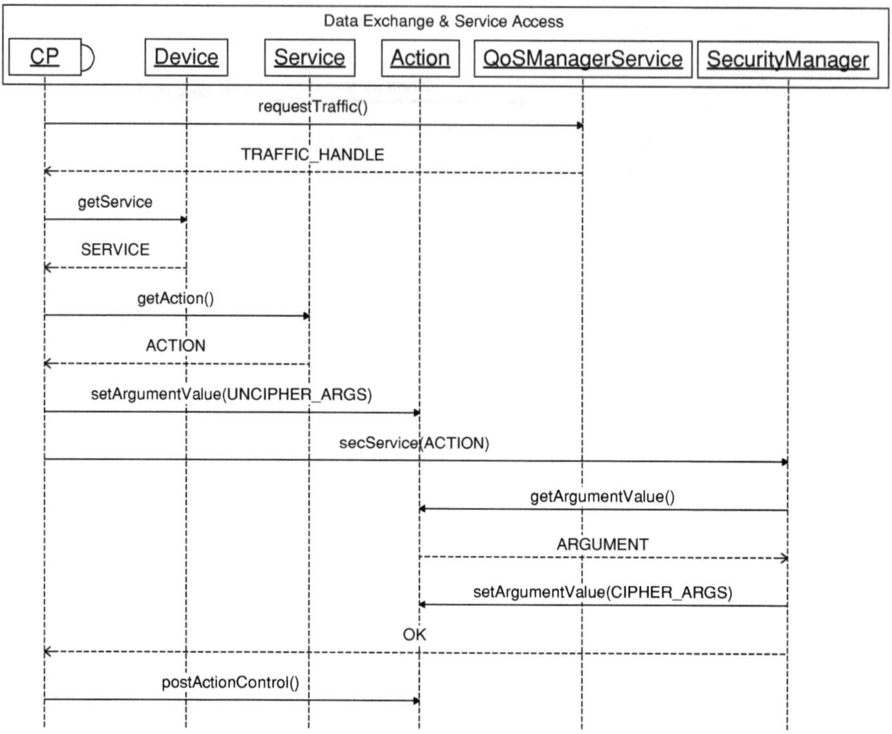

Fig. 13.2 The control point manages the QoS and Security

needed are specified, the control point tries to reserve it along the path from the source (the car-gateway) to the destination. To do this, the control point asks the QoS subsystem for the establishment of a flow described in the traffic descriptor. Then, the control point receives an action from the object structure representing the UPnP device in the local system and sets the arguments values (`setArgumentValue`). Finally, the control point requests a secure service to the security subsystem and sends the call description to the remote device. When the security subsystem receives a secure service request it repeats the process of setting the argument values into the action, but in this case, the argument values are previously ciphered. Thus, when the control point posts an action no malicious user can decipher the transmitted information.

13.4.2 Access Modules

An additional service enables the direct asynchronous and the transparent synchronous interoperability models. Respectively, the Event Manager and the UPnP-OSGi bridge described below are in charge to provide these functionalities.

1. *Event Manager*: The Event Manager follows the Java event model based on *listeners* and *notifiers* and adapts to the OSGi framework. In other words, the middleware bundles interested in the UPnP events must implement the listener interface that the Event Manager registers into the Services Registry. Then, they subscribe to events through the Subscription Manager interface. This interface allows configuring subscriptions. The Subscription Manager bundle maintains a list of Subscription Profiles where the event listeners and their preferences are configured. The Notification Manager bundle looks up this Subscription Profile list in order to notify events the CP dispatches. The Notification bundle subscribes every event since service initialization. The CP forwards any event from the UPnP network to their subscribers that is, to the Notification bundle. Before the Notification bundle notifies an event, it preselects a subset from the Subscription Profile list only containing the listeners of which preferences matches to the event description (eventing device, eventing service, event type, etc.).
2. *The Cross-platform Bridge*: The Cross-platform Bridge imports UPnP services into the OSGi framework so the CGW middleware can interoperate with them, and exports OSGi services to the UPnP network. Other remote CP in the vehicle network can invoke the CGW public services. The Cross-platform Bridge organizes these two functions in two modules: the Importer and the Exporter. They are joined together by a Service Inventory. The Service Inventory is a registry for both the CGW and the vehicle network services. The Inventory uses the OSGi (org.osgi.service.upnp) standard definition of the UPnP services as a transitional representation of any service. The Importer waits for further UPnP device registration events that the CP of the NSA notifies. When such an event arrives, the Importer extracts the services from the device description. Then, it creates one device in the OSGi representation as a wrapper for each service. The new device is registered within the Inventory. The Exporter listens to the OSGi Service Registry waiting for a REGISTERED event from the OSGi framework. When such event arrives, the Exporter analyzes the further registration structure that is, the interfaces the new service implements. In case the registered service was compliant to the bridge, the Exporter analyzes its interface declaration and dynamically builds a device in the OSGi representation. This device maps the middleware service's methods to UPnP actions. Then, the Exporter queries the UPnP stack for the publication of the farther UPnP device and finally registers the further started device in the Inventory.

13.5 Security in the NSA

Security in UPnP networks is an open issue. Some studies [16] state the necessity of providing security in the service oriented architectures applied to ubiquitous computing. The proposal is to provide a basic UPnP security service to be deployed over general purpose networks based on the UPnP protocol. Specifically, in the automotive network, the service described adds security features to the negotiation of parameters between the car-gateway and the remote UPnP devices. One typical security application in a car-gateway is the eCall. Such application opens a voice channel between the driver and the emergency services usually operating from a Public-safety Answering Point (PSAP). Often, the CGW and the PSAP server do not have the same streaming processing capabilities, so the audio connection parameters need to be negotiated to ensure compatibility. Also the URL where the audio streaming is emitting must be provided by both the car-gateway and the PSAP. These parameters are sufficient for anyone to listen to the conversation between the emergency services and the driver, so protecting this information can be considered a critical issue in such safety applications.

The main objective of the security subsystem is to cipher the communication between the control point and the UPnP devices trough an encryption system. Table 13.2 shows, only the security of the communication belonging to the control phase of the UPnP protocol will be encrypted. The discovery and the description phases put forward the device to any control point attached to the network so that are public phases which do not need to be encrypted.

The security service belongs to the add-on services class. This means that any UPnP compliant device can publish them independently from what kind of device it was. This approach is different from the quite recently published UPnP Security Service [17]. Although, the standard service is much more extensive, it is also too complex for vehicular environments purposes.

The proposal is based on two entities called the Security Manager and the Security Agent (Fig. 13.3).

These entities form a complementary peer where the Security Manager represents a service consumer and the Security Agent represents the service provider (Target Device). When a Security Agent meets a Security Manger, it sends its own security certificate to the manager containing the agent's public key (1: send certificate in Fig. 13.3) that can be generated using a public-key schema as RSA. Observing the same cryptographic schema, the manager now ciphers the service invocation (3: request in Fig. 13.3). The Security Agent deciphers the service invocation to be dispatched to the target device. This step must be done using a secure channel. Once the requested operation is finished, the agent may return the result to the manager. In this case, the Security Manager must previously share its own public-key with the agent. This allows a secure return answer.

Table 13.2 Security issues covered by the data exchange and service access layer

	Discovery	Description	Control	Eventing
Key management	Yes	–	No	–
Confidentiality	–	–	Yes	–
Non-repudiation	–	–	–	–
Integrity	–	–	Yes	–
Authentication	–	–	–	–
Authorization	–	–	–	–

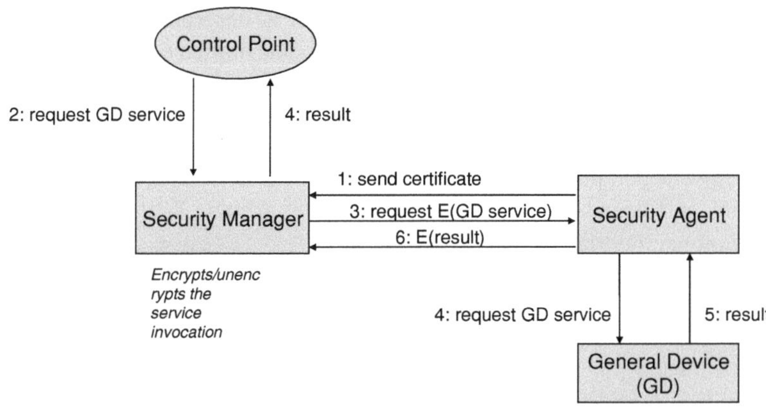

Fig. 13.3 UPnP entities collaborating to complete a secure service invocation

13.6 Security Subsystem Prototype

As a result of this work, the Security Manager and Security Agent prototypes have been developed. Both prototypes are described below.

13.6.1 Security Manager

The Security Manager is a CP that only interacts with the UPnP Security Agent devices. Its function consists of translating an UPnP service invocation into a secure UPnP service invocation. A secure UPnP invocation encrypts the contents of the arguments received by the specific action. The Security Manager implements a public key encryption algorithm. For each remote security agent, the Security Manger stores their correspondent public keys. Thus, the Security Manager ensures that any other Security Agent can reverse the encryption, so the information contained into the invocation never will be revealed to distrusting users.

13.6.2 Security Agent

This UPnP service is directly associated with a target device. The target device must publish a UPnP service. The Security Agent acts as a proxy between a Security Manager and the target services. The Security Manager deciphers the argument values from an action call and invokes the service via any local interface. The Security Agent and the target device must be located in the same physical device and communicate through a safe channel.

The implementation of the security subsystem is described in Fig. 13.4. The *ControlPoint* represents a generic UPnP control point implementation. It would be provided by the Cyberlink UPnP open libraries [18]. The class *SecurityManager* implements the security manager entity of the UPnP security architecture. This class inherits the functions of a standard CP so it is able to discover UPnP devices and filter those which publish a security agent service. The *SecurityManager* stores the security certificates provided by the security agents. Table 13.3 shows the *SecurityManager* interface.

The class *Certificate* represents a security certificate. The fields of a certificate are the `URL` and the Universal Unique Identifier (`UUID`) to identify the owner UPnP device, and the public `Key` which will be used to cipher the communications.

Finally, the *SecurityAgent* class represents the security agent service of the UPnP Security architecture. The *SecurityAgent* implements the algorithm to decipher messages from a security manager. It also manages the security certificate for this UPnP device. Table 13.4 shows the interface.

The security subsystem theory of operation (see Fig. 13.5) is as follows. The middleware calls the service `someService` with the argument values `PARAMS`. The control point asks the security manager for a secure service (`secService`) and supplies the target device identifier.

The security manager checks that the target device has an associated security agent. The security manager calculates ciphered values with a cryptographic function `Epk(PARAMS)`. The security manager invokes `someService` on the remote security agent. The security agent deciphers the argument values calculating the inverse function `E'pk(PARAMS)`. Then, the security agent invokes `someService` on the target device. In return, the device provides a response `RES`. The security subsystem propagates the response to the invoker middleware following symmetrically the previous steps.

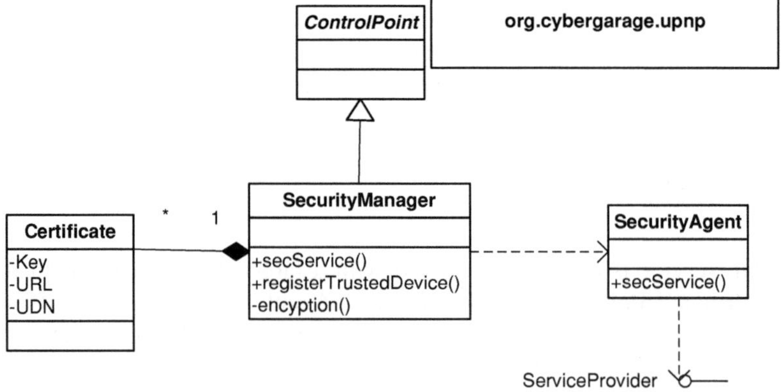

Fig. 13.4 The Security subsystem software architecture

Table 13.3 The SecurityManager interface

Methods	Description
secService	Request a secure service
registerTrustedDevice	Add a UPnP device to the trusted list
unregisterTrustedDevice	Remove a UPnP device from the trusted list
encryption	Cipher the arguments values of an action

Table 13.4 The SecurityAgent interface

Methods	Description
secService	Implements the UPnP secService UPnP action
dechiper	Dechiper the arguments values of the secService
getCertificate	Implements the getCertificate UPnP action. Return this security agent's public key

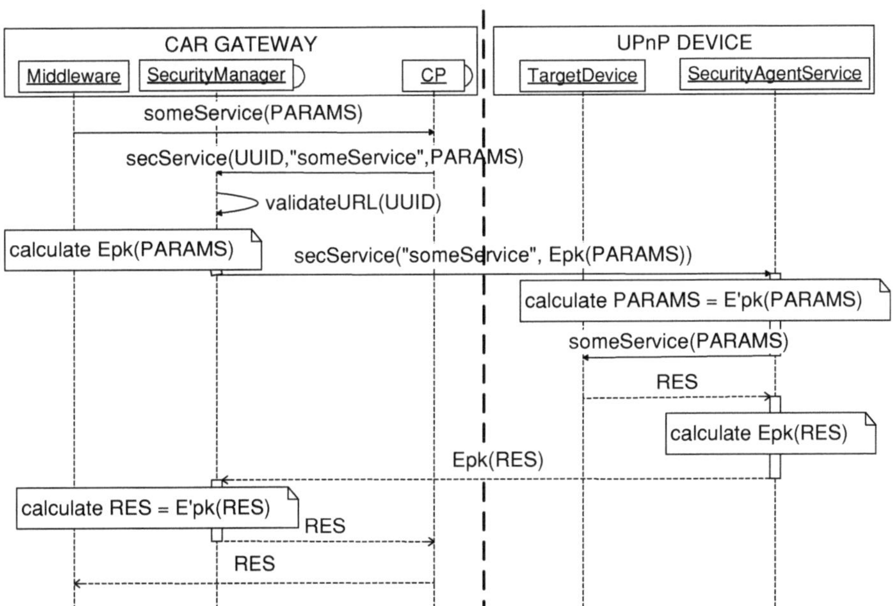

Fig. 13.5 Internal and the external (UPnP) message exchange

13.7 Event Manager Prototype

The event manager supports the direct asynchronous access to the NSA. This means that this module publishes an interface towards the middleware for subscribing to events which the UPnP network generates. The event manager architecture is a variant of the observer design pattern specialized on managing the asynchronous events of the UPnP network.

There are two event classes the event manager can dispatch: The events coming from the UPnP devices; the UPnP core catches such events from the network and propagates them up to the event manager. Secondly, the events the NSA generates. The control point is the source of such events which are dispatched whenever a device or a service connect, update or disconnect from the network. By gathering together these two event classes, the event manager continuously supplies updated information about the current state of the UPnP services and their availability.

The UPnP Events module is built from the Subscription Manager and the Notification bundles. These two modules share libraries and the common software architecture in the UML diagram (Fig. 13.6).

13.7.1 Subscription Manager

The *EventSubscriptionService* interface allows any *EventListener* either subscribing or unsubscribing the UPnP events. By default, when an *EventListener* subscribes to the *SubscriptionManager* this will be notified whenever the NSA control point listens an UPnP event. But the listeners can also configure their subscription preferences. This should be done through the `configureSubscription` method, which allows specifying a regular expression describing the family, type or name of the devices a subscriber is interested in. In this way, a subscriber can easily select, for example, the events coming from any QoS device, with the next call.

The *SubscriptionManager* class implements a *EventSubscriptionService* interface. It stores and updates a list of subscription profiles where the subscription

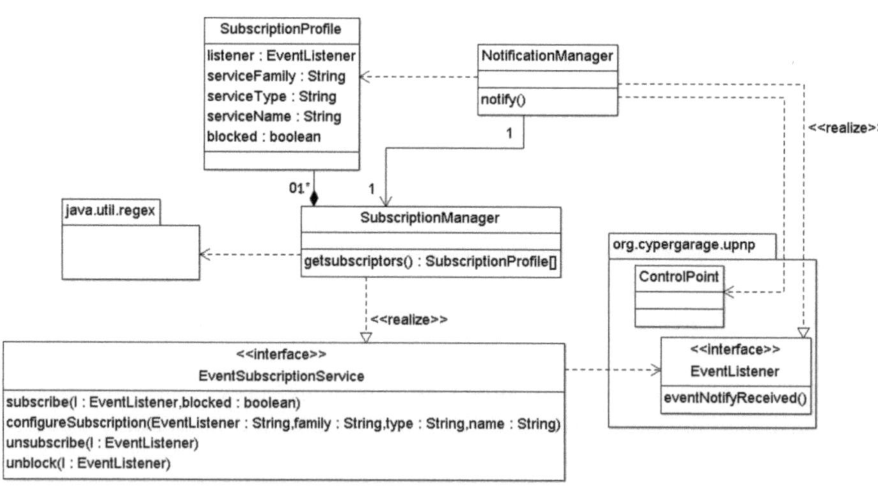

Fig. 13.6 The UPnP Event Manager software architecture

preferences for each *EventListener* are specified. The subscription preferences are stored in a *SubscriptionProfile* object. Attributes within a profile store the following parameters.

- *ServiceFamily*: This parameter can be used to describe all the services belonging to a UPnP standard, as for example the AV architecture, the QoS architecture, etc.
- *ServiceType*: The service type refers to one specific UPnP device, as for example the QoS Manager, the QoS Policy Holder or the QoS Device, all of them belonging to the QoS architecture family.
- *ServiceName*: The service name refers to the friendly name of a specific UPnP device. This parameter usually identifies only one device in a network. The Unique Device Name (UDN) can be used rather than the friendly name.

A subscription profile stores an array of device identifiers matching configuration preferences after the subscription manager parses the regular expression.

13.7.2 Notification Manager

A notification manager gets a list of event listeners from a subscription manager instance and notifies the subscribers. The notification manager acts as an event listener subscribed to all the UPnP event sources. While subscribing directly from the middleware modules and applications to the UPnP stack may cause the congestion of the middleware platform and extra processing time in the subscribers, the notification manager filters the events received and forwards them only to the interested subscribers.

13.8 Conclusions

A Network Service Access layer enables a car-gateway middleware to share internal services towards a VAN which has been described in detail. Such a layer allows interoperability between the OSGi framework and the UPnP service protocol in both directions. Furthermore, it provides the ontology that connects OSGi to UPnP concepts. This new approach shows that quality of service and security issues are inherent to any mobile environment but specially in vehicular networks. In particular, a new security schema for UPnP networks is provided.

Since the NSA provides a platform for fast prototyping of mobile distributed applications over services networks, the contributions of this approach are significant for middleware developers as well as for any on-board embedded systems industry. A more complex security architecture dealing with all UPnP protocol phases is forecasted for the near future. Also it is planned to upgrade to the UPnP Security standard.

References

1. Jiang D, Delgrossi L (2008) IEEE 802.11p: towards an international standard for wireless access in vehicular environments. In: Proceedings of vehicular technology conference (VTC), May, pp 2036–2040
2. Open Service Gateway Initiative (OSGi) Alliance (2010) http://www.osgi.org, March
3. UPnP-QoS Architecture:3 (2009) http://upnp.org/specs/qos/UPnP-qos-Architecture-v3.pdf
4. Buschmann F, Meunier R, Rohnert H, Sommerlad P, Stal M (1996) Pattern-oriented software architecture—a system of patterns. Wiley, Chichester
5. Santana JMS, Petrova M, Mahonen P (2006) UPnP service discovery for heterogeneous networks. In: IEEE 17th international symposium on personal, indoor and mobile radio communications, 11–14 Sept 2006, pp 1–5
6. Hong SG, Lee JW, Choi WS (2005) Open platform test framework for telematics terminal platform. In: Proceedings of the IEEE 62nd vehicular technology conference, vol 4, pp 2745–2748
7. Kang DO, Kang K, Choi S, Lee J (2005) UPnP AV architectural multimedia system with a home gateway powered by the OSGi platform. IEEE Trans Consum Electron 51(1):87–93
8. Sun Y, Huang WL, Tang SM, Qiao X, Wang FY (2007) Design of an OSEK/VDX and OSGi-based embedded software platform for vehicular applications. In: Proceedings of the IEEE international conference on vehicular electronics and safety, ICVES
9. Ai Y, Sun Y, Huang W, Qiao X (2007) OSGi based integrated service platform for automotive telematics. In: Proceedings of the IEEE international conference on vehicular electronics and safety, ICVES
10. Li Y, Wang F, He F, Li Z (2005) OSGi-based service gateway architecture for intelligent automobiles. In: Proceedings of IEEE intelligent vehicle symposium, May, pp 861–865
11. Zhang D, Wang XH, Hackbarth K (2004) OSGi based service infrastructure for context aware automotive telematics. In: Proceedings of the IEEE 59th vehicular technology conference, May, vol 5, pp 2957–2961
12. Seepold R, Martimez Madrid N, Gómez-Escalonilla JS, Reina A (2009) An embedded software platform for distributed automotive environment management. EURASIP J Embed Syst, vol 2009, pp 1–10. Article ID 856962 ISSN: 1687–3955
13. Dobrev P, Famolari D, Kurzke C (2002) Device and service discovery in home networks with OSGi. Commun Mag IEEE Commun Soc 40(8):86–92, New York, August
14. UPnP Base Driver (2010) http://domoware.isti.cnr.it/documentation.html, March
15. Open Service Gateway Initiative (OSGi) Alliance (2010) Javadoc. http://www.osgi.org/javadoc/r4v401/, March
16. Cotroneo D, Graziano A, Russo S (2004) Security requirements in service oriented architectures for ubiquitous computing. In: Proceedings of the 2nd workshop on middleware for pervasive and ad-hoc computing, October, pp 172–177
17. UPnP Security Ceremonies v1.0 (2009) http://upnp.org/download/standardizeddcps/UPnPSecurityCeremonies_1_0secure.pdf
18. Konno S (2009) Cyberlink development package for UPnP devices for Java. http://cgupnpjava.sourceforge.net/, May

Part IV
Software and System Architecture

Chapter 14
Applying Bayesian Networks for Intelligent Adaptable Printing Systems

Arjen Hommersom, Peter J. F. Lucas, René Waarsing and Pieter Koopman

14.1 Introduction

Many complex systems such as printers are required to make dynamic in-product trade-offs between various qualities of operation at the system level, which can be viewed as the capability to adapt. While many definitions of adaptability have appeared in literature (see [3] for a summary), we here define *adaptability* to be such system-wide trade-offs. In printing systems, system-wide qualities include the power division, the speed of printing, the power consumption, etc. Such trade-offs heavily depend on the system's environment, e.g., humidity, temperature, available power, etc. Failure to adapt adequately to the environment might result in faults or suboptimal behaviour.

The area of adaptive control has a long tradition of over 50 years. Several approaches in this field exist. First, model-reference adaptive control (MRAC) uses a reference model that reflects the desired behaviour of the system. On the

This paper originally appeared in the Proceedings of the Seventh Workshop on Intelligent Solutions in Embedded Systems (WISES 2009)[9].

A. Hommersom (✉) · P. J. F. Lucas · P. Koopman
Institute for Computing and Information Sciences, Radboud University Nijmegen, Nijmegen, The Netherlands
e-mail: arjenh@cs.ru.nl

P. J. F. Lucas
e-mail: peterl@cs.ru.nl

P. Koopman
e-mail: pieter@cs.ru.nl

R. Waarsing
Océ Technologies BV, Venlo, The Netherlands
e-mail: rene.waarsing@oce.com

M. Conti et al. (eds.), *Solutions on Embedded Systems*,
Lecture Notes in Electrical Engineering, 81, DOI: 10.1007/978-94-007-0638-5_14,

basis of the observed output and of the reference model, the system is tuned. The second type of adaptive controllers are so called self-tuning controllers (STC), which estimate the correct parameters of the system based on observations and tunes the control accordingly. In the last few decades, techniques from the area of artificial intelligence (AI), such as rule-based systems, fuzzy logic, neural networks, evolutionary algorithms, etc. have been used in order to determine optimal values for control parameters (see e.g. [6]).

The problem of adaptability as defined above has two typical characteristics. First, decisions are typically required at a low frequency, i.e., it is not necessary and not even desirable to change the speed or energy usage many times per second. Second, there is a lot of uncertainty involved when making decisions, in particular about the environment, the state of the machine, but also about the exact dynamics of the system. Complex systems usually cannot be modelled accurately, whereas adaptability requires one to make system-wide, complex, decisions. In order to deal with this uncertainty, techniques where probability distributions are learned from available data seem therefore appropriate. In this paper, we explore the use of Bayesian networks [23] to tune setpoints of local controllers of the system. The block diagram in Fig. 14.1 offers an overview of this approach.

One advantage of Bayesian networks is that they contain a qualitative part, which can be constructed using expert knowledge, normally yielding an understandable, white-box model. Moreover, the quantitative parameters of a Bayesian network can be learned from data. Other AI learning techniques, such as neural networks, resist providing insight into why the machine changes its behaviour, as they are black-box models. Furthermore, rules—possibly fuzzy—are difficult to obtain and require extensive testing in order to check whether they handle all the relevant situations.

The purpose of the present paper is to convey some of our experience in building Bayesian-network based controllers in the area of adaptive printing systems, which can be looked upon as special stochastic controllers. In our view, as systems get more and more complex, the embedded software will need to be equipped with such reasoning capabilities for making sound decisions. The paper is organised as follows. In the next section, we will introduce the necessary preliminaries with respect to Bayesian networks. In Sect. 14.3, we will look at a

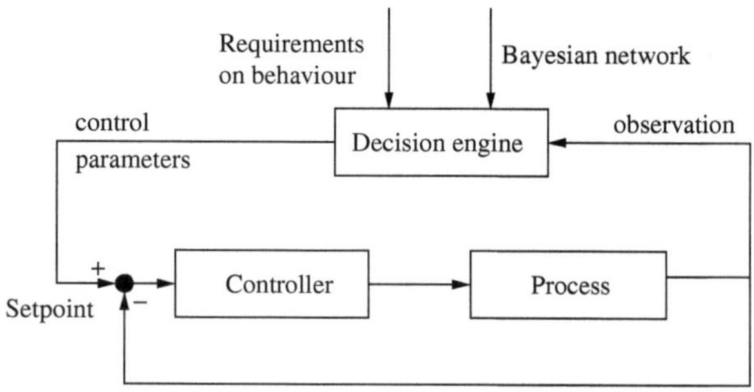

Fig. 14.1 Block diagram of an adaptive controller using a Bayesian network

specific case study where we would like to estimate the optimal setpoint under uncertainty. This example shows that some of the logic that might be needed in a rule-based system is implicitly encoded in the probability distribution. Another case is considered in Sect. 14.4; here the goal is to optimise the velocity of the engine. Both cases are compared to a traditional controller. In Sect. 14.5 the results obtained are compared to related approaches.

14.2 Preliminaries

A *Bayesian network* $B = (G, P)$ consists of a directed acyclic graph $G = (V, E)$, where V is a set of vertices and $E \subseteq V \times V$ is a set of directed arcs; with the probability distribution P is associated a set X of random variables that correspond one-to-one to the vertices of G, i.e., each vertex v corresponds exactly to one random variable X_v and vice versa. As the joint probability distribution P of the set of random variables X is factored in accordance to the structure of the graph G:

$$P(X) = \prod_{v \in V} P(X_v \mid X_{\pi(v)}),$$

where $\pi(v)$ is the set of parents of v, P can also be defined as a family of local conditional probability distributions $P(X_v \mid X_{\pi(v)})$, for each vertex $v \in V$. Bayesian networks can encode various probability distributions. Most often the variables are either all discrete or all continuous. Hybrid Bayesian networks, however, contain both discrete and continuous conditional probability distributions. A commonly used type of hybrid Bayesian network is the conditional linear Gaussian model (see [4, 12]). Efficient exact and approximate algorithms have been developed to infer probabilities from such networks (e.g., [2, 10, 14, 15]). Also important in the context of embedded systems is the fact that real-time inference can be done using Bayesian networks, i.e., produce an approximate probability at any time (cf. [8] for a comprehensive overview).

A Bayesian network can be constructed with the help of one or more domain experts. However, building Bayesian networks using expert knowledge, although by now known to be feasible for some domains, can be very tedious and time consuming. Learning a Bayesian network from data is also possible, a task which can be separated into two subtasks: (1) structure learning, i.e., identifying the topology of the network, and (2) parameter learning, i.e., determining the associated joint probability distribution, P, for a given network topology. In this paper, we employ parameter learning. This is typically done by computing the maximum likelihood estimates of the parameters, i.e., the conditional probability distributions, associated to the networks structure given data [13].

Temporal Bayesian networks are Bayesian network where the vertices of the graph are indexed with (discrete) time. All vertices with the same time index form a so-called *time slice*. Each time slice consists of a static Bayesian network and the time slices are linked to represent the relationships between states in time. If the structure and parameters of the static Bayesian network are the same at every time slice (with the exception of the first), one speaks of a *dynamic* Bayesian network, as such networks can be unrolled (cf. [21] for an overview).

14.3 Setpoint Estimation

14.3.1 Description of the Problem

For the type of printing system under consideration, various temperatures during the printing process play an important role. Low-level controllers make sure that the measurable temperatures are kept on setpoint. Due to design issues and considerations with respect to the cost price, it is not possible to place sensors at all places of interest; therefore, estimations have to be made.

In this section, we use a Bayesian network to estimate the appropriate setpoint for a heating component with the goal to influence the paper temperature when we can only measure the temperature of media (paper) that has passed this heating component. The temperature of the paper is influenced by uncertain aspects, such as the environmental temperature, the speed, the humidity of the paper, and the type of paper. In this case we focus on the latter aspect and assume that the other aspects are constant.

14.3.2 Experimental Setup

The qualitative structure of the domain was elicited from the domain experts. For the purpose of this paper, we focus on certain relevant parts of the complete network dealing with the specific problem of determining the correct setpoint of the heater. The structure of the domain consisting of two time slices is presented in Fig. 14.2.

The associated random variables for this network have been modelled as discrete variables by discretising the values to typical values that are used in the simulation. The setpoint variables have a domain size of 12; media temperature has a domain size of 16 and we consider three paper types: $80, 120$, and $160\,\mathrm{g/m^2}$ paper.

In order to acquire data and to test the system, a physical model of the system was created using Simulink [18]. The data that was generated was used to learn the conditional distributions of the model by calculating the parameters associated to the qualitative structure of the Bayesian network.

The adaptive control was implemented by a low-level PID controller (see e.g., [22]) that controls the temperature of the heater and a Bayesian network to manipulate the setpoint of this controller.

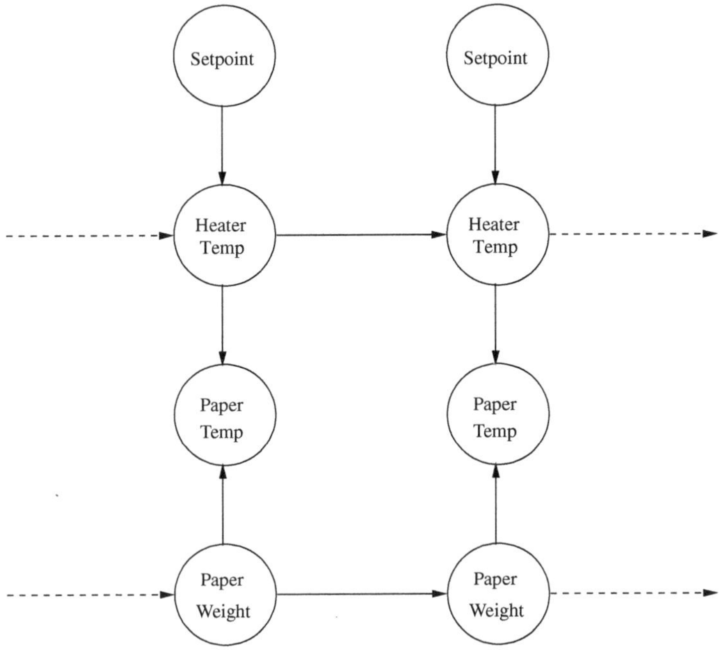

Fig. 14.2 Simplified Bayesian network of the print domain

14.3.3 Case 1: Keep Paper Temperature on Setpoint

First, we use a Bayesian network to choose the next setpoint such that the temperature of the paper will be at a setpoint T_{set} based on observations of the paper temperature and the setpoint at time t. Let $\text{OBS}_t = \{\text{PaperTemp}_t = T, \text{Setpoint}_t = \text{SP}\}$. We then calculate:

$$\text{SP}^* = \underset{\text{SP}' \in \text{Setpoint}}{\text{argmax}} \; P(\text{PaperTemp}_{t+1} = T_{\text{set}} \mid \text{OBS}_t, \text{Setpoint}_{t+1} = \text{SP}')$$

and adapt the setpoint of the heater controller to SP^*. The Bayesian network is simplified, in particular by forgetting about the history, except for the immediate history when making decisions (first-order Markov assumption). Due to this simplification, sometimes the interpretation of the measurements can be misleading. There are several solutions to this problem. For example, we may extend the model to incorporate additional evidence of earlier states, or we may sample less, i.e., by waiting to the system returns to a steady situation. One simple heuristic that proved to be successful in this situation is avoid making decisions when the interpretation is highly uncertain, e.g., when:

$$P(\text{PaperWeight}_t = w \mid \text{OBS}_t) < k$$

for all paperweights w and where k is some tuning constant less than 1. Of course, such a controller can also be implemented well using a standard PID controller that controls directly based on the measurement of the paper temperature. The results of such a PID controller is compared to the Bayesian network approach (with $k = 0.9$) and is presented in Fig. 14.3. The PID controller seems smoother, which is most likely due to the fact that we have discrete value for the setpoint in the adaptive controlling setting. However, for the most part, the behaviour of the adaptive controller is similar to the PID control.

For this control task, Bayesian networks do not provide much benefit over PID controllers, and we can only hope not to do worse than these standard, well-understood, PID controllers. We are therefore aiming at more complex controllers, where traditional control theory starts to become more difficult. One example is discussed in the next section.

14.3.4 Case 2: Avoid Faulty Temperatures

As mentioned earlier, in order to get high quality prints, it is of importance to have a lower threshold for some temperatures. Figure 14.3 shows that if we try to keep the paper at the T_{set} temperature, temperatures may drop below this value when the media changes. This could lead to a system fault. One solution is to put the setpoint at a higher temperature which provides a buffer for the media changes; however, if it is unnecessarily high, energy is lost and it may also cause problems at other parts of the printing process.

The advantage of Bayesian networks is that various probabilistic constraints can be put on the control signal. In this case, we are interested in the lowest temperature that ensures that we avoid dropping below T_{set}. Formally, to decide on the next setpoint, we calculate the minimal SP′ such that

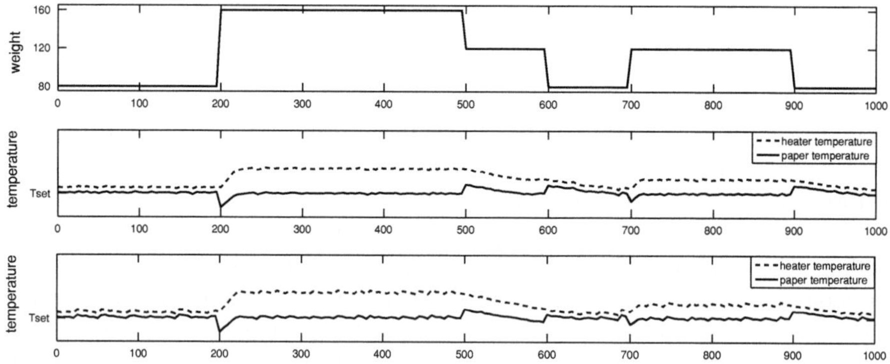

Fig. 14.3 *Top figure*: weight of paper that is being printed on, which changes dynamically. *Middle figure*: paper temperature controlled by a PID controller. *Bottom figure*: paper temperature controlled by an adaptive controller using a Bayesian network

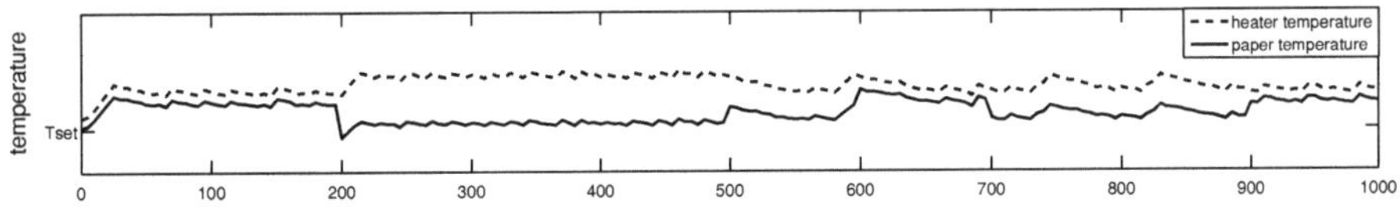

Fig. 14.4 Faults are avoided using an adaptive controller. Paper weight changes are the same as in Fig. 14.3

$$P(\text{PaperTemp}_{t+1} < T_{\text{set}} \mid \text{OBS}_t, \text{Setpoint}_{t+1} = \text{SP}') < \epsilon$$

i.e., the probability that the resulting temperature will be lower than T_{set} will be less than some threshold ϵ. The result can be found in Fig. 14.4 (with $\epsilon = 0.01$). What is interesting here is that the heater temperature is relatively high when the paper weight is lower. This is because the system anticipates on paper that might arrive with a high paper weight as this high paper weight causes a sudden large drop in temperature. This type of logic could be modelled by any system; however, it is interesting to see here that this is implicit in the probability distribution that has been learned from data.

14.4 Dynamic Speed Adjustment

14.4.1 Description of the Problem

The productivity of printers is limited to the amount of power available, in particular in environments which depend on weak mains. If there is insufficient power available, then temperature setpoints cannot be reached, which causes bad print quality. To overcome this problem, it is either possible to decide to always print at lower speeds or to adapt to the available power dynamically. In the section, we explore the latter option by a dynamic speed adjustment using a Bayesian network.

14.4.2 Modelling

The structure of the fragment of the model at each time slice is shown in Fig. 14.5. The requested power available is an observable variable that depends on low-level controllers that aim at maintaining the right setpoint for reaching a good print quality. The error variable models the deviation of the actual temperature from the ideal temperature, which can be established in a laboratory situation, but not during run-time. If this exceeds a certain threshold, then the print quality will be below a norm that has been determined by the printer manufacturer.

Both velocity and available power influence the power that is or can be requested by the low-level controllers. Furthermore, the combination of the

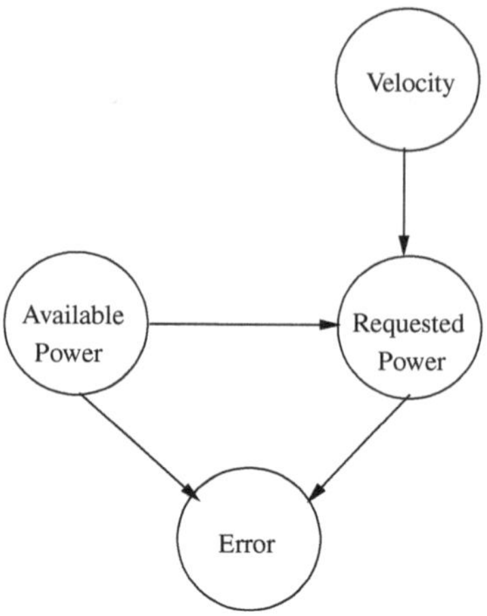

Fig. 14.5 Structure of the Bayesian network of each time slice

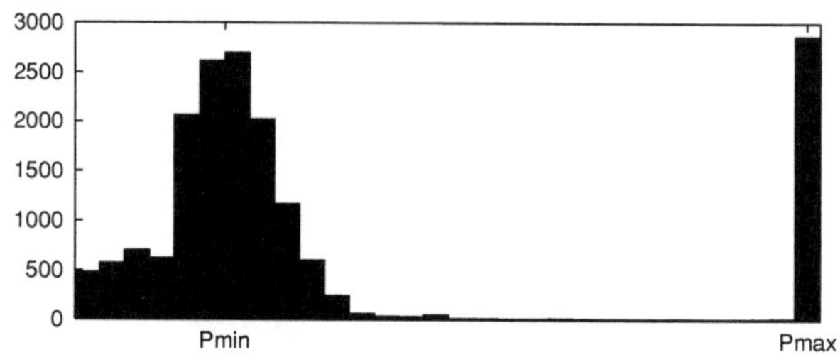

Fig. 14.6 Distribution of requested power

available power and the requested power is a good predictor of the error according to the domain experts.

For our experiments we again use two time slices with the interconnections between the available power—which models that the power supply on different time slices is not independent—and requested power, which models the state of the machine that influences the requested power.

In order to choose the family of distributions, we can consider to model the variables as Gaussian variables. This is reasonable as most variables are normally distributed, except for the available power (see Fig. 14.6). Fitting a Gaussian distribution to such a distribution will typically lead to insufficient performance. However, it can be interpreted as a mixture of two Gaussian distribution, one with mean P_{low} (Watt) and one with mean P_{high} (Watt) with a small variance. Such a distribution can be modelled using a hybrid network as follows. The network is augmented with an additional (binary) parent node S with values 'high' and 'low' for the requested power variable. For both states of this node, a Gaussian model is associated to this variable. The marginal distribution of requested power is obtained by basic probability theory as

$$P(\mathrm{P}_{\mathrm{req}}) = \sum_{S} P(\mathrm{P}_{\mathrm{req}} \mid S)P(S).$$

14.4.3 Classification

One of the main reasoning tasks of the network is to estimate the error, i.e., the deviation from the ideal temperature, given a certain velocity and a certain observations. We could consider this a classification performance, i.e., the print quality is *bad* or *good*. This provides means to compare different models and see how well it performs at distinguishing between these two possibilities. A standard way to visualise and quantify this is by means of a Receiver Operating Characteristic (ROC) curve, which shows the relation between the false positive ratio and the true positive ratio (sensitivity). The area under the curve is a measure for its classification performance.

We have compared three models, i.e., a discrete model, a fully continuous model and a hybrid model for modelling the distribution of the requested power with two Gaussians. The classification performance is outlined in Fig. 14.7. As expected, the fully continuous model performs worse, whereas the hybrid and discrete show a similar trend. The advantage of the discrete version is that the probability distribution can easily be inspected and it has no underlying assumptions about the distribution, which makes it easier to use in practice. The hybrid version however allows for more efficient computation as we need a large number of discrete to describe the conditional distributions. For this reason, we have used the hybrid version in the in following experiments.

14.4.4 Simulation of the Bayesian Controller

As the error information is not available during runtime, the marginal probability distribution of the error in the next time slice is computed using the information about the power available and power requested. This error is a Gaussian random variable with mean μ and standard deviation σ. Given a maximum error that we allow, denoted by $E_{\max}$, we pick the highest velocity v such that the marginal

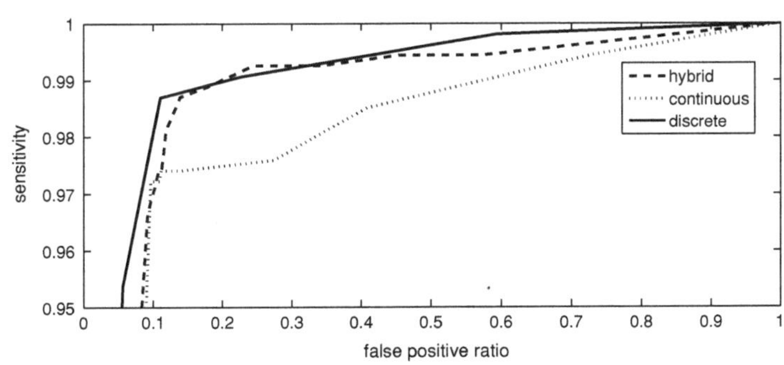

Fig. 14.7 ROC curves of the three Bayesian networks. The hybrid and discrete versions show the best classification performance

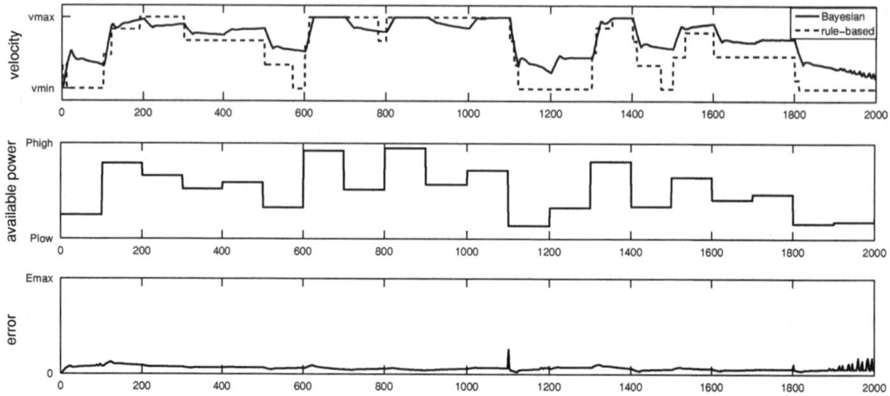

Fig. 14.8 In the centre figure, the available power is plotted, which is fluctuating. At the *top*, we compare the velocity of the engine which is controlled by a rule-based system and by a Bayesian network. *Below*, we present the error that the controller based on the Bayesian network yields, which is within the required limits

probability distribution of $P(\text{Error}_{t+1})$ is such that $\mu + k\sigma < E_{\max}$, where k is a constant. Different values of k correspond to different points on the ROC curve as depicted in Fig. 14.7. For a Gaussian variable, more than 99.73% of the real value of the error will be within three standard deviations of the mean, so for example $k = 3$ would imply that $P(\text{Error}_{t+1} < E_{\max}) > 99.87\%$. However, the sensitivity, i.e., the chance that the paper is warm enough, can be chosen arbitrarily, by also increasing the false positive ratio, i.e., by reducing the overall productivity.

In order to evaluate the approach, we compared the productivity of the resulting network with a rule-based approach that incorporates some heuristics for choosing the right velocity. The productivity is defined here simply as $\int_0^\tau v(t)dt$, where τ is the simulation time.

In order to smooth the signal that the network produces, we employ a FIR (Finite Impulse Response) filter in which we average the decisions of the last 10 s. The resulting behaviour was simulated and is presented in Fig. 14.8 (with $k = 3$). Compared to the rule-based approach, we improve roughly 9% in productivity while keeping the error within an acceptable range. While it could certainly be the case that the rules could be improved and optimised, again, the point is that the logic underlying the controller does not have to be designed. What is required is a qualitative model, data, and a probabilistic criterion that can be inferred.

14.5 Discussion and Conclusions

So far, adaptive controllers based on explicit Bayesian networks have not been extensively investigated. The most closely related work is by Deventer [5], who investigates the use of dynamic Bayesian networks for controlling linear and nonlinear systems. The main difference with his work is that he estimates parameters from a Bayesian network using a given model of the system. In contrast, we aim at using models that were learned from data. Such data can come from measurements during design time or during run-time of the system.

Bayesian inference is well-known for trying to infer a hidden state in a dynamic model. Typical applications are filtering, i.e., trying to infer the current hidden state given the observations in the past and smoothing where past states are inferred. For example, the Kalman filter [11] is well-known in stochastic control theory (see e.g., [1] for an overview) and is a special case of a dynamic Bayesian networks, where the model is the linear Gaussian variant of a hidden Markov model, i.e., it describes a Markov process with noise parameters on the input and output variables. Non-linear variants, such as the extended Kalman filter or the unscented Kalman filter (see e.g., [19]) are approximate inference algorithms for non-linear Gaussian models by linearisation of the model. More recently, particle filters [16], also known as sequential Monte Carlo methods, have been proposed as an alternative, which relies on sampling to approximate the posterior distribution.

The difference with these filtering approaches is that for Bayesian networks there is an underlying domain model which is understandable. As Bayesian networks are general formalisms, they could also be used or re-used for diagnostic purposes, where it is typically required that a diagnosis can be represented in a human-understandable way so that proper action can be taken. Furthermore, it is well-known that the structure of the graphical part of a Bayesian network facilitates the assessment of probabilities, even to the extent that reliable probabilistic information can be obtained from experts (see [17] in the medical domain). One other advantage compared to black-box models is that the modelled probability distribution can be exploited for decision theory. This is particularly important if one wants to make real trade-offs such as between productivity and energy consumption. This is another direction that will be explored in the future.

With respect to the choice of the underlying distribution that is associated to a Bayesian network, several choices can be made. For exact inference, conditional linear Gaussian models are the standard way to deal with continuous variables in this area as exact inference can be used; however, they are restricted to modelling linear systems. Discrete distributions can then be used and have been successfully applied in, for example, medicine. Another option is to model the distribution on the basis of a mixture of truncated exponentials (MTE) [20], which allow one to model non-linear systems and, furthermore, allows for exact inference of required probabilities. This is a direction which we will explore in the near future. Of course, in particle filters, sampling methods can be employed for approximating the posterior from a wide range of non-linear distributions.

Bayesian networks have drawn attention in many different research areas, such as AI, mathematics and statistics. In this paper, we have explored the use of Bayesian networks for designing an adaptable printing systems. We have shown that the approach is feasible and can help to design an intelligent system. We believe that these techniques can have a wide application in the engineering sciences in particular for control and fault detection. The latter has been investigated before, e.g., [7], in which Bayesian networks were applied for both consistency-based as well as abductive diagnosis. Results of this paper provide evidence that explicit Bayesian networks can also be useful for the development of adaptive control systems.

Acknowledgements This work has been carried out as part of the OCTOPUS project under the responsibility of the Embedded Systems Institute. This project is partially supported by the Netherlands Ministry of Economic Affairs under the Embedded Systems Institute program. We would like to thank the anonymous reviewers and the members of the OCTOPUS project for their helpful suggestions and feedback. We also thank Marcel van Gerven for making his Bayesian network toolbox available.

References

1. Åström KJ (1970) Introduction to stochastic control theory. Academic Press, New York
2. Casella G, Robert C (1999) Monte Carlo statistical methods. Springer, New York
3. Chmarra MK, Arts L, Tomiyama T (2008) Towards adaptable architecture. In: ASME 2008 international design engineering technical conferences DETC2008-49971
4. Cowell RG, Dawid AP, Lauritzen SL, Spiegelhalter DJ (1999) Probabilistic networks and expert systems. Springer, New York
5. Deventer R (2004) Modeling and control of static and dynamic systems with Bayesian networks. PhD thesis, University Erlangen-Nürnberg, Chair for Pattern recognition
6. Farrell JA, Polycarpou MM (2006) Adaptive approximation based control: unifying neural, fuzzy and traditional adaptive approximation approaches. Adaptive and learning systems for signal processing, communications and control series. Wiley-Interscience, Hoboken
7. Flesch I (2008) On the use of independence relations in Bayesian networks. PhD thesis, University of Nijmegen
8. Guo H, Hsu WH (2002) A survey of algorithms for real-time Bayesian network inference. In: Darwiche A, Friedman N (eds) AAAI/KDD/UAI02 joint workshop on real-time decision support and diagnosis systems, Edmonton, Canada
9. Hommersom A, Lucas PJF, Waarsing R, Koopman P (2009) Applying Bayesian Networks for Intelligent Adaptable Printing Systems. In: Proceedings of the IEEE Seventh International Workshop on Intelligent Solutions in Embedded Systems WISES09, Ancona, Italy, June 25-26 2009, pp 127–133
10. Jordan MI, Ghahramani Z, Jaakkola T, Saul LK (1999) An introduction to variational methods for graphical models. Mach Learn 37(2):183–233
11. Kalman RE (1960) A new approach to linear filtering and prediction problems. J Basic Eng 82(1):35–45
12. Lauritzen SL (1992) Propagation of probabilities, means and variances in mixed graphical association models. J Am Stat Assoc 87:1098–1108
13. Lauritzen SL (1995) The EM algorithm for graphical association models with missing data. Comput Stat Anal 19:191–201
14. Lauritzen SL, Spiegelhalter DJ (1988) Local computations with probabilities on graphical structures and their application to expert systems. J R Stat Soc 50:157–224
15. Lerner U, Parr R (2001) Inference in hybrid networks: theoretical limits and practical algorithms. In: Breese J, Koller D (eds) Uncertainty in artificial intelligence, vol 17. Morgan Kaufmann, San Francisco, pp 310–318
16. Liu JS, Chen R (1998) Sequential Monte Carlo methods for dynamic systems. J Am Stat Assoc 93:1032–1044

17. Lucas PJF, Boot H, Taal BG (1998) Computer-based decision-support in the management of primary gastric non-Hodgkin lymphoma. Meth Inform Med 37:206–219
18. MATLAB (2008) The MathWorks Inc, version R2008A
19. Maybeck PS (1979) Stochastic models, estimation, and control. Academic Press, New York
20. Moral S, Rumí R, Samarón A (2001) Mixtures of truncated exponentials in hybrid Bayesian networks. In: Sixth European conference on symbolic and quantitative approaches to reasoning with uncertainty, vol 2143 of LNAI, pp 156–167
21. Murphy KP (2002) Dynamic Bayesian networks: representation, inference and learning. PhD thesis, UC Berkeley
22. Ogata K (2002) Modern control engineering, 4th edn. Prentice-Hall, Inc, Upper Saddle River
23. Pearl J (1988) Probabilistic reasoning in intelligent systems: networks of plausible inference. Morgan Kaufmann, San Mateo

Chapter 15
Applicability of Virtualization to Embedded Systems

Tackling Complexity by "Divide and Conquer"

Robert Kaiser

15.1 Introduction

The performance of embedded system platforms has grown steadily during the recent years. In a complex, distributed system of embedded control units (e.g. an aircraft or a car), one single embedded computer is now powerful enough to take on loads that previously had to be handled by multiple dedicated nodes [1, 2]. Taking advantage of this potential, there is a new tendency to reduce the total number of embedded control units. While this trend is only in its beginnings in the car industry [3], modern aircraft already use the "IMA"[1] concept, where a network of uniform embedded computers is used to execute a multitude of different applications [4, 5].

Besides reducing cost and increasing performance, this also promises less complex hardware and thus better system reliability due to fewer components. Software complexity, however, increases: A multitude of more or less independent programs that previously lived each on its own dedicated node now have to share a common machine. Thus, the system infrastructure must provide mechanisms to prevent a program malfunction from affecting other programs in the system.

Moreover, the programs that are to be hosted by one machine may have very diverse requirements for their operating system interfaces: Some may have to ensure timely operation,[2] some may require a rich set of operating system services,

R. Kaiser (✉)
Bingen University of Applied Sciences, Bingen, Germany
e-mail: kaiser@fh-bingen.de

[1] Integrated Modular Avionics.

[2] Where the definition of "timely" as well as the degree to which it has to be ensured may vary largely.

M. Conti et al. (eds.), *Solutions on Embedded Systems*,
Lecture Notes in Electrical Engineering, 81, DOI: 10.1007/978-94-007-0638-5_15,

some may need to minimize their trusted code base for reasons of safety or security. It is impossible for a single operating system to cover all these conflicting requirements, however, traditionally, there exists only one operating system interface per machine.

These problems are similar to those that once led to the consolidation effort in the server field. There, virtualization was a successful solution: Virtual machines can be instantiated multiple times. Thus, it is possible to have multiple independent logical subsystems within a single physical machine. Programs which exist in different subsystems can not interfere with each other, except through interfaces which are provided (and controlled) by the virtualization environment. Also, any resources used by the subsystems are controlled by the virtualization environment, so a program can never exceed the resource limitations imposed on it by the environment. This strict control over resources and communication interfaces is the key to the safe and secure isolation by means of virtualization: Complex system software can be broken down into manageable components, each of which can decide for itself, which other components it depends upon.

It seems logical to apply virtualization to embedded systems now. However, this raises the question whether existing virtualization technology can simply be re-used "as is", or if any conceptual changes must be made to make it applicable to embedded systems.

The rest of this paper is organized as follows: In Sect. 15.2, a brief overview of the relevant virtualization techniques and their inner workings is given. Section 15.3 compares the requirements for embedded virtualization against the functionalities offered by the different approaches. Special attention is taken with respect to the ability of virtualization environments to support real-time applications, as this requirement does not exist in the space of server systems, but is frequently needed for embedded systems.

15.2 Virtualization Techniques

Virtualization is a broad term in computer science: Memory can be virtual, as can be storage, networks or even desktops. In the context of this paper, virtualization refers to the provision of environments (called *virtual machines* (VMs)), in which programs can execute.

15.2.1 Virtual Machines

Probably the most common example of such an environment is a Java Virtual Machine (JVM) [6]. Basically, a JVM is a program which emulates the behavior of a hypothetical machine by interpreting byte code. The benefit of this approach is the ability to run the same byte code program on different architectures without having to recompile it. This platform independence is generally considered to be a significant enough improvement to justify the loss of efficiency due to the interpretation of byte code.

Technically, a real machine can be emulated in much the same way as a hypothetical one. This is exemplified by the "Bochs" program [7] which emulates a complete IA-32 machine to sufficient detail so as to allow execution of a full-fledged operating system. However, compared to native code execution by a physical processor, Boch's binary code interpretation is very inefficient.[3]

If the underlying physical (host) machine features the same instruction set as the machine that is being emulated, and if that instruction set also satisfies certain formal criteria [8], then, instead of interpreting code, the host processor can execute most of the instructions directly. Only a few "sensitive" instructions (see below) remain to be emulated. With this *native* virtualization approach, virtual machines can be nearly as efficient as physical ones. This form of virtualization is therefore examined in more detail in the following text. It will be referred to as "virtualization" in the rest of this text.

15.2.2 Virtual Machine Monitors

Virtualization was invented by IBM in the late 1960s [9]. At the time it was motivated by the desire to improve the utilization of (expensive) mainframes by making them sharable between several users and/or applications. The software component that implemented virtualization was referred to as a *Virtual Machine Monitor* (VMM).

IBM's VMM offered multiple virtual machines which were each exact copies of the underlying host machine. It could host multiple independent operating systems along with their applications in different VMs within a single physical machine. It was a requirement that a virtual machine be indistinguishable programmatically from a real one. Consequently, neither the operating systems nor their applications needed any adaptation to the VM environment: All code that had originally been written for the physical machine could be reused without changes.

In [8], Popek and Goldberg state formal requirements for a computer architecture to be virtualizable in this way: They define the set of *sensitive instructions* as the subset of the machine's instruction set that interrogate or modify the resource configuration of the virtual machine. For example, a machine instruction manipulating the processor status word (PSW) would obviously qualify as sensitive, but even just reading the PSW is a sensitive operation too as it allows a program to make decisions based on the state of the physical machine.

[3] Bochs is primarily intended as a tool for test and debug purposes, but not for productive systems.

For virtualization to be transparent, all sensitive instructions must be emulated by the VMM, while all non-sensitive instructions can be executed directly by the processor. To make this possible, the VMM is installed as a trap handler and the virtualized program is executed in non-privileged mode. Provided that all sensitive instructions are also privileged instructions, any attempt by the program to execute one of them in non-privileged mode generates a trap that invokes the VMM. The VMM then emulates the instruction transparently. Therefore, according to Popek and Goldberg, for a machine architecture to be virtualizable, the set of sensitive instructions must be equal to the set of privileged instructions or a subset thereof.

This virtualization method is also referred to as "full" virtualization to distinguish it from the method of paravirtualization which will be described next. Full virtualization has the advantage of being able to execute any binary code without changes. However, it is only applicable to machine architectures that fulfill Popek and Goldberg's virtualizability criteria. The prevalent IA-32 architecture, for example, is not virtualizable according to these criteria: its instruction set comprises 17 instructions which are sensitive but do not generate a trap [10].

15.2.3 Paravirtualization

The term *paravirtualization* was coined by Gribble et al. when they introduced the "Denali" VMM [11]. Paravirtualization gives up the requirement of binary compatibility, at least for operating system code. The code of guest operating systems is adapted to the virtual machine environment by replacing sensitive instructions with appropriate system calls[4] into the VMM. The adaptation can be done manually or even automatically, e.g. by using special compilation tools [12]. It only needs to be done for programs that expect to run in privileged mode on a physical machine (i.e. operating system kernels), whereas application programs can be kept unchanged. Usually, the source code of an operating system kernel must be available for it to be "paravirtualized" in this way. Therefore, closed source operating systems can normally not be adapted. Today, the Xen Hypervisor [13] is probably the most prominent VMM using paravirtualization.

As stated earlier, the IA-32 architecture is not virtualizable according to the criteria given in [8]. However, by "creative programming" it was nonetheless possible to virtualize IA-32 machines. The first virtual machine monitor to demonstrate this was VMware [14]. The method it uses is referred to as *scan before execution*: when a piece of code has been loaded into memory and before it is executed, the program's code is scanned and all instances of the aforementioned 17 sensitive instructions are replaced by appropriate hypercalls. Thus, although VMware is generally considered to be a full virtualization environment

[4] So-called *hypercalls*.

(because operating systems are able to run in its virtual machines unchanged), the method it uses would probably be described more suitably as "on the fly paravirtualization".

Obviously, this dynamic re-writing of code leads to additional overhead, so the approach is less efficient than the normal, "static" paravirtualization. Furthermore, hypercalls can offer a higher level of abstraction than emulation at the machine instruction level (i.e. the amount of useful work done by one hypercall is usually more than that of a single emulated machine instruction), therefore, the run-time efficiency of paravirtualization is generally better than that of full virtualization.

15.2.4 Microkernel-Based Paravirtualization

Microkernels originate from an entirely different line of thought than VMMs: Their goal is to reduce the complexity of operating systems by reducing the concepts that are implemented as privileged code to a minimum. In [15], Liedtke demands that "a concept be tolerated inside the kernel only if moving it outside the kernel... would prevent the implementation of the system's functionality". Following this principle, a microkernel should only provide a small set of mechanisms and ideally no policies at all. Policies are to be implemented by user-level software, utilizing the kernel-provided mechanisms as necessary. In this way, different, alternate policies can coexist in a common system. For example, operating system services such as I/O or memory management, which would have been implemented by the kernel in a classical monolithic operating system, are provided by user-level servers. The role of the microkernel is merely to establish an infrastructure (i.e. inter process communication, or *IPC*) for applications to communicate with their servers.

IPC is the one central service provided by a microkernel. It is not only used for direct client–server communication: Hardware-generated events such as interrupts or traps are translated by the microkernel into IPC messages which are then sent to a responsible handler (a user level process). Thus, interrupts "look" like processes sending messages when an associated event has happened. In this way, the "policy" portion of interrupt or trap handling can be implemented outside the kernel, only the "mechanism" part remains in the kernel.

It has been demonstrated that a microkernel infrastructure such as L4 [16] can be used as a VMM to support paravirtualized operating systems [17, 18] in much the same way as was described in the previous subsection. Given the different roots of VMMs and microkernels, it is remarkable to see the similarities between the outcomes of the two lines of development. There is an ongoing academic debate about wether or not VMMs are in fact "microkernels done right" (see [19, 20]). The one point that the opponents in this debate seem to agree upon is that VMMs share so many similarities with microkernels that they can in fact be considered a specific form of a microkernel.

15.3 Requirements (and Non-requirements) for an Embedded VMM

We now look at the functionalities of the different virtualization approaches. These functionalities are inherited from their previous use in the server or workstation area, so a re-evaluation in the light of the embedded system field is necessary.

15.3.1 Isolation and Communication

The need for isolation was the initial motivation for bringing up the subject of virtualization for embedded systems. All virtualization approaches that were introduced in the previous section do feature strong isolation between virtual machines. Any faults happening in one VM remain confined to that VM, they can not spread throughout the system. For server consolidation, this strong isolation is usually all that is needed, i.e. the individual VMs can be treated like individual physical machines and there is no compelling need for them to interact with one another in ways any different from real machines, i.e. via a network. In contrast, in embedded systems, multiple subsystems may contribute to the overall functionality of the embedded device. Therefore, an effective, secure mechanism for inter-VM communication is frequently required in addition to isolation.

The inter-VM communication facilities provided by most VMMs are typically "retrofitted" ones (e.g. pseudo network interfaces), thus communication with other VMs or the outside world may encompasses significant protocol overhead. Microkernels, on the other hand, are traditionally well-equipped in this respect: The early approaches to microkernel design (e.g. the Mach mikrokernel [21]) and their failure have shown the importance of a high-bandwidth, secure IPC mechanism. Second generation microkernels such as L4 have learned from these lessons: L4 was constructed from the ground up to deliver very efficient IPC.

15.3.2 Architecture Independence

The server and desktop markets are largely dominated by the IA-32 processor family today. Therefore, virtualization approaches (such as VMware) have been specifically tailored for the properties of the IA-32 architecture and have gone to great lengths to compensate for its shortcomings. In the embedded market, there is no single dominant processor architecture and some of the more prevalent embedded processor architectures also do not meet the virtualizability criteria in [8]. Furthermore, the major advantage that full virtualization has over paravirtualization, i.e. the ability to execute kernel code without recompilation, loses much if its attraction in the embedded space as it is common for embedded system developers to have access to the full source code to their operating systems anyway. Thus, from the perspective of embedded systems use, paravirtualization is preferable to full virtualization: the disadvantage of requiring an adapted OS is not an issue, plus it performs better.

Java byte code programs are platform agnostic by definition, so a JVM-based approach offers even better architecture independence. This level of independence comes at a price though: it takes significantly more resources to provide a given functionality in byte code form rather than as native code, and the byte code's performance is generally lower. Given that embedded systems are usually designed to provide their functionality with minimal resources, the increased resource requirements of JVM-based virtualization has hindered a wide adoption of such approaches to date. Some notable exceptions exist, though, in cases where the ability to install and uninstall services dynamically without knowing the details of the target platform is valued higher than the cost of resources [22]. In this context, approaches have been proposed which also leverage the isolation features of the JVM in much the same way as the native virtualization methods we are mainly considering in this text [23, 24]. Nevertheless, their extensive memory consumption and their lack of real-time capabilities (see below) limits these approaches to a small niche within the embedded systems market.

15.3.3 Size of Trusted Code Based

Any safety or security related application must consider the amount of code upon whose correctness it relies. This *trusted code base* includes all privileged code (i.e. the kernel) as well as any software modules the kernel relies upon. Depending on the criticality of the application's function, all trusted code may have to undergo rigorous testing or it may even have to be mathematically proven correct. The effort involved in these tests depends directly on the amount of trusted code, so there is a strong need to keep the trusted code base of critical applications minimal.

Looking—for example—at the Xen hypervisor, its code base itself is already quite large (about 100,000 lines of code), but it also requires (an relies upon) a fully fledged Linux system in the privileged "Domain 0". This results in a total trusted code base of several million lines of code for which an exhaustive test, let alone a proof of correctness is out of the question. Some microkernels (e.g. Mach) are known to be similar in size, however, newer approaches are typically in the 10,000 lines of code range. So, again, microkernel-based virtualization would seem like a preferable approach.

Similar to Xen, JVM-based approaches, besides their JVM, rely on the correctness of the host operating system which is needed to support them. The (prototype) implementations presented in [23, 24] employ Linux as host OS, so the size of trusted code again amounts to several million lines of code. However, it has been shown that it is possible to implement a JVM on the basis of a much smaller set of OS functionality, e.g. that of a microkernel [25].

15.3.4 Superfluous Functionality

The size of the Xen and similar hypervisors is at least partly due to the implementation of certain policies inside the privileged code which will be of little use for embedded systems. For example, Xen allows to change the resource allocation of its virtual machines at run time as well as to do live migrations of virtual machines. Since they are implemented inside the kernel, these policies contribute in full to the memory footprint of the hypervisor and they also increase the trusted code base of all applications unnecessarily.

In contrast, microkernel development efforts have generally observed the principle of policy/mechanism separation more stringently: In a microkernel-based system, all policies are provided by servers. Thus, each application can individually select the set of servers it wants to communicate with. Each application can therefore make a fine-grained selection of its trusted code base based on the services it requires.

15.3.5 Real-Time Capabilities

Deterministic timing behavior is a frequent requirement for many, but not all embedded systems. A computing system which has to interact with a physical or technical system must match that system's timing.

It is difficult to guarantee deterministic timing for programs which are executed in a generic JVM environment. Unpredictable delays due to garbage collection are a major problem and running the JVM on top of a non-real-time, general purpose operating system such as Linux makes matters even worse. Consequently, [23, 24] do not discuss real-time behavior at all. This is acceptable for their specific application domain (i.e. residential gateways). Nevertheless, deterministic timing is an important requirement for a large portion of the embedded space and these approaches fail to address it.

Regarding native virtualization, Popek and Goldberg postulate in [8], that a virtual machine shows exactly the same behavior as a physical machine, however, they explicitly exclude the machine's temporal behavior from their considerations. The difference in timing has several reasons: trapping and emulating sensitive instructions is obviously more expensive than direct execution of same and replacing instances of these instructions with hypercalls during prescan as VMware does also requires additional effort. Nevertheless, these overheads are predictable, i.e. they occur in the same way and amount whenever a program is executed by the VM. Therefore it is possible for a process scheduler to take these overheads into account.

However, if multiple VMs exist in a single system, the VMM must switch the physical machine between the contexts of its virtual machines. Thus, besides several other responsibilities, a VMM also acts as a scheduler, allocating portions of the overall CPU capacity to its VMs. The guest operating systems hosted by the VMs schedule their own, private set of processes. Thus, there is a two-level hierarchy[5] of schedulers in which the VMM's scheduler forms the first level.

Most VMMs use some form of proportional share scheduling, because it reflects the concept of a virtual machine: VMs receive a share (i.e. a percentage) of the CPU's computational resources. Ideally, they should all receive their CPU shares in parallel, so each of them should be able to execute at any time, where the speed of their virtual CPUs would be portions of the real CPU's speed. But this is technically not possible because a CPU is not sharable: it can only be allocated to one activity at a time. Therefore, proportional share schedulers approximate the idealized behavior by switching between their virtual machines cyclically, running each of them for a finite *quantum* of time. The quality of this approximation improves as the quantum is made smaller, but at the same time, context switching cost increases.

Figure 15.1 shows the estimated context switching overhead of a system of three VMs, as a function of quantum size. These values were obtained from a simulation of a proportional share scheduler, where previously measured context switch delays were taken into account (see [26]). For quantum sizes below roughly 1 ms, the overhead increases dramatically, reaching more than 90% in the worst case. In order to stay within acceptable bounds (e.g. 10%), the quantum size would have to be in the 1–10 ms range. Figure 15.1 also shows the delay for which a VM may be blocked in the worst case. This delay directly adds to the jitter of real-time programs that are hosted by virtual machines. It is proportional[6] to the number of virtual machines and the quantum size. For any quantum value yielding an acceptable switch overhead (i.e. 1 ms), the corresponding delay is a multiple of this quantum, i.e. several milliseconds.

There are many practical real-time applications which can accept a temporal variation of this magnitude. Such applications can thus be executed by a virtual machine without problems. There are, however, also applications that require better timing predictability. For these applications, alternative scheduling approaches at the VMM level are needed. In order to cover the whole range of timing constraints (e.g. from microseconds to minutes) that a single embedded system may have to support, the VMM must apply different scheduling policies depending on the timing requirements of the particular application, or, more precisely, the virtual machine hosting the application (see [27]).

Here again, many VMMs are problematic because the scheduling policy is built into the VMM and thus difficult to replace.[7] But flexible scheduling is also the Achilles heel of microkernel-based solutions: L4 and its derivatives include a simple, priority-based scheduler. The plan is to implement external, user-level

---

[5] In principle, any of the processes hosted by a guest operating system could recursively host multiple processes, so the hierarchy could in fact have any number of levels.

[6] Note that the quantum axis in Fig. 15.1 is scaled logarithmically.

[7] Xen, however, does feature a pluggable scheduler architecture.

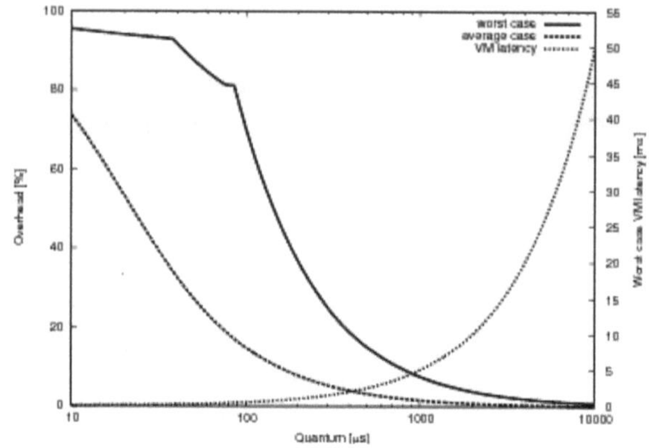

Fig. 15.1 Proportional share switch overhead and worst-case virtual machine delay vs. quantum (*source*: [26])

scheduling policies that dynamically modify the kernel-level task priorities to put through their own scheduling policies. However, such user-level schedulers are currently a research topic, no solutions are readily available at this time.

The "PikeOS" microkernel, which is a commercial development based on L4 concepts, features a scheduling policy that combines priority-driven and time driven scheduling [28]. Here, the time driven schedule is in fact controlled by a user-level scheduler, while priority ranges are assigned to tasks statically. The time-driven part of the scheduler can be used to implement cyclic executives with very low jitter. Also, time partitioning as required by avionics standards (i.e. [29]) can be implemented with this approach. The PikeOS scheduling approach has the ability to run non-real-time, general purpose operating systems alongside with hard real-time ones.

15.4 Conclusion and Outlook

Virtualization is a promising technology to deal with the problems of upcoming complex embedded systems. By creating multiple isolated virtual machines hosting different subsystems, the overall system becomes manageable. However, most of the virtual machine environments available today were developed for use in server or desktop systems. They are not very well suited for use in embedded systems in their current form, because they lack certain important capabilities on the one hand and because they contain functionalities that are not needed in the embedded space on the other hand. However, this is clearly not a limitation of the concept of virtualization as such: In many cases, virtual machine monitors have incorporated policies which were suited for the use case at hand. These policies are now hard to remove/replace. Microkernels, however, have been built according to the principle of separation between policy and mechanism and they can also serve as a foundation for paravirtualization. Due to their lack of policies, they stand a better chance of being adaptable for the needs of embedded virtualization.

Some approaches in this regard already exist, and we expect to see an increasing number of use cases for these approaches in future complex embedded systems.

References

1. Greene R, Lownes G (1994) Embedded CPU target migration, doing more with less. In: TRI Ada '94: Proceedings of the conference on TRI-Ada '94. ACM Press, New York, pp 429–436
2. Stepner D, Rajan N, Hui D (1999) Embedded application design using a real-time os. In: DAC '99: Proceedings of the 36th ACM/IEEE conference on design automation. ACM Press, New York, pp 151–156
3. Broy M (2006) Challenges in automotive software engineering. In: ICSE '06: Proceeding of the 28th international conference on software engineering. ACM Press, New York, pp 33–42
4. Sanchez-Puebla MA, Carretero J (2003) A new approach for distributed computing in avionics systems. In: ISICT '03: Proceedings symposium on Information and communication technologies. Trinity College Dublin, pp 579–584
5. Bate I, Kelly T (2003) Architectural considerations in the certification of modular systems. Reliab Eng Syst Saf 81:303–324
6. Lindholm T, Yellin F (1999) Java virtual machine specification. Addison-Wesley Longman Publishing Co., Inc, Boston
7. Lawton KP (1996) Bochs: a portable PC emulator for Unix/X," LINUX J, vol 1996, no 29es, p 7, sep 1996
8. Popek GJ, Goldberg RP (1974) Formal requirements for virtualizable third generation architectures. Commun ACM 7(7):412–421
9. Creasy RJ (1981) The origin of the VM/370 time-sharing system. IBM J Res Dev 25(5): 483–490
10. Robin J, Irvine C (2000) Analysis of the Intel Pentium's ability to support a secure virtual machine monitor [Online]. http://citeseer.ist.psu.edu/robin00analysis.html
11. Whitaker A, Shaw M, Gribble S (2002) Denali: lightweight virtual machines for distributed and networked applications. University of Washington technical report 02-02-01
12. LeVasseur J, Uhlig V, Chapman M, Chubb P, Leslie B, Heiser G (2005) Pre-virtualization: slashing the cost of virtualization. Technical report PA005520, NICTA, October
13. Barham P, Dragovic B, Fraser K, Hand S, Harris T, Ho A, Neugebauer R, Pratt I, Warfield A (2003) Xen and the art of virtualization. In: ACM symposium on operating systems principles, Bolton Landing, NY, USA
14. Muller A, Wilson S, Happe D, Humphrey GJ (2005) Virtualization with VMware ESX server. Syngress Publishing, Inc, Rockland
15. Liedtke J (1995) On-kernel construction. In: ACM symposium on operating systems principles, Copper Mountain Resort, CO, USA, pp 237–250
16. Liedtke J (1996) L4 reference manual—486, Pentium, Pentium Pro. http://os.inf.tu-dresden.de/L4/l4refx86.ps.gz
17. Hartig H, Hohmuth M, Liedtke J, Schnberg S, Wolter J (1997) The performance of μ-kernel-based systems. In: 16th ACM symposium on operating system principles (SOSP), pp 66–77
18. Hartig H, Baumgartl R, Borriss M, Hamann C-J, Hohmuth M, Mehnert F, Reuther L, Schönberg S, Wolter J (1998) DROPS: OS support for distributed multimedia applications. In: Proceedings of the eighth ACM SIGOPS
19. Hand S, Warfield A, Fraser K, Kotsovinos E, Magenheimer D (2005) Are virtual machine monitors microkernels done right? In: HOTOS'05: Proceedings of the 10th conference on hot topics in operating systems. USENIX Association, Berkeley, p 1
20. Heiser G, Uhlig V, LeVasseur J (2006) Are virtual-machine monitors microkernels done right? SIGOPS Oper Syst Rev 40(1):95–99

21. Accetta M, Baron R, Bolosky W, Golub D, Rashid R, Tavanian A, Young M (1986) Mach: a new kernel foundation for UNIX development. USENIX Summer 93–112
22. OSGi "Open Service Gateway Initiative Alliance" (2009) [Online]. www.osgi.org
23. Ibanez M, Martinez Madrid N, Seepold R (2007) Virtualization of residential gateways. In: International workshop on intelligent solutions in embedded systems (WISES07), June
24. Royon Y, Frenot S, Mouel FL (2006) Virtualization of service gateways in multi-provider environments. In: University, Vasteras, Sweden, pp 385–392
25. Hoffmann A (2005) Portierung und Validierung der Java 2 Micro Edition fur das Mikrokern-Betriebssystem PikeOS. Diploma Thesis, Wiesbaden University of Applied Sciences, Dept. DCSM, September
26. Kaiser R (2008) Empirische Ermittlung Cache-bedingter Umschaltverluste. GI/ITG Fachgruppentreffen, Wiesbaden, March
27. Kaiser R (2008) Alternatives for scheduling virtual machines in real-time embedded systems. In: IIES '08: Proceedings of the 1st workshop on isolation and integration in embedded systems. ACM, New York, April, pp 5–10
28. Kaiser R, Wagner S (2007) Evolution of the PikeOS Microkernel. MIKES 2007, Sydney, Australia, National ICT Australia, Sydney NSW 2052, Australia, Tech Rep, January
29. Kaiser R (2009) Complex embedded systems—a case for virtualization. WISES 2009, Ancona, Italy, June

Chapter 16
Distributed Trading Architecture with Sensors Support for a Secure Decision Making

Javier Martínez Fernández, Ralf Seepold and Natividad Martínez Madrid

16.1 Introduction

In these days, trading is under pressure due to unstable stock markets. So, traders require to overcome own emotions during decision making. In most of the cases stress or panic provokes wrong decisions and this should be avoided.

Therefore, it would be desirable to avoid making decisions in time of panic or stress, decisions that are not taken in a rational way. The stock exchange system is the representation of money; and many systems have been set up to analyze all types of data. For example: future markets, derivatives, index, or shares. Systems such as MetaTrader [1] or Metastock [2] provide to users all types of data and even risks hedging using derivatives.

Besides, there are developments of expert systems for stock trading as Brown et al. [3] developed in prolog for the buying and selling of futures in Chicago's market, or for stocks' recommendation on the Spanish market, Martínez and Heyn [4]. It is possible to see the creation of complex statistical models that try to find a pattern in the market as Wolberg [5] or Ang and Quek [6] and even research groups of trading systems as Cao et al. [7] and Tradingsys [8].

There are a great variety of programs that collect stock information and expert systems and research studies that treat this information. This article tries to explain the need to know "when" these systems should work and "when" these systems

J. Martínez Fernández
Universidad Carlos III de Madrid, Avda. de la Universidad 30, Leganes 28911, Spain

R. Seepold (✉)
University of Applied Sciences Konstanz, Brauneggerstr. 55, Konstanz 78462, Germany
e-mail: ralf.seepold@htwg-konstanz.de

N. Martínez Madrid
Reutlingen University, Alteburgstraße 150, Reutlingen 72762, Germany
e-mail: Natividad.martinez@reutlingen-university.de

M. Conti et al. (eds.), *Solutions on Embedded Systems*,
Lecture Notes in Electrical Engineering, 81, DOI: 10.1007/978-94-007-0638-5_16,

are indeed effective. It will be described a trading architecture that tracks behavior by means of sensors to detect the moment of entering in an unsafe decision making mode cause of stress. A platform connecting to sensors and to trading parameters will be used to embed an expert system ready to support a trading decision. It is crucial to detect the appropriate moment to activate the expert system and to support the trader.

The next sections will explain how it is possible to measure the stress of a trader and the proposal trading architecture taking advance of this acknowledgment in more detail. Finally, conclusions are presented and some indication of the future research work is given.

16.2 Measuring a Trader's Stress

There are several studies of the human behavior in the short term operating and about what variables can have influence on it. All these studies are experimental. For example, researchers Coates and Herbert [9] detected visually whether a financial intermediary (adviser) is qualified to deal with the vicissitudes of the market just with their intuition.

The article reviewed dates at the end of 2006 and aims to clarify the role of the endocrine system to assume risks in the financial field. The study focuses on a group of male traders in the City of London and the steroid hormones. The testosterone and the UFC (urine free cortisol) are known catalysts for cognitive responses and behavioral changes. It is intended to measure changes in levels of both steroids, the influence on them of the events which traders face up in their daily life and the effect caused on the norms of behavior and on the decision making. Finally, it was planned to determine whether the responses and decisions made in moments of strain and psychological stress are significantly influenced by the levels of these hormones. We can see an unsatisfactory male chauvinist approach, probably because the testosterone is a predominantly male hormone (also produced by women though at lower levels), but there are no accused sexual differences in cortisol.

Beginning with the testosterone, this hormone appears in sexual behavior and in the competitive confrontation. When someone is in a competition, its levels grow in the winner and lower in the loser. This androgenic award is known as the winner effect. Ultimately, besides other implications on the secondary sexual characteristics, testosterone causes aggressive behavior. Their highest levels in blood were found after adolescence, they remain for about 15 years, to be gradually diminishing.

Cortisol, is produced by the adrenal glands and plays a central role in the behavioral and physiological response to physical or psychological stress. It is very sensitive to situations where we do not have the control, to news and uncertainties. When a stressful situation comes up, after the emergency, hormone levels and the physiological processes return to normal. The problem emerges when stress is prolonged (persistent fall in the stock market). In this case, cortisol levels in the blood are fired, altering the performance and interpretation cognitive, besides other physiological disorders that result in a distortion of sensory perception. The experiment runs for about 8 days and is done on the premises of a broker, by measuring the levels of these hormones while operators perform their daily work. They found that daily testosterone levels were significantly higher on days when the trader wins over the daily average of the last month. In the case of cortisol, there was not a relationship between the loss and increased levels of this steroid. It was studied if there was any correlation between their levels and the risk. It was found that the higher volatility (standard deviation of change in the value of a financial instrument with a specific time horizon) of the daily trader's profits and losses, the highest daily levels of cortisol and the standard deviation of the daily levels. This suggests that the individual levels of cortisol are not related to the rate of economic return, such as testosterone, but it is related with the variable nature of the returns. At this stage of scientific progress, nobody is surprised when it says, quite firmly, that the response to everyday events and to uncertainty and stress moments are measured by hormonal cascades that after millions of years of evolution have improved responses to increase survival. It will take many generations for any kind of adaptation to this changing environment will be improved. Because of this approach, in the stock sphere, managers for more than 25 years are trying to deposit the responsibility of its clients' portfolio management in the quantitative management, mathematics and probability theory.

If testosterone is responsible for competitive fighting, its maximum level is reached after adolescence and is stable for 15 or 20 years, so the explosive combination of youth—capital management must be taken in mind. This, for example, will affect the platform depending on the market in which we work, as the winner effect—the youth encourages a manager to increase exposure to risk, this can go well when the market is bullish but when the markets are falling is very dangerous. Moving from euphoria to despair often leads to hasty and intuitive decisions, built in spikes or valleys hormonal, usually in the worst of times. The problem with these studies is that there is no way to measure in real time the hormone levels of traders and translate it to suitable stress data for the trading process.

Johnston [10] Biopsychology professor at the University of New México after a hormonal study says that the hands are testimony of the hormonal flow in the fetal stage and through the measurement of the index and the ring fingers; it is possible to have a degree indicator of exposure to fetal testosterone. Increased exposure is reflected in a ring finger longer than index one. The conclusion that tries to provide, without experimental support, is to exclude advisors with a ring finger longer than index finger (because of its negatives implications to control moments of panic or euphoria) and rely more on the minor asymmetry finger present, especially if their financial intermediary is continually abusing of the intuition (whether fundamental or technical). However, in this case, we can rule out many traders without giving their one chance in the trading process.

It is important to obtain a stress data in the right way to allow feedback for the trader about the stress level in real time, and biometric sensors have the answer. According to [11] when you are experiencing stressful emotions, whether you are

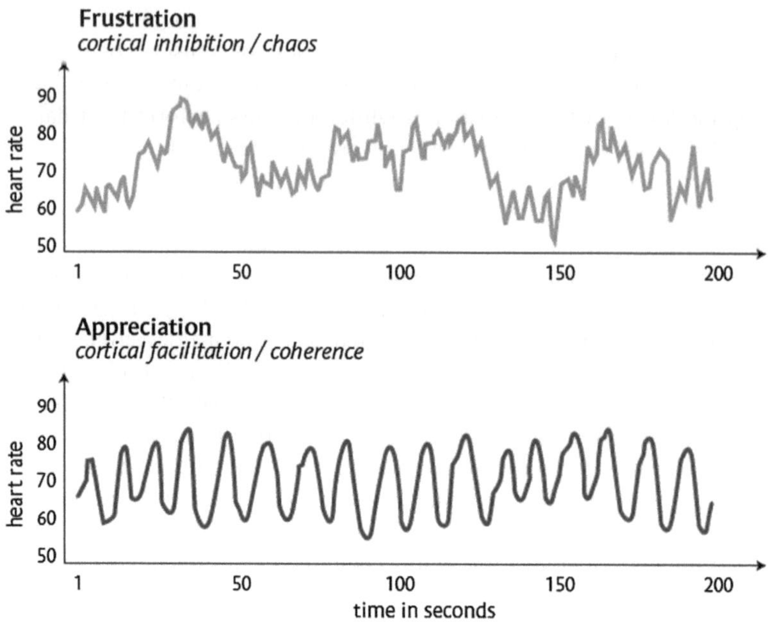

Fig. 16.1 Different heart rhythms (coherence/no stress and chaos/stress)

conscious of them or not, higher brain processes become seriously compromised. This phenomenon is called *cortical inhibition*. In the same study, this cortical inhibition is tested with the impact of stress on the cardiovascular system in real time. The more stable the frequency and shape of the waveform of the heart rate, the more it reflects a coherent system. In physiological terms, coherence describes the degree to which respiration and heart rate oscillate at the same frequency. When physiological coherence occurs, the brain associates it with feelings of security and well-being. Figure 16.1 extracted from this study shows the difference of the heart wave in coherence and chaos situations.

It therefore, seems to be clear that our behavior can vary depending on certain levels, and that our objective of moving away from operating at any given time according to our constants is not unreasonable. Sensors can support the real time stress information, and in the next section we can see the power of this information in a trading architecture.

16.3 Trading Architecture with Sensors Support

The architecture is designed to act in case that the user's behaviour indicates an outstanding situation where the stress on the person increases significantly. Furthermore, it focuses on the profile of a short-term investor and a market daily worker "trader", where emotions can actually have a great impact. Obviously, the behavior of a trader in the futures market probably is distinct of the behavior of one trader in the stock market or in the bund market or in the commodities' market etc. Therefore, depending on the market, rules will be interpreted stricter or more relaxed, because each market has an operating mode reflected by a distinct margin of error and risk.

The current implementation is designed to support a trader that is working at home. This is quite common in the USA, but it is not restricted to that since today's devices can be mobile. The core of the architecture is a residential gateway running an OSGi [12] framework. The application is designed to be fully compatible with other users working at the same time (in parallel) with the gateway. So, there is no need to have exclusive access or dedication to devices. The following modules (bundles) will be provided on the residential gateway:

1. *Biometric Module*: The user's capabilities will be monitored by a Stress Monitor of HeartMath's (with emWave software) [13]: A heart rate sensor put in the ear of the person informs in real time of the stress level, named in the software coherence level. This sensor is connected by USB pen drive where the information is processed and sent it to the computer. Besides, we can see the data on real time through log files.
2. *My Profile Module*: This module is responsible for measuring the risk profile of the user depending on his age and experience in financial markets since the biometric responses are also conditioned by these rules.
3. *Stock Information Module*: This module connects to the Internet, and it will provide stock information in real time. An example of this module would be Infobolsa PowerStation belonging to Infobolsa [14]. This module gathers the next sub-modules:

   - *Storage Module*: Various databases are needed to store stock information that will be used later by the expert system for decision making.
   - *Update Databases Module*: This module will be activated at the end of each session. It stores tracking data of the stock for a proper function of the expert system.
   - *Warning Module*: It is responsible for detecting the right moment to trigger the expert system. The process is continuously running and collecting data from the stress sensor. It will activate the expert system module when the user enters into an unsafe decision making zone, and it may even block the current operation (depending on configuration preferences).

4. *Expert System Module*: This module comes into operation when the alert module activates it, and it is responsible for the operation. In the section on expert system, we will look in more detail. The architecture is presented in Fig. 16.2.

Regular operation works in the following way:

A. The trader uploads his profile (age, experience, type of market in that will operate, etc.)

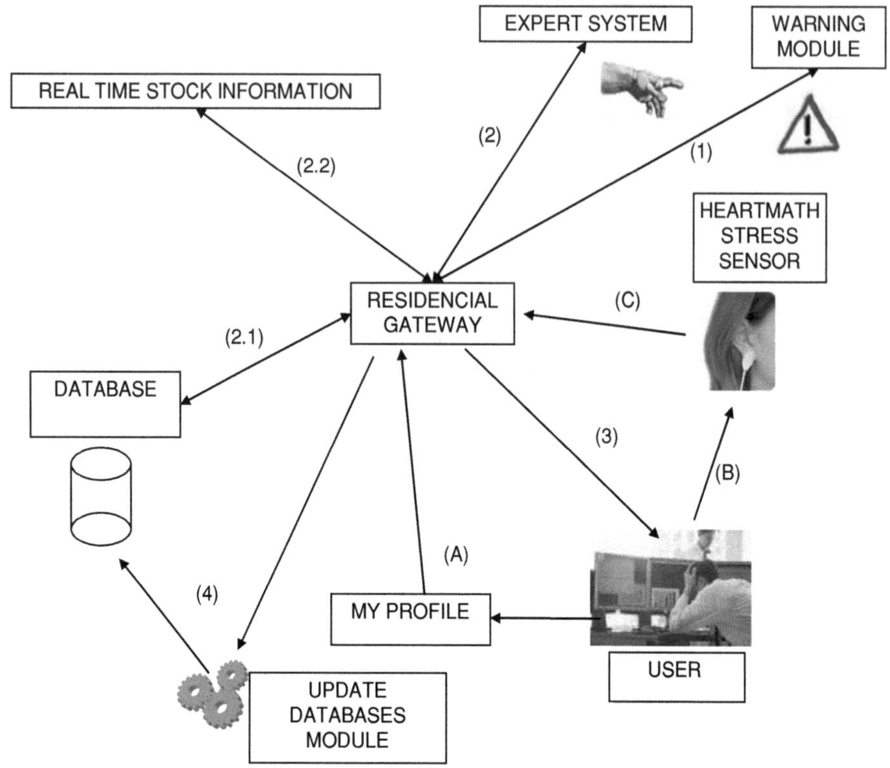

Fig. 16.2 Distributed trading architecture with sensors support

B. The trader connects the stress sensor
C. The stress sensor collects information that is continuously revised by the warning module

In case the alert module triggers and the user enters in non secure decision making and the expert system is triggered and the operation mode will be the following:

1. Module that is running in the residential gateway activates the expert system module and blocks the user operation.
2. The expert system retrieves data for an analysis of the situation.
   2.1. Stored data is requested from the database.
   2.2. Real time data are required from the real time operating system.
3. Once the database manager provides the requested information to the expert system, the expert system sends the results to the user, who must approve the operation.
4. The update module will be launched automatically at closing time.

16.4 Expert System

Common sense becomes the enemy on the stock market, and undoubtedly the influence of external conditions, such as news, global macroeconomic structure and even the feelings can result in an incorrect operative.

One of the inconveniences of the expert systems [15] is the lack of emotions, quality only attributable to human reason, but in the trading process, that problem of expert systems translates into a virtue in stressful situations, because the emotions never interfere in the Expert System, an automatic system that complies with rigor, with the knowledge of an expert, and that cannot be influenced by these external agents. This knowledge will be implemented through rules of inference based on sources as Dussauchoy and Nchatain [16], Gómez and Montes [17] and Amat Salas [18] at the time of architecture design. The system implements the following components of an expert system:

- The Knowledge Base: An expert system contains knowledge of the facts and the experiences of experts in a certain domain. In the next section, this point will be explained in detail.
- The inference mechanism of an expert system that can simulate the solving strategy of an expert.
- The explanatory component or explanation engine, explains to the user the solution strategy.
- The User Interface is used to enable queries in a natural language. The user interface consists on web pages.
- Necessary databases.

The structure of the expert system follows the MYCIN model by Buchanan and Shortliffe [19], because its understanding is very simple and its efficiency very high, keeping high degree of parallelism with the expert system to develop.

MYCIN is an expert system used in clinic diagnostics, initiated by Ed Feigenbaum and subsequently developed by E. Shortliffe and his colleagues. Its function is to advise the doctors in the investigation and determination of diagnoses in the field of infectious diseases. The MYCIN system, when it is consulted by the doctor first asks general information about the patient: name, age, symptoms, etc. Once this information is known by the system, the expert system establishes a hypothesis. First, it checks the accuracy of the premises of the rule to verify the hypothesis. This is done by searching for the corresponding statements in the knowledge base. Furthermore, it is done by certain questions to the user. Here we find questions like: Have you practiced in the patient some form of surgical intervention? With responses received, the MYCIN verifies or rejects the hypothesis. A series of test has demonstrated that MYCIN works as well as a doctor. The expert system is based on the MYCIN model, so the intrinsic operation is the same as described above. Figure 16.3 shows how to relate the following entities.

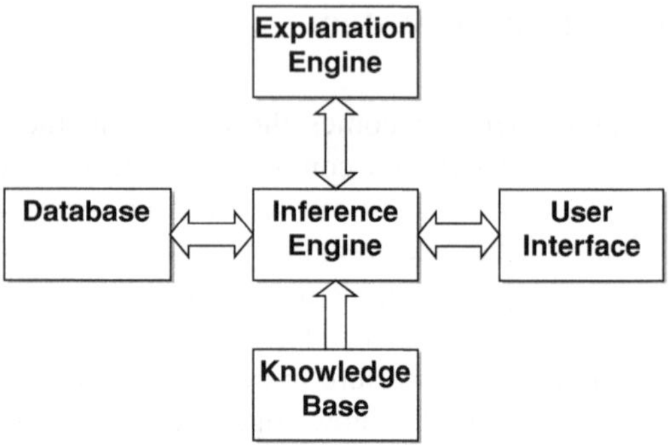

Fig. 16.3 Entities of an expert system

16.5 Knowledge Base Structure

The Expert System builds the rules of the knowledge base based in different analysis. A brief explication of each one is given in this section. Obviously, with a major number of indicators, the expert system will be able to create better rules. The intention of this point is to explain the indicators that generate the rules with major weighting, inside each kind of analysis. It is important to take in mind that each user configures the expert system to give the preference between the rules.

16.5.1 Technical Analysis

It is based on derived rules of a creation of charts thought to several numerical parameters.

1. *RSI: Relative Strong Index*: This indicator measures the strong of the offer and demand on the market. To calculate this indicator the system applies the next formula:

$$\mathrm{RSI} = 100 - (100/(1 + \mathrm{RS}))$$

where RS is the quotient between the summation of bullish close prizes (SA) and summation of bearish close prizes (SB) at 14 days:

$$\mathrm{RS} = \mathrm{SA}/\mathrm{SB}$$

with this indicator, it is possible to detect an overbought or an oversold in a share in this way: if the indicator is lower than 30 the share is oversold, and it is possible that indicates a possible buy. On the other hand, if the indicator is higher than 70, the share is overbought, and it is possible that indicates a possible sell. In Figs. 16.4 and 16.5 is shown how this indicator acts. In the expert system, the rule will have more relevance if the lecture of RSI is more extreme as follows:

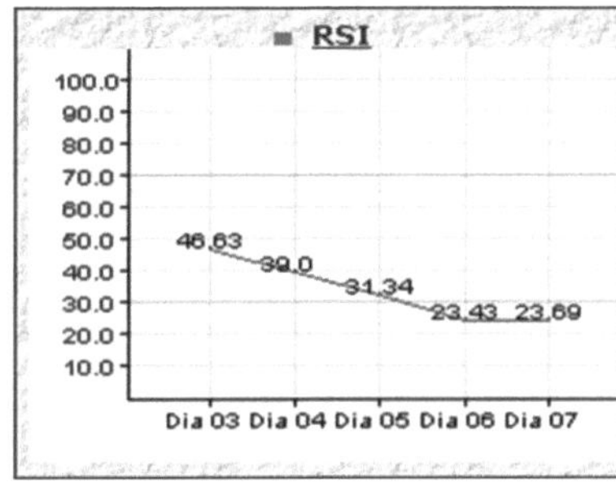

Fig. 16.4 RSI with buy signal

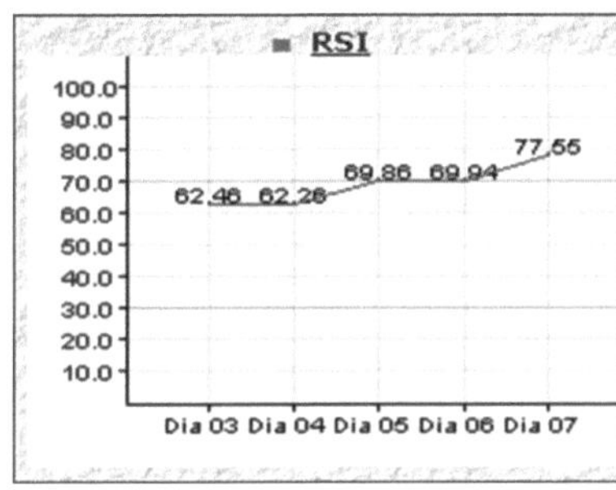

Fig. 16.5 RSI with sell signal

$$\text{Weighting} = (100 - (100 - \text{RSI}))/100$$

So with a RSI of 70 the Weighting will be 0.7, in case of 90 will be 0.9 and the rule will have more relevance.

2. *MACD*: This indicator uses two exponential mobile averages to create a chart with a softly tendency that avoids the abrupt movement of the shares in some moments that far to the user of seeing the truth tendency. For example, Fig. 16.6 shows the prize of one share and its mobile average (the solid line of the truth tendency).

The indicator MACD is composed of two lines, the first is named MACD, and it is the subtraction of two exponential mobile averages, one of 12 days (Mx(c, n)) and other of 26 days (Mx(c, m)):

$$\text{MACD} = \text{Mx}(c, n) - \text{Mx}(c, m)$$

The second is named SIGN and is the arithmetical average of 9 days of MACD:

$$\text{SIGN} = \text{Mx}(\text{MACD}, 9)$$

A buy signal is produced when the line MACD crosses the line SIGN and a sell signal is shown in the Figs. 16.7 (MACD line starts at -1.36) and 16.8 (MACD line grows up to 0.17).

Fig. 16.6 Mobile average against prize

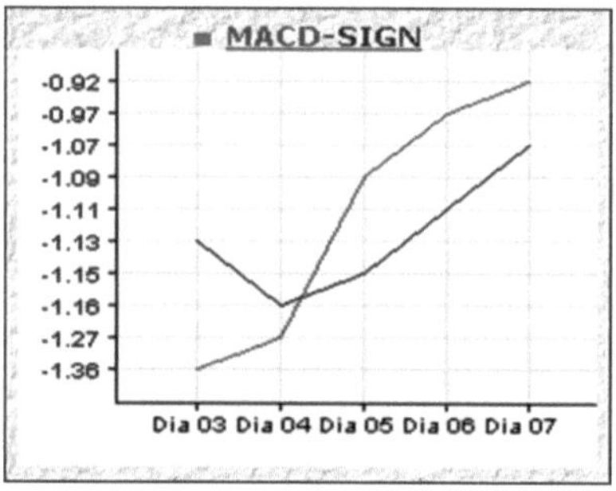

Fig. 16.7 MACD with buy signal

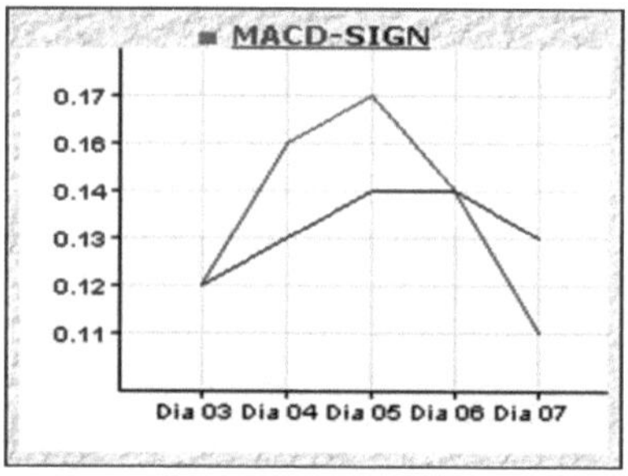

Fig. 16.8 MACD with sell signal

The user specifies the system's preference for this rule.

16.5.2 Fundamental Analysis

The fundamental analysis is based on the theoretical value of the company as a function of the balance of the company and its book value. The variation of these indicators is very slow (normally each 6 month when the company presents the results), so, in the expert system the weighting for the short term operative should be low and it depends on the user. Some parameters that use the Expert System are:

1. *PER*: Price Earning Ratio. PER is a simple but effective indicator that indicates the number of times that the profit per share is included in the prize of the share. So, a low PER is indicating an undervalued share and a recommendation of buy for this one. The user should assign the priority for this indicator in the Expert System. The formula is:

$$\text{PER} = \text{Prize of Share}/\text{Profit per Share}$$

2. *BPA*: Profit per Share (Beneficio por Acción—in Spanish): This indicator measures the payoff in shares. It is calculated as shown

$$\text{BPA} = \text{Net Profit}/\text{Number of Shares}$$

In this case, a great BPA indicates a major profit. For the Expert System it consists in add this parameter for calculation of the final recommendation for the user.

16.5.3 Feeling Market Analysis

Feeling Market Analysis is the most controversial analysis because it measures the feeling of the people with the market. A pessimistic feeling on the market indicates that the market is next to turn to bullish and vice versa.

This feeling of the market is measured with statistics that are published each certain time, these are the direct entries for the Expert System, but with this point, there is a problem: The veracity of a statistics; people reply to the questions to influence statistics by their operatives and their desires, so the question is: Is there any possibility to obtain a truth feeling of these statistics? Perhaps the reply to this is *no*, but the Trading Architecture with Sensors Support fixes this problem in the next point, where at least the user knows how is your real feeling, and this is most important for his operative.

16.5.4 Feeling Sense of the User (with the Sensors)

While the user is operating under a great stress, and his emotions are his enemies, obviously the stress is his major enemy. However, the track of these feelings is measured for the HeartMath sensors connected to the user. When the stress level measurements values detect an unsafe area, our Trading Architecture shows an alert. The decision of how the trader must manage this information depends on the context of the trader and the Trading Architecture configuration. For example, if the trader is a private trader, the action of stopping operative and Expert System starts to manage the trading process or to continue in a manual way could be managed by the own trader. However, if the context is an inversion bank where the trader is an employee of this entity, the action could be to stop the operative and Expert system takes the control. Anyway, in both cases, independently how the operative is managed, now we have a stress aware trader. Therefore we have a secure decision making.

16.6 Conclusions and Future Work

We have born our natural predisposition to dualism and theology predispose us to believe, without the need for contrast, without questioning the habits, values, and lessons of our adults. This predisposition leads us during adulthood to keep believing and trusting on concepts and behaviors that are wrong, even though evidences continually alert us. We are masters at denying the evidence and maintain unhealthy habits, and that in times of stress is really dangerous, even more if our work is to be a stock trader.

The paper begins with the clear objective to avoid decisions or wrong attitudes driven by moments of stress in which we are typically unable to make right decisions despite perhaps having the necessary tools. The question to resolve is "when" is the right moment to receive support. If it is possible to determine when such moments occur, then it is possible to avoid bad decisions. In stress situations our body change, and as we have seen, our hormone levels are altered. Biometric parameters like heart rate variability show with suitable sensors (like HeartMath stress sensor) our stress level. The platform has been described above can alert us when we are facing a moment of stress. In this sense with a well-defined expert system, we can certainly reduce the risk we face in those times when our reasoning is interfered by our emotions.

Our system is focused to a stock trader, but it can enter in any domain. The union of the concepts of Expert System and Telematics (networks of sensors and monitoring) can detect this critical moment and most important, it allows to the user to be aware of this sensitive moment and it avoids an unsafe decision making.

References

1. Metatrader (2010) http://www.metaquotes.net/metatrader
2. Metastock (2010) http://www.equis.com
3. Brown CA (1991) Holland, I y Mesch, R. "TRADER: an expert system for trading commodities futures", Artificial intelligence on wall street, 1991. Proceedings, ISBN: 0-8186-2240-7

4. Martínez J, Heyn A (2002) Sistema Experto para la Recomendación de Valores en el Ibex 35. Master Thesis, Universidad Politécnica de Madrid, Spain
5. Wolberg JJ (2000) Expert trading systems: modeling financial markets with kernel regression. Wiley, ISBN-10: 0471345083
6. Ang KK, Quek C (2006) Stock trading using RSPOP: a novel rough set-based neuro-fuzzy approach. IEEE Trans Neural Netw 17(5):1301–1315
7. Cao L, Luo C, Zhang C (2007) Developing actionable trading strategies for trading agents. In: IEEE/WIC/ACM international conference on intelligent agent technology
8. Independent Reflections on Trading Systems, (Tradingsys) (2010) http://www.tradingsys.org
9. Coates JM, Herbert J (2008) Endogenous steroids and financial risk taking on a London trading floor. In: Proceedings of the national academy of sciences, 2008—national acad sciences
10. Johnston VS (2000) Why we feel, "the science of human emotions", ISBN 0738203165
11. Cryer B, McCraty R, Childre D (2003) Pull the plug on stress. Harv Bus Rev 81(7):102–107
12. Open Service Gateway initiative (OSGi) (2010) http://www.osgi.org
13. Stress Monitor, Health Reviser (2010) http://www.healthreviser.com/content/stress-monitor
14. Infobolsa (2010) http://www.infobolsa.es
15. Sánchez y Beltrán JP (1998) Sistemas Expertos. "Una metodología de programación. Editorial RA-MA", ISBN 84-86381-48-7
16. Dussauchoy A, Nchatain J (1988) Sistemas Expertos. Métodos y Herramientas. Editorial Paraninfo, 84-283-1599-X
17. Gómez A, Montes C (1997) Ingeniería del Conocimiento". Editorial Centro de Estudios Ramón Areces, S.A, 978-84-8004-269-7
18. Amat Salas O (2004) La bolsa. "Funcionamiento y técnicas para invertir". Editorial Deusto S.A, ISBN: 200978-84-234-2240-1
19. Buchanan BG, Shortliffe EH (1984) Rule-based expert systems: the Mycin experiments of the Standford Heuristic Programming Project". Addison Wesley Publishing Company, pp 233–262, ISBN 9780201101720

Chapter 17
Migrating from a Proprietary RTOS to the OSEK Standard Using a Wrapper

A Feasibility Study

Joachim Denil, Serge Demeyer, Paul De Meulenaere, Kurt Maudens and Kris Van Stechelman

17.1 Introduction

With the introduction of the AUTOSAR standard, Automotive Open System Architecture [1], many automotive companies insist that their suppliers use standard components like an OSEK [2] (open systems and their corresponding interfaces for automotive electronics) compliant operating system in their products. OSEK is a joint project of the German and French automotive industry that defines an open software architecture for automotive control units in vehicles. According to a study by Schoof and Wybo [3], the migration from OSEK to AUTOSAR is easier then directly adopting AUTOSAR. Gaining knowledge and expertise with OSEK helps understanding the philosophy behind AUTOSAR.

Despite the success of OSEK, a lot of suppliers are still using their own proprietary real-time operating system (RTOS) [4]. The "not invented here syndrome" certainly plays a role in this choice, however suppliers have often good reasons not to adopt third party components. First of all because they have a lot more experience with their proprietary software; secondly, because the internal design process and tool-chain is often highly dependent on the proprietary operating system, and thirdly, because the real-time operating system is tuned and optimized for the underlying hardware platform. This implies that switching to third party components comes with a great risk and a considerable cost.

J. Denil (✉) · S. Demeyer
Lab on Reengineering, University of Antwerp, Antwerp, Belgium
e-mail: Joachim.Denil@kdg.be

J. Denil · P. De Meulenaere
TERA-Labs, Karel de Grote University College, Antwerp, Belgium

K. Maudens · K. Van Stechelman
Spicer Off-Highway Products Division Belgium, Dana Corporation, Brugge, Belgium

M. Conti et al. (eds.), *Solutions on Embedded Systems*,
Lecture Notes in Electrical Engineering, 81, DOI: 10.1007/978-94-007-0638-5_17,

Hence suppliers are reluctant to introduce a new OSEK compliant operating system in their software.

A possible way out of such a catch-22 situation is to wrap the proprietary software with an interface that complies to the standard specification. Such a *wrapper* (also known as a wrapper-façade [5] or adapter [6]) is—when feasible—indeed a cheap way to integrate an existing component in a system that expects a different interface. Moreover, it allows for an incremental migration strategy (the wrapped software component can later be replaced with another component that adheres to the interface), which reduces risk. However, introducing a wrapper also has a performance impact as it triggers extra processor cycles and consumes more memory. Performance is a crucial aspect of a real-time operating system, thus must be assessed carefully.

To investigate this trade-off, we conducted a feasibility study adapting a proprietary real-time operating system from an off-highway automotive company to the current OSEK OS specifications. Afterwards, one of the applications running on the proprietary RTOS was migrated to the interface of the OSEK OS. Our study was guided by the following research questions:

- *RQ1—Is the construction of the wrapper feasible?* What are the technical implications when building such a wrapper?
- *RQ2—What is the performance impact of the wrapper?* Can we quantify the extra computation time and memory consumption caused by introducing the wrapper?
- *RQ3—How can the application be reengineered?* What adaptations are necessary to the application to make it compliant with the OSEK OS interface while preserving the correct behavior?

As part of this feasibility study, we must show that the wrapped operating system is indeed a valid OSEK implementation. In our study, we have used the MODISTARC specifications [7] which are designed to test and assure that an implementation conforms to the OSEK specification.

17.2 Related Work

Part of this work was previously published, this a major revision of the work published in [8].

Since the release of the OSEK standard, a lot of commercial and academic implementations have been realized. One academic initiative is Trampoline [4]. It offers a full OSEK operating system, communication and network management layer for several hardware platforms. A configuration tool is provided to generate an optimal trampoline kernel. Another academic initiative is the EMARALDS-OSEK [9]. It uses several memory and performance optimizations for efficiency. Implementation details of both these operating systems were used during the case study. Some vendors offer an OSEK compliant wrapper for their own commercial RTOS.

An example of this is the Micrium μC/OS-II OSEK extension layer [10] that provides a certified OSEK wrapper for the μC/OS-II kernel. To the best of our knowledge, currently few examples of wrappers for proprietary operating systems and their impact on performance are documented in the literature.

17.3 Migration Strategy

OSEK provides a standard software architecture for distributed control units in vehicles. It meets two stringent automotive requirements: real-time support and small memory footprints. OSEK actually consists of a set of standards: (a) OSEK OS specifies the behavior and APIs for a real-time operating system; (b) OSEK COM describes an interface for transferring data between applications (through local communication or by use of a network); (c) OSEK NM provides system wide management functions; and (d) OSEK OIL proposes a language to configure the OSEK system. In the scope of this feasibility study, we restrict ourselves to the operating system because that was the primary focus for the off-highway company requesting the feasibility study.

In Fig. 17.1, the migration mechanism is shown. The first major step is constructing a wrapper around the proprietary RTOS. This OSEK-wrapper is used to implement OSEK behavior on top of the RTOS. The wrapper allows creating a valid OSEK implementation without any significant changes to the proprietary RTOS, as the MODISTARC specification says: "*All what behaves like OSEK is OSEK*" [11].

After the OSEK wrapped RTOS is confirmed to be a valid OSEK implementation, the application can gradually be migrated towards this new RTOS. This process is shown in Fig. 17.2 where during the process, some tasks still use services from the proprietary RTOS and others use the services offered by the OSEK-wrapped RTOS.

17.4 Construction of the Wrapper

Within OSEK, some different options can be taken: OSEK OS defines two major conformance classes. The basic class only allows basic tasks while the extended class offers a basic and extended task. Details of these types of tasks will be discussed in the next section. The choice however is highly dependent on the proprietary operating system, some RTOSs can be wrapped with an extended class

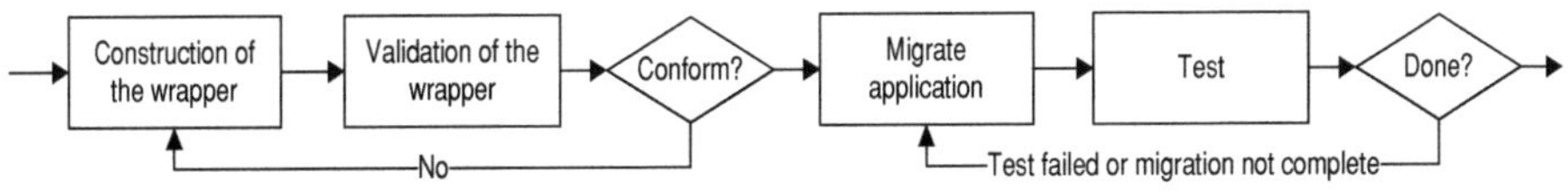

Fig. 17.1 Proposed migration method to the OSEK-OS

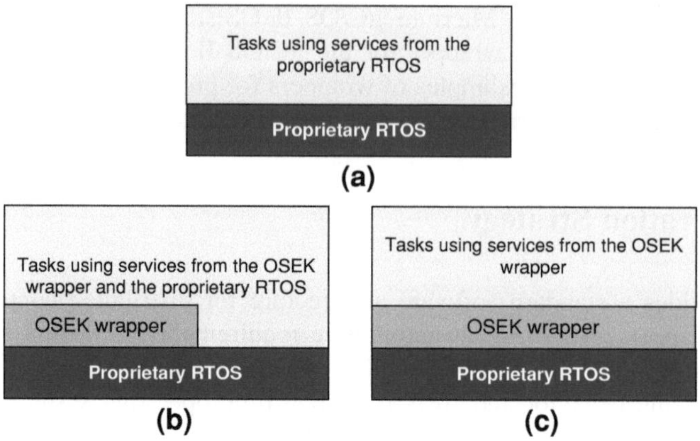

Fig. 17.2 Gradual migration of the application. **a** Initial state; **b** while porting the application, some services of the proprietary RTOS are still used; **c** result when the migration has been completed

OSEK-wrapper while others are limited to the basic conformance class. This has no impact on whether the application can be migrated, since the reengineered application won't need services in the wrapper that were not available in the proprietary RTOS.

We split up the construction of the wrapper in the following isolated subparts: (a) migration of the task model, (b) migration of the scheduler, (c) migration of the event management, (d) migration of the resource management, (e) migration of the interrupt services and (f) migration of the alarm functionality.

For each of these subparts of the migration, we compare the detailed mechanisms from the proprietary operating system to the OSEK-specifications. If there are major differences, the OSEK-wrapper often needs to keep state- or context variables to translate between both OSs. If on the other hand there are only minor differences, the OSEK wrapper can often translate between the interfaces without the need to keep context variables.

17.4.1 Task Model

17.4.1.1 Requirements

Depending on the chosen conformance class, an OSEK compliant operating system must allow for two sorts of tasks: the *basic task* and the *extended task*. The basic task is the simplest type of task, switching between (a) the READY state (indicating that the task is ready to run); (b) the RUNNING state (indicating that the task is currently being executed) and (c) the SUSPENDED state (indicating that a task is terminated and can be restarted later the beginning). The extended

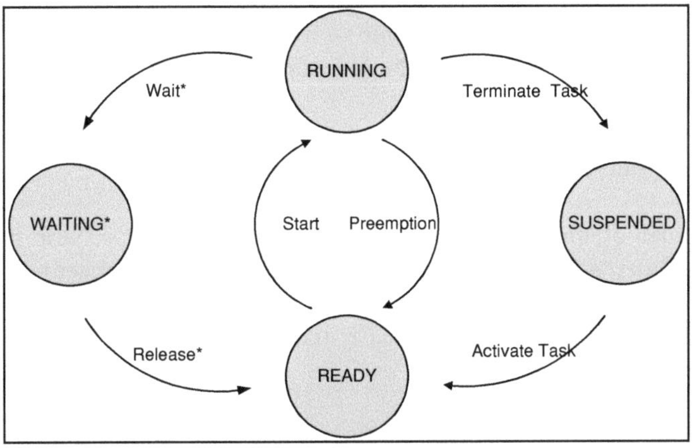

Fig. 17.3 The OSEK task model, states and transitions with a * are only available to extended tasks

task adds an extra WAITING state, used when a task is paused until it is released to the READY state by means of a system call. The state-machines specifying the legal sequences of state transitions for both the basic and extended tasks are shown in Fig. 17.3. The *Start* and *Preemption* transitions are the responsibility of the scheduler. The others are available as system calls in the OSEK operating system.

17.4.1.2 Wrapper Implications

If the state machine of the proprietary operating system does not match with the state machine of the OSEK tasks, the OSEK-wrapper has to adapt the state-machine of the proprietary operating system to the OSEK state machine. The wrapper should: (a) introduce and/or exclude states that do not match the OSEK model and implement the state transitions to and from these states. Implementing a WAITING state in the wrapper, when no WAITING state is available in the proprietary operating system seems unlikely. In this case, the wrapper can still be targeted to a basic conformance class; (b) record and synchronize the state of the defined tasks in the wrapper; (c) implement and translate the interface for this new behavior. If on the other hand a one-to-one match between the OSEK state machine and the state machine of the proprietary operating system is possible, only the interface for manipulating the task state has to be translated.

17.4.2 The Scheduler

17.4.2.1 Requirements

The scheduler of an OSEK operating system must obey to a fixed priority scheme. Therefore, all tasks are assigned a fixed priority at system configuration. Tasks that are in the READY state are given time to execute on the processor. The higher priority tasks are processed before the lower priority tasks.

17.4.2.2 Wrapper Implications

If the scheduler of the proprietary operating system is not a static priority pre-emptive scheduler, a new scheduler must be constructed. The scheduler and all the OSEK-tasks run within one task of the proprietary RTOS. The mechanisms for context switching are coded in the OSEK-wrapper.

17.4.3 Events

17.4.3.1 Requirements

The event-architecture is the primary synchronization mechanism in an OSEK operating system. Events are only available to extended tasks and are used to initiate the transitions to and from the WAITING state. Events are not independent objects but are assigned to extended tasks. If the required conformance class of the OSEK-wrapped operating system aims at an extended conformance class, the event mechanism must be available.

17.4.3.2 Wrapper Implications

The proprietary operating system should also feature some kind of event-architecture, otherwise it is most unlikely that a wrapper interface can be constructed. If such an event mechanism is present, the wrapper should (a) map the event data structure from the wrapped operating system to the OSEK task records; (b) translate the interface manipulating the events. If there is no event-architecture available, the OSEK-wrapped operating system can be targeted to a basic conformance class.

17.4.4 Resources

17.4.4.1 Requirements

Protection of shared resources is done by the resource mechanism. This mechanism must work according to the OSEK-PCP protocol (priority ceiling protocol). The OSEK-PCP protocol is a variant of the original priority ceiling protocol presented in [12]. When a task acquires a resource, the priority of the task is temporarily raised to the priority of the highest priority task that will ever use the resource. This is called the ceiling priority of the resource. This mechanism ensures that, while a task occupies a certain resource, other tasks that share the same resource cannot get scheduled.

17.4.4.2 Wrapper Implications

If the proprietary operating system has a resource protection mechanism available with a priority ceiling protocol, the wrapper should translate the interface that manipulates these resources. Otherwise the resource mechanism must be constructed within the wrapper. The method is to delay the activation of tasks (i.e. the transition of a task from SUSPENDED to READY, or from WAITING to READY) until the resource is released.

17.4.5 Interrupts

17.4.5.1 Requirements

OSEK defines that a separate stack must be used for interrupt service routines (ISR) that use OSEK system calls. The separate stack is introduced for reducing the memory requirements of the OSEK system. If this stack is not available, all run-time stacks of the tasks should be big enough to allow storage of multiple ISR-frames (context that is saved before executing an ISR). Besides the separate stack, an interface is defined to allow the manipulation of interrupt flags in the hardware.

17.4.5.2 Wrapper Implications

When the proprietary operating system does not provide a separate stack for handling interrupts, a stack in the OSEK-wrapper can be used. Implementation pointers for a stack in the wrapper can be found in [6]. The interface for manipulating the interrupt flags has to be provided if these are not available in the proprietary operating system.

17.4.6 Counters and Alarms

17.4.6.1 Requirements

The handling of recurring events in OSEK is done by a mechanism of counters and alarms. An alarm is attached to a counter. If this counter reaches a specific value, the alarm is triggered. The alarm can activate a task, set an event or call a special routine. Alarms can be set on an absolute value or relative to the current counter value and can be cyclic or single shot.

17.4.6.2 Wrapper Implications

If no such mechanism is available in the RTOS it can easily be constructed using a sorted linked list. The counter has access to this list using a next pointer. This pointer always points to the alarm nearest to the current counter value. More implementation details can be found in [4].

17.4.7 Compliance of the Wrapper

When a part of the OSEK-wrapper has been written, it can be tested using the MODISTARC specifications [5]. The MODISTARC test suite is also used to assure that the whole system is an OSEK compliant system. The test procedure is a black-box test. For this, the verification of the separate stack used by the interrupt service is not validated.

17.5 Migrating the Application

Migrating an application, to make it compliant with the OSEK interface, is done in two steps. The first step is to identify which functions or modules need to be reengineered (reverse engineering). In a second step, the identified modules or functions are redesigned such that they use the services of the OSEK-wrapper instead of the services of the proprietary RTOS. Reengineering patterns, as described in [13], are used to structure the process.

When migrating the application it is tempting to adapt the application in a single step. However this approach could introduce a lot of errors. To gain confidence in the process and to detect errors earlier in the migration cycle, it is recommended to use a gradual approach. Still it is important to keep the wrapper in a consistent state. This can be achieved by adapting all tasks that are dependent on each other (use shared data, activate or set events of each other) in one step. Afterwards the changes must be tested using the integration and system tests of the application.

Another pitfall is the introduction of new OSEK-services in the application. The architecture of the application has been designed with hardware resources and timing behavior in mind. Introducing new services, such as restarting a task that runs in an infinite loop or introducing other events could change the architecture of the entire application. However, sometimes it is inevitable to use these new services.

17.6 Experimental Set-Up

Since OSEK is designed to run on small embedded systems, speed and memory consumption are key issues in a real-time operating system. Introducing the OSEK-wrapper around the operating system causes overhead. In the next sections the experimental set-up for measuring this impact is described.

17.6.1 Worst Case Impact of the Wrapper

In this section the overhead of the services offered by the OSEK-wrapper are compared to the services of the proprietary RTOS. With the results of these measurements, an assessment can be made whether these impacts are acceptable for the application.

17.6.1.1 Response Time Overhead of the Task Model

To assess the influence of context switching with the wrapper, we compare the execution time of a typical configuration of tasks executing on the bare proprietary operating system against the same configuration on the wrapped version. Therefore two or more tasks are executed to obtain the execution times: one, the highest priority task records the current time and then terminates. Another, the lowest priority task calculates the difference in time. Extra care has been taken so that the recording and output of the time does not influence the measurements. Tasks with intermediate priorities, that only terminate, are added to see the impact of the number of tasks.

17.6.1.2 Response Time Overhead of the Event Mechanism

The impact on the event mechanism can be measured when the wrapper is designed for an extended conformance class. Therefore two tasks are defined. The highest priority task waits for an event, while the lowest priority task sets this event. This functionality is executed in a loop that loops for a predefined amount of time. When a wait-release cycle is completed, a counter is incremented. To assess the influence of the wrapper, we compare the number of wait-release cycles in the proprietary operating system and the OSEK-wrapped operating system.

17.6.1.3 Response Time Overhead of the Resource Mechanism

Measuring the impact of the adapted resource mechanism is based on the method proposed in [14]. Only a single task is used. This task gets and releases a resource in a loop that lasts for a predefined amount of time. When a get-release cycle is completed, a counter is incremented. The difference in get-release cycles between the proprietary operating system and the OSEK-wrapped operating system is an indication of the performance loss introduced by the wrapper.

17.6.1.4 Memory Impact of the OSEK-Wrapper

ROM consumption is measured by compiling the target without any defined tasks and resources. The compiler optimizations are turned off, so no dead code in the wrapper and RTOS is removed. This is done for both the proprietary RTOS and the OSEK-wrapped RTOS. RAM consumption is quantified similar. The source code is compiled with 0–5 tasks and 0–5 resources for both the proprietary RTOS and the OSEK-wrapped RTOS.

17.6.2 Impact of the Wrapper on the System

The effects of the wrapper must also be assessed when porting the application.

17.6.2.1 Execution Time of Tasks

For all tasks in the original system and in the migrated system, the time is recorded when activating the task and when the task is terminated. The difference between the termination and the activation gives the total execution time of a task. From these values, the best case, worst case and averages can be used to evaluate the impact of the wrapper.

17.6.2.2 Number of Activations of a Task

In a strict interval, the number of activations of the task are recorded. This can be used to see whether the behavior of the application remains the same. This is a critical measurement for time-triggered tasks.

17.6.2.3 Memory Impact of the Wrapper

ROM and RAM consumption are measured for both the original system (application with the proprietary RTOS) and the ported system (application, wrapper and proprietary RTOS). Compiler optimizations are used as in the specifications of the original system.

17.7 Case Study

The main prerequisites when selecting the case were: (a) the whole system should be built on a proprietary RTOS, (b) there should be differences in the behavior of the OSEK-OS and the proprietary RTOS, (c) there should be different tasks in the application and these tasks should use different services of the proprietary RTOS.

Our selected case study used to verify this method adheres to these prerequisites. It is a control application for a transmission unit in off-highway vehicles. The proprietary RTOS differs from the OSEK-OS in task model and event mechanism. It does not have any support for alarms and counters. The application built on top of this RTOS consists of several tasks. Some of these tasks run in an endless loop and yield the processor to wait for an event. The others are time-triggered tasks that are activated by the scheduler on predefined intervals.

17.7.1 Construction

- *Task model*: Two types of tasks are available in the proprietary RTOS. On the one hand it has simple tasks that do not have a WAITING state available. The proprietary operating system activates these tasks automatically in a time triggered way. On the other hand there are tasks that allow a WAITING state but run in a continuous loop; (they do not have a SUSPENDED state). Both the standard and extended tasks were created using the second type of tasks. When terminating the task in the OSEK-wrapper, the stack pointer and program counter of the task is reset. A waiting state is used to emulate the OSEK SUSPENDED state.
- *Event mechanism*: The operating system in our case study has a mechanism to put tasks in and out of the WAITING state. By multiplexing and demultiplexing these calls, the OSEK event mechanism is created in the OSEK-wrapper.
- *Alarms and counters*: A general counter and alarm mechanism was created in the OSEK-wrapper following the proposed guidelines. The counter for these alarms is a time-triggered task in the wrapper since all hardware counters were used by the application or the proprietary RTOS.
- *Other parts*: The OSEK-wrapper translates the interfaces of all other services and keeps the state information synchronized.

The OSEK-wrapper designed in this case study, was verified to be an extended OSEK compliant implementation.

17.7.2 Worst Case Impact of the Wrapper

The worst case impact of the wrapper was measured using the experimental setup of Sect. 17.6.2.

- *Impact on context switching*: The context switching with the OSEK-wrapper is 3.8 times slower compared to the original case due to the reset of the program counter and stack pointer, putting the task in a WAITING state and keeping the OSEK-wrapper in a consistent state. Though the overhead introduced is linear.
- *Impact of the resource mechanism*: Speed of the event mechanism is decreased by 37%. This is the overhead caused by the added functionality.
- *Impact of the event mechanism*: Due to keeping the OSEK wrapper in a consistent state, the overhead for taking and releasing a resource is 40% extra.
- *Memory impact*: In the case study, the ROM increased by 21%. The RAM-usage increased by less than 1% without the resources necessary for each task structure and resource structure needed for the bookkeeping in the OSEK-wrapper.

Figure 17.4 compares the impact of the event and resource mechanisms. We evaluated that this impact is acceptable for our case and started the reengineering process of the application.

17.7.3 Impact of the Wrapper on the System

The application was gradually migrated to the OSEK-wrapped RTOS. Scenario-based tests were used after every change to confirm that the changes had no effect on the behavior of the application.

The impact on the timing behavior of individual tasks in the OSEK-wrapped RTOS is proportional with the arrival events of the higher priority tasks. This is can be observed since lower priority tasks are preempted by the higher priority tasks whose running time is slightly affected by the individual overheads of the mechanisms. This effect is apparent on a medium or low priority task that needs a lot of computation time and is preempted by a higher priority task.

Fig. 17.4 Impact of the OSEK-wrapper on the event and resource mechanisms

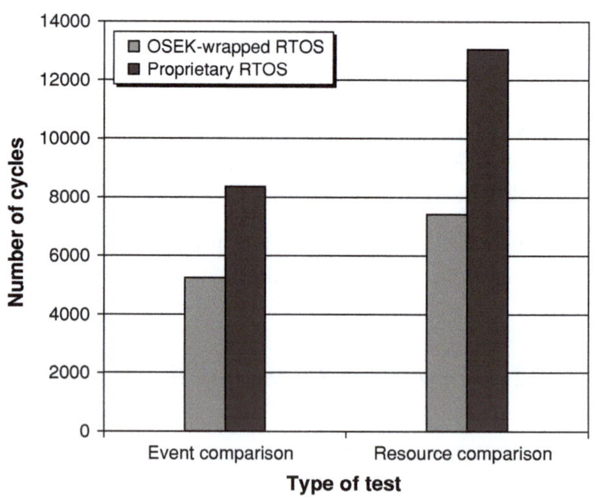

Table 17.1 Comparison of the average execution times (in scheduler ticks) of three tasks between the original system and the ported system

System	HPT	MPT	LPT
Original system	132	2944	42
Ported system	132	2949	43

We illustrate the above with three tasks shown in Table 17.1. The HPT is a cyclic high priority task with few computations. The MPT is a cyclic medium priority task that requires much computation time. The LPT is the lowest priority task with almost no computations. In the application, the HPT is scheduled four times more than the MPT and 20 times more than the LPT. The effects of the above are apparent on the tasks shown in the table.

Still, it is confirmed that after porting the application behaves according to the requirements of the system. In this particular case, the relative response time overhead of all the tasks is acceptable.

The whole system (application, OSEK-wrapper and proprietary RTOS) consumes less than 1% more ROM and only 0.1% more RAM.

17.7.4 Threats to Validity

17.7.4.1 Internal Validity

The migration of the proprietary RTOS is highly dependent on the developer who conducted the case-study. Moreover, knowledge of the proprietary RTOS is key to the success of the migration. In this case study, the proprietary RTOS is treated as a black-box, the developer had no insights into the internals of the RTOS. This could affect the performance and outcome of the case-study since other constructs might have been possible with less effort and better performance characteristics.

17.7.4.2 External Validity

Though this study is based on one case study, the methods proposed can be generalized to other proprietary real-time operating systems, since most RTOSs have a similar functionality. The guidelines proposed for reengineering the application can be used outside the scope of this case study. The measured performance impacts are highly dependent on the case-study.

17.8 Conclusion

In this paper we have demonstrated that it is feasible to build an OSEK compliant wrapper for a proprietary real-time operating system. No modifications to the hardware, the original code-base nor the tools are necessary with this approach.

We documented the steps and necessary implications a software designer must take into account in order to construct such an OSEK-compliant wrapper and to port the application to this new RTOS.

The OSEK-wrapper implies a significant increase in computation time and consumes a considerable amount of extra memory. This is caused by defining redundant information for tasks, events and resources in the wrapper, necessary to synchronize with the information in the proprietary real-time operating system. In this particular case, this overhead was acceptable, but this decision depends of course on the particular services built on top of the operating system and on the available hardware resources.

Consequently, we conclude that, when faced with the requirement of adopting an OSEK compliant operating system, the construction of a wrapper is a viable alternative compared to integrating third party components. The wrapper indeed causes a certain performance penalty, but is quite cheap to implement and reduces the inherent risks and costs associated with adopting third party components.

References

1. Fennel H et al (2006) Achievements and exploitation of the AUTOSAR development partnership [online]. www.autosar.org
2. Zahir A, Palmieri P (1998) OSEK/VDX-operating systems for automotive applications. IEE seminar OSEK/VDX open systems in automotive networks
3. Schoof J, Wybo D (2006) No detour needed: getting to autosar via OSEK. SAE in-vehicle software and hardware systems
4. Bechennec J, Briday M, Faucou S, Trinquet Y (2006) Trampoline, an open source implementation of the OSEK/VDX RTOS. In: Proceeding of the eleventh IEEE international conference on emerging technologies and factory automation (ETFA06)
5. Schmidt D, Stal M, Rohnert H, Buschmann F (2000) Pattern-oriented software architecture volume 2: patterns for concurrent and networked objects. Wiley, New York
6. Gamma E, Helm R, Johnson R, Vlissides J (1995) Design patterns: elements of reusable object-oriented software. Addison Wesley, Reading
7. OS test procedure [online] (1999) www.osek-vdx.org, OSEK Std., Rev. 2.0
8. Denil J, Demeyer S, Demeulenaere P, Maudens K, Vanstechelmans K (2009) Wrapping a real-time operating system with an OSEK compliant interface—a feasibility study. Intelligent solutions in embedded systems, 2009 seventh workshop on, pp 157–164, 25–26 June 2009
9. Zuberi KM, Pillai P, Shin KG (1999) EMERALDS-OSEK: a small real-time operating system for automotive control and monitoring. In: Proceedings of SAE international congress and exhibition
10. Micrium (2009) μc/os-II rtos osek layer [online]. www.micrium.com
11. Conformance Testing Methodology [online] (1999) www.osek-vdx.org, OSEK Std., Rev. 2.0
12. Sha L, Rajkumar R, Lehoczky J (1990) Priority inheritance protocols: an approach to real-time synchronization. IEEE Trans Comput 39(9):1175–1185
13. Lamie W (2007) To find the RTOS with the best real-time performance, you've got to do an apples-to-apples comparison. DSP Design Line
14. Demeyer S, Ducasse S, Nierstrasz O (2008) Object-oriented reengineering patterns. Square Bracket Associates, Kehrsatz

Part V
Power Aware Design

Chapter 18
A Sigma–Delta Controlled Power Converter for Energy Harvesting Applications

Rocco d'Aparo, Simone Orcioni and Massimo Conti

18.1 Introduction

In recent years mobile technology has become an important part of today's life. From smart-phones to PDAs, from notebooks to entertainment multimedia devices, a great share of people do massive use of electronic technology. One of the main target in the implementation of portable devices is the improvement of battery life time. To this aim, semiconductor technology evolution like the transition from CMOS to very low voltage CMOS has allowed a reduction in power consumption and then an addition in endurance.

The evolution of batteries allowed also an improvement in autonomy. Although today's battery are small if compared to that of some decade ago, their capacity in term of A/h has grown. Nowadays, technology allows to make devices that can run using a very little amount of energy. For example, modern music players do not make use of any electric motor like they did some years ago. Other example is about the active RFIDs that can run for years without needing any battery maintenance. Since the energy consumption of modern mobile devices is becoming very low, recently some researchers are working to replace or to combine battery with other kinds of sources taking energy from environment. The environment can be considered like a big energy container, where energy exists in a great deal and in many different types of physical signals: from light to electromagnetic noise, from temperature to wind, from mechanical vibrations to acoustical noise. Energy harvesting is becoming an important target to improve battery life time or even to remove batteries in mobile applications. Some decades ago, the main technique

R. d'Aparo, S. Orcioni and M. Conti (✉)
Dipartimento di Ingegneria Biomedica, Elettronica e Telecomunicazioni,
Università Politecnica delle Marche, 60131 Ancona, Italy
e-mail: m.conti@univpm.it

M. Conti et al. (eds.), *Solutions on Embedded Systems*,
Lecture Notes in Electrical Engineering, 81, DOI: 10.1007/978-94-007-0638-5_18,

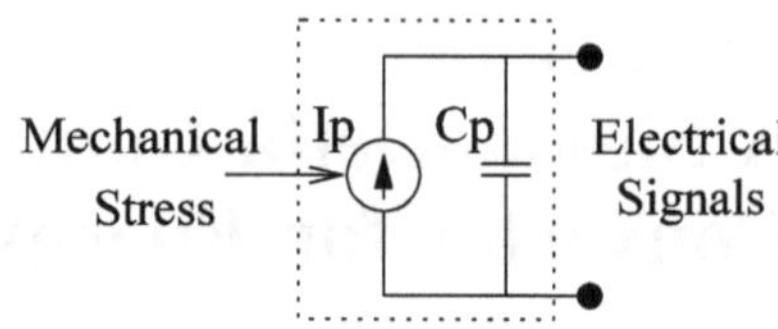

Fig. 18.1 Electric current generator based piezoelectric resonant model

used to scavenge energy from the environment made use of photovoltaic effect. The most common application was in pocket calculators.

In [1] a technique that can be used to harvest energy from electromagnetic noise is shown. In [2] a mechanical vibrating source is used to recovery energy, while in [3, 4] the body motion is used.

In this work, partially presented in [5], a full digital controlled switching power supply connected to a piezoelectric transducer is proposed, to store and to manage energy. The digital controller is based on Sigma Delta ($\Sigma\Delta$) modulation, and since all system is battery free, it stores or consumes energy depending on load energy requirement and stored energy availability. Load alternates Power-ON status during which it can consume more energy than that the transducer can produce, and Stand-by status during which load power consumption is low and system can save energy. When energy level reaches a sufficient value, load is turned ON. During this time, the stored energy of the system can decrease. When the load finishes its task or even the stock of energy reaches a minimum level, load is forced in Stand-by. During this time, system saves energy. When its level reaches a sufficient value, the load is turned ON again and the routine can restart.

Piezoelectricity is a phenomena characterising some crystals that, when subjected to a mechanical stress, produce an electric field. This is a reversible phenomena: such crystals, when subjected to an electric field, change their geometry like they would be under a mechanical stress.

Modelling a piezoelectric material is not simple. Its behaviour depends upon a great number of parameters like crystal characteristics, stress characteristics, system geometry. Piezoelectric phenomena can be described by the so called piezoelectric constitutive equations. Electrical models of piezoelectric generators have been presented in different works [2, 6]. Here we will use the model like depicted in Fig. 18.1, characterised by a current generator and a parallel capacitor. The amplitude of input current as well as its shape depend on crystal physical and geometric characteristics and stress acceleration. The value of the capacitor depends only on physical and geometric characteristics.

18.2 System Specifications and Design

The proposed system stores or uses energy depending on stored energy availability and load requirements. The system stores the energy in a tank capacitor, called in Fig. 18.2C_{rect}.

When the reserve of energy is sufficiently high, a DC–DC switching converter begins to give power to the load. During this time, since the power produced by the generator can be less than the power consumed by the load, system energy reserve can decrease. If it becomes too small, load is forced in Stand-by status during which the stored energy into the tank capacitor C_{rect} grows. When the stored energy reaches a sufficient level to finish the execution of load task or it reaches the maximum allowable level, the load is turned ON.

All system is composed by two stages. The first stage, after the piezoelectric generator, is an AC–DC converter. The other is a digital controlled switching DC–DC converter. To design and modelling the system, we have assumed the piezoelectric transducer working under a stationary mechanical stimulus. We have also assumed piezoelectric generator working at the resonance. This means that current generator produces a sinusoidal fixed frequency signal. The complete block diagram of the proposed system is shown in Fig. 18.3.

Since the power produced by the piezoelectric depends on the load, to find the working point where piezoelectric generator produces the maximum power, the AC–DC converter has been connected to a pure variable resistive load as shown in Fig. 18.2.

Figure 18.4 reports the voltage across C_{rect} (V_{rect}) and the value of the piezoelectric power production in steady state condition assuming C_{rect} big enough so that V_{rect} is constant. In our piezoelectric model, current generator is sinusoidal with an amplitude of 2.6 mA and a frequency of 50 Hz. The maximum production of energy occurs when $R_L = 25$ kΩ. Maximum power production for the proposed model is 17 mW when $V_{rect} = 20$ V.

In the proposed system, the output of the AC–DC stage is connected to the input of a switching power converter. It cannot be assumed as a resistive load neither its value is 25 kΩ. To emulate a maximum efficiency load, the system is provided with management blocks, depicted in dark gray in Fig. 18.3, that hold the rectified voltage close to 20 V as possible. In this condition, the piezoelectric generator works with the maximum efficiency and generate about 17 mW.

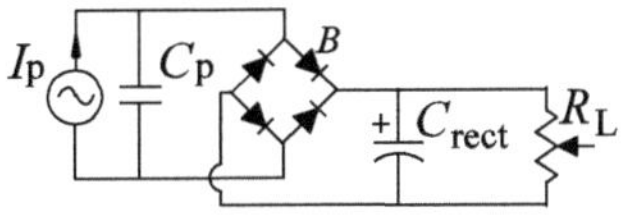

Fig. 18.2 Test circuit used to characterize piezoelectric power production

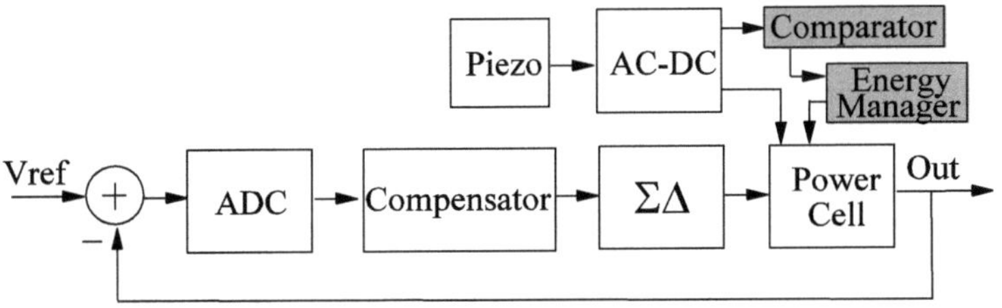

Fig. 18.3 Block diagram of overall system

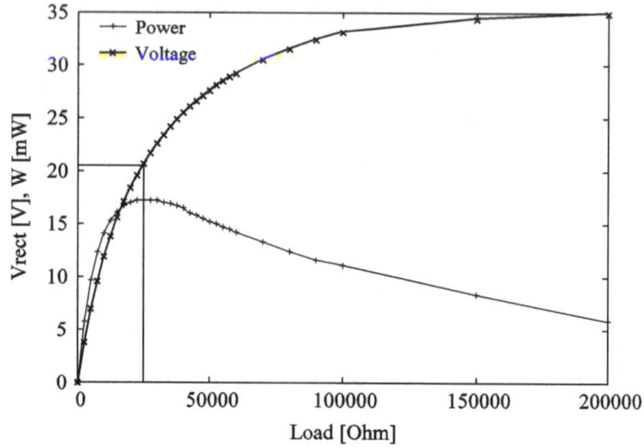

Fig. 18.4 Piezoelectric power production and voltage at the rectifier output for a pure resistive load. Current generator is sinusoidal with amplitude 2.6 mA and frequency 50 Hz

Table 18.1 System specifications

V_0	3.3 V
Δ_{ADCmax}	25 mV
P_{0ON}	30 mW
P_{0OFF}	5 mW
V_{rppmax}	1 mV
T_{ON}	50 ms

We have also assumed that the load has two states: Power-ON and Stand-by. In Power-ON all system requires a power of 30 mW, instead when in Stand-by, it requires only 5 mW. Moreover, we have assumed that the load is a wireless sensor. During the Power-ON, the sensor samples its quantity and then sends it to a server in 50 ms. After the transmission, wireless sensor is forced in Stand-by until the stored energy is high enough to allow the repetition of the routine. Table 18.1 reports all system specifications that have been used in the project.

18.3 AC–DC Stage

The first stage connected to the piezoelectric generator is a classical passive AC–DC converter consisting in a bridge rectifier (B) and a tank capacitor (C_{rect}) like reported in Fig. 18.5.

Since the maximum efficiency is reached when V_{rect} is about 20 V, we have designed a device that measures V_{rect} and controls that its value is as close as possible to such a value.

We have assumed that in steady state V_{rect} can vary between 20 V $\pm$ Δ.

Calling V_i the value of V_{rect}, E_i the energy stored in C_{rect} before the load is turned ON and V_f, E_f the values of same quantities after the load task has been executed, assuming $V_i - V_f = 2{\cdot}\Delta$ and $(V_i + V_f){\cdot}½ = 20$ V, since during Power-ON

Fig. 18.5 AC–DC stage

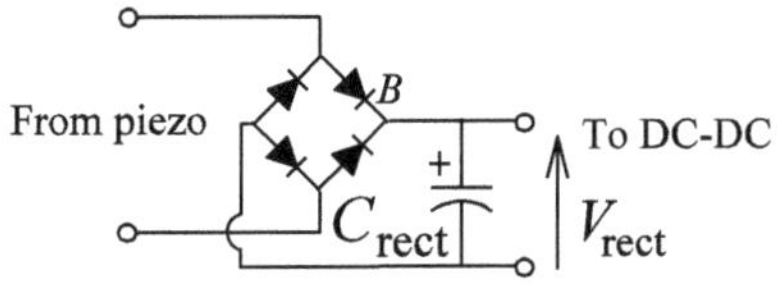

the wireless sensor requires more energy respect that the piezoelectric generator can produce, we will have:

$$E_f = E_i + P_I T_{ON} - P_{0ON} \cong E_I + P_M T_{ON} - P_{0ON} T_{ON} \tag{18.1}$$

where P_I is the power produced by the piezoelectric generator, P_M is its maximum efficiency steady state value, T_{ON} is the time interval during which the load task is executed and P_{0ON} is the power consumption of load during T_{ON}. Since the energy stored into a capacitor is $E = ½\ CV^2$:

$$T_{ON} \cong \frac{E_I - E_F}{P_{0ON} - P_M} = \frac{C_{rect}}{2} \frac{(20+\Delta)^2 - (20-\Delta)^2}{P_{0ON} - P_M} \tag{18.2}$$

then:

$$T_{ON} \cong \frac{C_{rect}}{2} \frac{4\Delta 20}{P_{0ON} - P_M} \tag{18.3}$$

After the load has accomplished its task, it goes in Standby, and stand there for T_{OFF} seconds. In this state, power consumption is about 5 mW. During this time, since the energy recovered by piezoelectric generator is greater than the energy used by load, the energy stored into C_{rect} increases. The value of T_{OFF} is:

$$T_{OFF} \cong \frac{E_I - E_F}{P_M - P_{0OFF}} = \frac{C_{rect}}{2} \frac{(20+\Delta)^2 - (20-\Delta)^2}{P_M - P_{0OFF}} \tag{18.4}$$

Then:

$$T_{OFF} \cong \frac{C_{rect}}{2} \frac{4\Delta 20}{P_M - P_{0OFF}} \tag{18.5}$$

Since $T_{ON} = 50$ ms, as reported in the system specifications of Table 18.1, fixing Δ so that the piezoelectric generator works with maximum efficiency, for example $\Delta = 2$ V:

$$C_{rect} = 8\ \mu F \tag{18.6}$$

With this choice T_{OFF} is about 53 ms. Power-ON and Stand-by states alternate themselves with a period of about 103 ms. Like it will be shown afterwards, this value is close the simulated results. To maintain V_{rect} between 18 V and 22 V a comparator has been used.

18.4 DC–DC Stage

Switching power supply is a common choice in today's applications because of their high efficiency. Such characteristic can be well suited to mobile applications where an increasing in efficiency means a battery life improvement.

In our project a step down buck converter, like that represented in Fig. 18.6, has been used to keep output voltage at a fixed level compatible with load specifications. The output voltage V_0 is maintained at 3.3 V, while V_{rect} changes between 18 V and 22 V.

To design the power cell, we have assumed that the power converter works in Continuous Conduction Mode (CCM) [7].

To this aim minimum value of inductance must be chosen

$$L_{MIN} = \frac{V_0}{2\overline{F_{SW}}I_0}\left(1 - \frac{V_0}{V_{rect}}\right) \tag{18.7}$$

where $\overline{F_{SW}}$ is the average switching frequency of the power cell, and I_0 is the load current. In standard PWM based power converter, power cell switching frequency is usually a system specification. The value of the inductor, as indicated in Eq. 18.7, decreases increasing the switching frequency.

In the present work, the power cell is controlled by means of a second order ΣΔ modulator. ΣΔ modulators cannot be characterised by a unique value of switching frequency. ΣΔ modulators produce a pseudo random impulsive signal, whose average switching frequency $\overline{F_{SW}}$ is a function of duty cycle δ and of the clock frequency F_{CLK} of the modulator [8, 9], as reported in Eq. 18.8.

$$\overline{F_{SW}} = \begin{cases} \delta F_{CLK} & \delta < 1/2 \\ (1-\delta)F_{CLK} & \delta \geq 1/2 \end{cases} \tag{18.8}$$

The duty cycle versus average switching frequency is reported in Fig. 18.7.

The minimum $\overline{F_{SW}}$ has been chosen equal to 500 kHz, meaning that the F_{CLK} must be

$$F_{CLK} \geq \frac{\overline{F_{SW}}}{\delta_{MIN}} = 3.3\,\text{MHz} \tag{18.9}$$

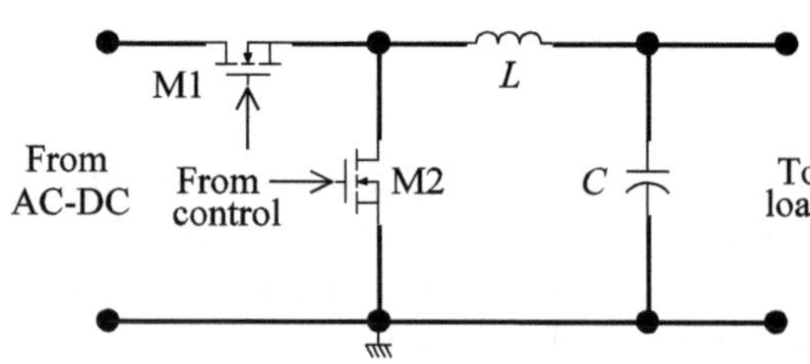

Fig. 18.6 DC–DC stage

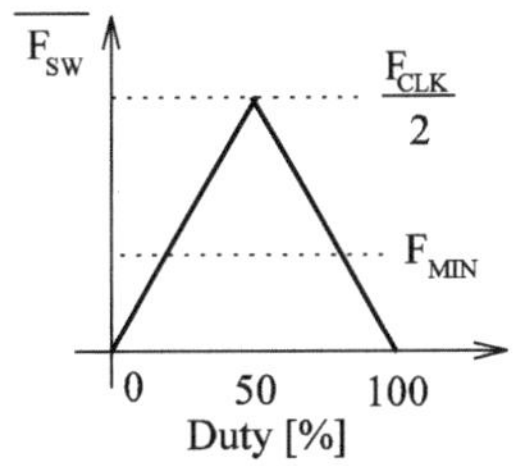

Fig. 18.7 Average switching frequency of power cell versus duty cycle

Since the maximum load value is $R_{0MAX} = V_0^2/P_{0OFF} = 2{,}178\ \Omega$, the inductor has been chosen $L = 5$ mH. Capacitor in the power cell is usually designed considering the maximum allowable voltage/current ripple on the load

$$C_{MIN} = \frac{V_0}{8L\overline{F_{SW}^2}V_{rpp}} \tag{18.10}$$

To limit load transition effect on V_0, the capacitor has been chosen greater than its minimum value, $C = 5\ \mu$F. Using the Space State Averaging (SSA) [10, 11] average transfer function of power cell can be assumed as

$$F(s) = \frac{v_0(s)}{\delta(s)} = \frac{V_{IN}}{LCs^2 + \frac{L}{R}s + 1} \tag{18.11}$$

Neglecting ADC delay, since linearized signal transfer function of modulator is unitary [9], the compensator network can be designed in order to stabilise the closed loop system. In this way, the closed loop bandwidth (*BW*) of the overall system can be defined. We have assumed that the maximum bandwidth of system is $BW = 50$ kHz.

One of the main problem of digital controlled switching power supply are limit cycles. They are high amplitude low frequency (lower than switching frequency) signals that can make the device unstable. To avoid this problem, three are the conditions in a unitary feedback digital controlled system that must be ensured [12]. Among these, modulator precision needs to be at least one bit more than the ADC precision. This means that modulator must be designed so that the quantisation step of ADC Δ_{ADC} must be at least double with respect to the modulator precision, here called ΔV_0. Like reported in system specifications, since the precision of ADC is $\Delta_{ADC} = 15$ mV, ΔV_0 must be equal or less than 7.5 mV. Usually the modulator precision is expressed in terms of number of bit. It can be also expressed in term of the ratio between the power of an input sinusoidal signal with maximum amplitude so that the quantizer does not saturate, and the power of quantisation noise, all expressed in dB. For a classical N bit quantizer

$$\text{SNR} = 1.76 + 6.02N \tag{18.12}$$

This means that one bit corresponds to about 6 dB in term of SNR. $\Sigma\Delta$ modulator used in power conversion, also if it uses a single bit quantizer, produces a greater SNR, since it makes use of two techniques known as oversampling and noise shaping [9].

For a second order single bit ΣΔ modulator the precision is:

$$\mathrm{SNR} = -5.12 + 50\log_{10}\left(\frac{F_{\mathrm{CLK}}}{2BW}\right) \qquad (18.13)$$

where $\frac{F_{\mathrm{CLK}}}{2BW}$ is called oversampling ratio (OSR). Since 6 dB correspond to about 1 bit in resolution, step regulation can be rewritten as:

$$\Delta V_0 \cong \frac{V_{\mathrm{IN\,max}}}{2^{\mathrm{SNR}/6.02}} \qquad (18.14)$$

In the present project, since $F_{\mathrm{CLK}} \geq 3.3\,\mathrm{MHz}$, as indicated in Eq. 18.9, the minimum OSR value is 3.3 MHz/2·50 kHz = 33. Since F_{CLK} has been chosen equal to 3.5 MHz, the OSR value is equal to 35. This value allows a SNR of about 72 dB corresponding to a precision greater than that reported in the system specifications.

18.5 Dynamic OSR ΣΔ Modulation and MPPT

As just reported above, the minimum allowable switching frequency has been taken 500 kHz. If the mechanical excitation on the piezoelectric transducer changes in frequency and/or in amplitude, under the assumption that the generator works in the resonance zone, the power versus load characteristic reported in Fig. 18.4 changes its shape.

In Fig. 18.8 is reported the Power, Vrec versus load when the mechanical stress has the same amplitude of that reported in Fig. 18.4, but a frequency of 100 Hz. If the rectified voltage is maintained close to 20 V as made when the stimulus was at 50 Hz,

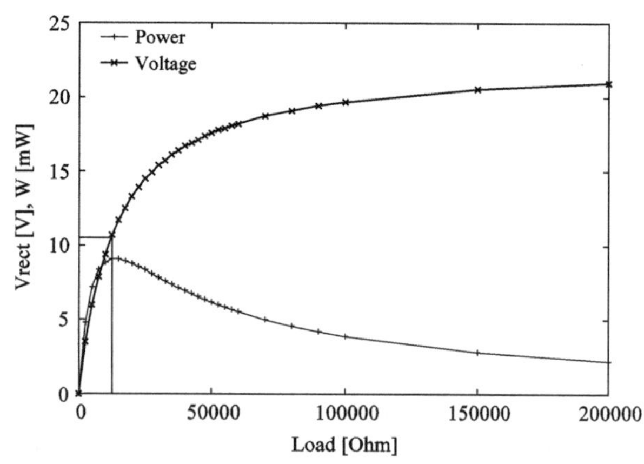

Fig. 18.8 Piezoelectric power production and voltage at the rectifier output for a pure resistive load. Current generator is sinusoidal with amplitude 2.6 mA and frequency 100 Hz

the power production of the piezoelectric sensor drops down under 5 mW. To improve its efficiency, a Maximum Power Point Tracker (MPPT) algorithm must be adopted to find the rectified voltage at which power production is the biggest.

In this case the maximum power production of 9 mW corresponds to a rectified voltage of 10.5 V. From Eq. 18.3, since C_{rect} was fixed to 8 μF and T_{ON} was fixed to 50 ms, then $\Delta = 6.5$ V so that the rectified voltage can change between 3.5 V and 16.5 V. This means a maximum duty cycle of $\delta_{MAX} = V_0/V_{rect\,min} = 0.94$. Unfortunately, the use of $F_{CLK} = 3.5$ MHz can produce a $\overline{F_{SW}}$ less than $\overline{F_{SW\,min}} =$ 500 kHz for which the power cell has been designed to work. This means that the SSA model used in the design of the loop control, could be not valid, since the power cell can work in Discontinuous Conduction Mode (DCM). A solution to a such problem is to ensure that under critical conditions, for example when $\delta = \delta_{min}$ ($\delta = \delta_{max}$), $\overline{F_{SW}}$ must be greater than $\overline{F_{SW\,min}}$.

In this project, since the more critical condition is $\delta = 94\%$, $\overline{F_{SWmin}}$ should be fixed at 500 kHz at least. Using Eq. 18.8 and accordingly to first condition on limit cycle absence [12], F_{CLK} can be chosen about 8.3 MHz.

Unluckily, driving the modulator with this signal, the system performances in term of efficiency can drop down. In fact, when δ is close to 0.5, $\overline{F_{SW}}$ is above 4 MHz, increasing dynamic power switching loss and compromising system efficiency. Good system performances can be obtained restricting $\overline{F_{SW}}$ in a suitable range $\left[\overline{F_{SW\,min}} : \overline{F_{SW\,max}}\right]$, for example into [0.5:1] MHz. To this aim Dynamic OSR ΣΔ modulation can be used [13].

Figure 18.9 reports the Dynamic OSR (DOSR) ΣΔ modulator. The LookUp Table (LUT) feeds ΣΔ modulator with duty cycle level and selects a suitable clock. To choose the clock frequency values to be associated to each dimming level, $\delta = 0.5$ has been associated to the maximum allowable commutation frequency $\overline{F_{SW}} = \overline{F_{SW\,max}} = 1$MHz.

From Eq. 18.8, the first clock frequency F_{CLK1} to drive the modulator has been chosen equal to 2 MHz. Such value should be used until $\overline{F_{SW}}$ is greater than $\overline{F_{SWmin}}$, then until $0.25 \le \delta \le 0.75$.

When duty cycle became less than 0.25 or bigger than 0.75, F_{CLK2} must be used. Fixing in $\delta = 0.75$ (or $\delta = 0.25$) $\overline{F_{SW}} = \overline{F_{SW\,max}}$,$F_{CLK2}$ can be fixed to 4 MHz. Similarly, F_{CLK2} can be used until $\overline{F_{SW}}$ is bigger than $\overline{F_{SW\,min}}$, when $0.125 \le \delta < 0.25$ and $0.75 < \delta \le 0.875$. Proceeding in this way, the entire range $[\delta_{min} : \delta_{max}]$ has been associated with appropriate F_{CLK} values.

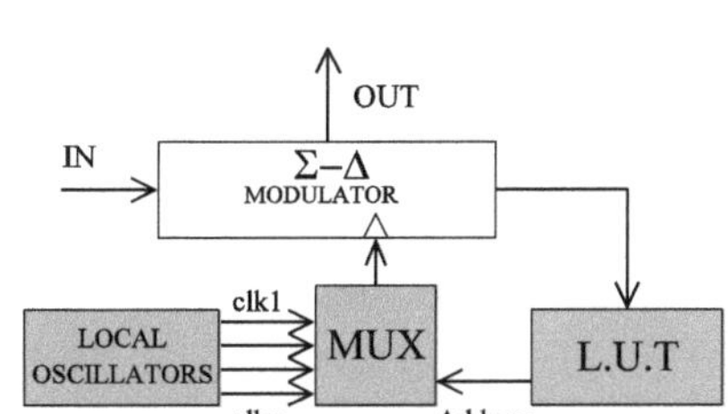

Fig. 18.9 Dynamic ΣΔ modulator

Table 18.2 reports the clock frequencies associated with the respective duty cycle ranges, and Fig. 18.10 reports the average switching frequency versus duty cycle for the proposed DOSR ΣΔ modulator.

18.6 System Modelling and Simulations

System modelling and simulations have been performed using SystemC-WMS environment [14, 15], a SystemC extension for mixed signals simulation. In SystemC-WMS, analogue blocks interact to each other exchanging energy wave quantities instead of using current and voltage signals. As an example, Fig. 18.11 reports a SystemC-WMS schematic representation of the interconnection between an LC filter and the load, modelled as an ideal switch with a non zero ON and OFF resistance.

Figure 18.12 shows some code lines modelling the circuit reported in Fig. 18.11. In Fig. 18.13 the SystemC-WMS code that declare and implement an LC filter is reported.

Table 18.2 LUT parameters for dynamic OSR modulation

Left range	Right range	Clock frequency (MHz)
$0.25 \le \delta < 0.5$	$0.5 < \delta \le 0.75$	$F_{CLK1} = 2$
$0.125 \le \delta < 0.25$	$0.75 < \delta \le 0.875$	$F_{CLK2} = 4$
$0.06 \le \delta < 0.125$	$0.875 < \delta \le 0.94$	$F_{CLK3} = 8$

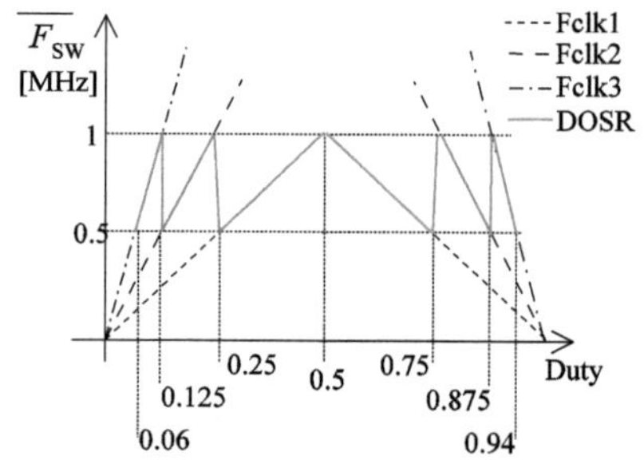

Fig. 18.10 Average switching frequency $\overline{F}_{SW}$ versus duty cycle magnitude δ for different clock frequency and with DOSR

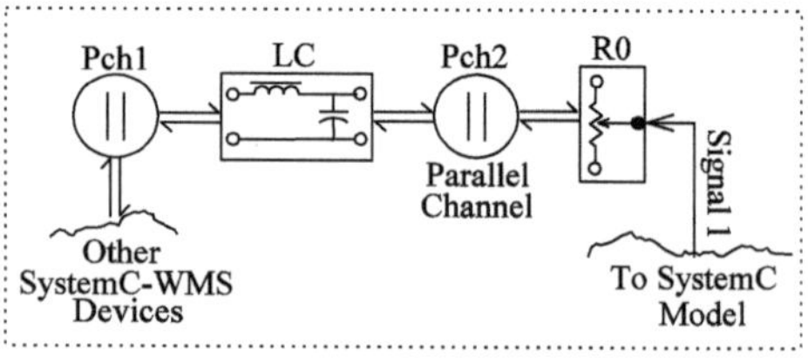

Fig. 18.11 A part of the power cell SystemC-WMS model

```
/********Implementation of the circuit**********/
//Channels declaration
ab_signal<electrical, parallel> Pch1, Pch2;
//Instance of LC module
LC_filter LC("LC", 5000 uH,  uF, 1e-7,1e-7);
//Connection of LC module
  LC(Pch1,Pch2);
//Istance of R0 module
switch_RonRoff R0("R0", 363 ohm, 2200 ohm);
//Connection of R0 module
  R0.control(Signal_1);
  R0(Pch2);
```

Fig. 18.12 Implementation of the circuit reported in Fig. 18.11

```
// Declaration of LC_filter two ports:
struct LC_filter : wave_module<2, electrical>, analog_module
  {
  SC_HAS_PROCESS(LC_filter);
  LC_filter (sc_core::sc_module_name name, double s_ind,
  double p_cap, double res_min, double res_max);
private:
  void calculus ();
  void field (double *var) const;
  const double C,L;
  };
// Implementation of class LC_filter two ports:
LC_filter::LC_filter(sc_core::sc_module_name name, double
  s_ind, double p_cap, double res_min, double res_max):
  analog_module(2,res_min,res_max),C(p_cap),L(s_ind) {
  SC_THREAD(calculus);
  }
void LsCp_ladder_rcc::field (double *var) const
  {
  const double sqrt_R1 = port[0]->get_normalization_sqrt();
  const double R1 = port[0]->get_normalization();
  const double sqrt_R2 = port[1]->get_normalization_sqrt();
  const double R2 = port[1]->get_normalization();
  double a1 = port[0]->read();
  double a2 = port[1]->read();
  var[0] = state[1]/L+(2*a2)/sqrt_R2-state[0]/(R2*C);
  var[1] =2*a1*sqrt_R1-(R1/L)*state[1]-(state[0])/C;
  }
  void LsCp_ladder_rcc::calculus ()
  {
  const double sqrt_R1 = port[0]->get_normalization_sqrt();
  const double sqrt_R2 = port[1]->get_normalization_sqrt();
  state[0]=0;
  state[1]=0;
  while(step()) {
    double a1 = port[0]->read();
    double a2 = port[1]->read();
    double b1=a1-(sqrt_R1/L)*state[1];
    double b2 =-a2+state[0]/(sqrt_R2*C);
    port[0]->write(b1);
    port[1]->write(b2);
      }
  }
;
}
```

Fig. 18.13 Declaration and implementation of an LC filter in SystemC-WMS

Figure 18.14 reports the SystemC-WMS schematic representation of the whole system. A piezoelectric device is modelled with its resonance electric equivalent scheme. The current generator produces a sinusoidal fixed frequency signal.

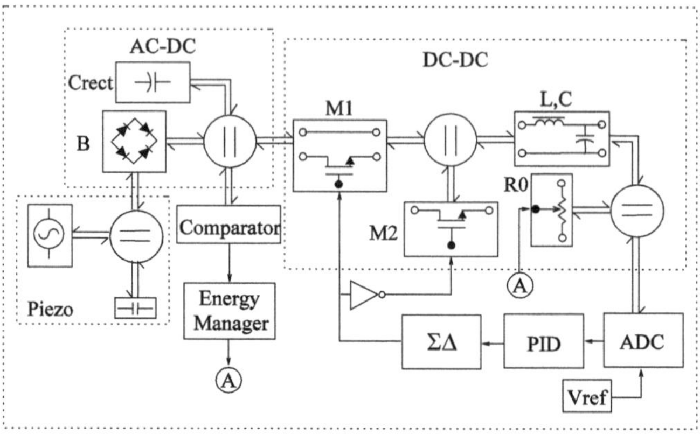

Fig. 18.14 SystemC-WMS mixed signals model of proposed device

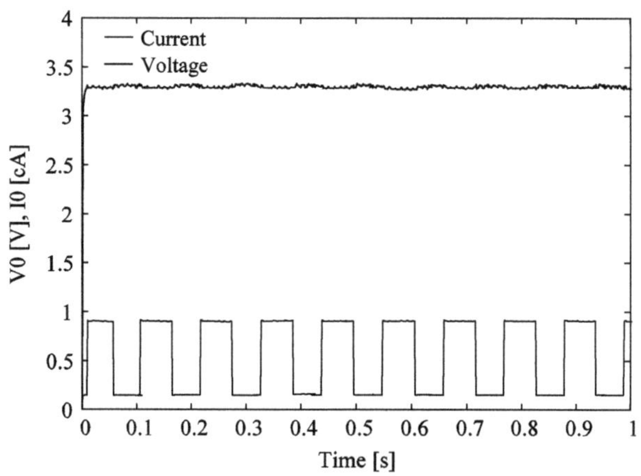

Fig. 18.15 Voltage across and current through wireless sensor supply pins when I_p is at 50 Hz

Its value depends on mechanical stress intensity and on frequency as well as on crystal physical and geometrical characteristics. Parallel capacitor C_p depends only on crystal physical and geometrical characteristics. In our model it is supposed $C_p = 0.2\ \mu F$.

The first stage connected with the piezoelectric generator is the AC–DC converter, composed by a bridge rectifier and by an output capacitor. The aim of this stage is storing energy into C_{rect}. Some management blocks are used to maintain the capacitor stored energy at a suitable level so that the piezoelectric generator can work close to the maximum efficiency point.

The final stage is a DC–DC step down converter. It holds output voltage at 3.3 V independently from load status.

In Fig. 18.15 the time diagram of the output voltage and of the output current are reported, for different Power-ON state and Standby state transition cycles of the wireless sensor.

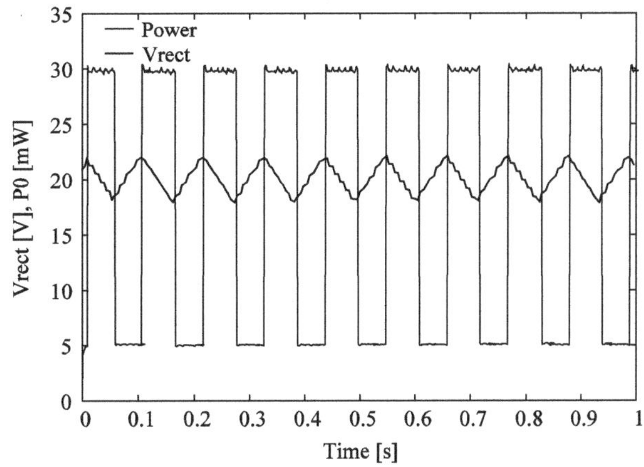

Fig. 18.16 Power used by the wireless sensors and the rectified voltage across C_{rect} when I_p is at 50 Hz

In Fig. 18.16 the rectified voltage V_{rect} and the power used by the load P_0 is reported for different Power-ON and Standby state transitions. When the load is in Power-ON and requires 30 mW, V_{rect} goes down, conversely when it is in Stand-by, V_{rect} grows.

18.7 Conclusion

In this work a digital controlled power supply has been proposed. It recovers environmental energy by means of a piezoelectric transducer working at its resonance frequency and under a stationary mechanical stress.

When the system is connected with an unknown piezoelectric device working at its resonance frequency, an MPPT algorithm together a DOSR ΣΔ modulator should be used. The proposed device allows to store or to supply energy depending on load power requirements and system energy availability. The system has been modelled at system level using SystemC-WMS. The system works without using any battery.

References

1. Sample A, Smith JR (2009) Experimental results with two wireless power transfer systems. www.techonline.com/learning/techpaper/2129020
2. Ottman G, Hofmann H, Bhatt A, Lesieutre G (2002) Adaptive piezoelectric energy harvesting circuit for wireless remote power supply. In: IEEE Transactions on Power Electronics, vol 17, no 5, pp 669–676, September 2002
3. Mateu L, Fonellosa F, Moll F (2003) Electrical characterization of a piezoelectric film-based power generator for autonomous wearable devices. In: XVIII Conference on Design of Circuits and Integrated Systems

4. Niu P, Chapman P, Riemer R, Zhang X (2004) Evaluation of motions and actuation methods for biomechanical energy harvesting. In: Power Electronics Specialists Conference, PESC 04, IEEE 35th, vol 3, June 2004, pp 2100–2106
5. d'Aparo R, Orcioni S, Conti M (2009) A digital controlled energy scavenger power converter. In: Proceedings of the IEEE Seventh International Workshop on Intelligent Solutions in Embedded Systems WISES09, 25–26 June 2009, Ancona, Italy, pp 165–170
6. Loreto Mateu NLMP, Codrea C, Spies P (2006) Energy harvesting for wireless communication systems using thermogenerators. In: Proceedings of the 21st Conference on Design of Circuits and Integrated Systems, Barcelona, Spain
7. Rashid MH (2010) Power electronics handbook. Butterworth Heinemann, ISBN:978-0-12-382036-5
8. Orcioni S, d'Aparo R, Conti M (2007) A switching mode power supply with digital pulse density modulation control. In: Circuit Theory and Design, ECCTD 2007, 18th European Conference, Seville, Spain, Aug 2007, pp 603–606
9. Aziz P, Sorensen H, van der Spiegel J (1996) An overview of sigma-delta converters. IEEE Signal Process Mag 13(1):61–84
10. Middlebrook RD, Cuk S (1976) A general unified approach to modelling switching-converter power stages. In: Power Electronics Specialists Conference, Cleveland, Ohio, pp 18–34
11. Moussa W, Morris J (1990) Comparison between state space averaging and PWM switch for switch mode power supply analysis. In: Southern Tier Technical Conference, Proceedings of the 1990 IEEE, Apr 1990, pp 15–21
12. Peterchev A, Sanders S (2001) Quantization resolution and limit cycling in digitally controlled PWM converters. In: Power Electronics Specialists Conference PESC 01 IEEE 32nd, vol 2, pp 465–471
13. Conti M, Orcioni S, d'Aparo R (2009) Dynamic OSR sigma delta controller for monolithic switching converters. SPIE EUROPE, Dresden, vol 7363, p 73630W
14. Orcioni S, Biagetti G, Conti M (2005) SystemC-WMS: a wave mixed signal simulator. In: Proceedings Forum on Specifications & Design Languages (FDL'05), Lausanne, CH, Sep 2005, pp 61–72
15. Orcioni S, Ballicchia M, Biagetti G, d'Aparo RD, Conti M (2008) System level modelling of RF IC in SystemC-wms. EURASIP J Embed Syst 2008:1–11. doi:10.1155/2008/371768 (Article ID 371768)

Chapter 19
Energy Efficient Data Transmission of On-Chip Serial Links

A Case Study

George Kornaros

19.1 Introduction

As CMOS technology scales down, the ratio between wire and gate capacitance is continuing to grow. In today's ultra deep submicron (UDSM) designs, the interconnect wires play a major role in the timing behavior of logic gates and in the power consumption of large Systems-on-Chip (SoC). The main reasons are: (i) the relative scaling of cell capacitances and, (ii) the thinner and more aggressive metal pitch of modern technologies causing the increase of inter-wire, or coupling capacitances. The latter effect, in particular, is increasingly important in technologies below 90 nm, in which coupling capacitances between adjacent wires become significantly larger than the capacitance between a wire and the substrate, known as the self capacitance.

Designers argue that coupling capacitances are very important (see [1, 2]) mostly because they affect wire delays: transitions to opposite values on adjacent wires will exhibit longer delays than for other types of transitions; in addition, the delayed transitions are often preceded by spurious spikes. These effects are commonly grouped under the term "*crosstalk*". Increased coupling capacitances, however, are also critical for power consumption, because (i) for some types of transitions will cause these large coupling capacitances to switch, and (ii) the spurious transitions will also dissipate unnecessary power. Coupling effects are particularly critical in long, cross-chip buses, because of the large capacitances due to their length, and also because conventional routing algorithms tend to keep bus wires close together, thus increasing the number of adjacent wires.

G. Kornaros (✉)
Electronics & Computer Engineering Department, Technical University of Crete, Chania, Crete, Greece
e-mail: kornaros@epp.teiher.gr, kornaros@gmail.com

M. Conti et al. (eds.), *Solutions on Embedded Systems*,
Lecture Notes in Electrical Engineering, 81, DOI: 10.1007/978-94-007-0638-5_19,

Table 19.1 Substrate capacitance (Cs) and coupling capacitance (Cc) of the victim wire due to transitions of the adjacent aggressor wires

Type	Ceff	Transition patterns			
1	Cs	(↑, ↑, ↑)	(↓, ↓, ↓)		
2	Cs + Cc	(–, ↑, ↑)	(–, ↓, ↓)	(↑, ↑,–)	(↓, ↓,–)
3	Cs + 2.Cc	(–, ↑,–)	(–, ↓,–)	(↓, ↑, ↑)	(↓, ↓, ↑)
		(↑, ↑, ↓)	(↑, ↓, ↓)		
4	Cs + 3.Cc	(–, ↑, ↓)	(–, ↓, ↑)	(↑, ↓,–)	(↓, ↑,–)
5	Cs + 4.Cc	(↑, ↓, ↑)	(↓, ↑, ↓)		

Table 19.1 categorizes the worst-case crosstalk patterns that may occur among three adjacent wires and cause the increase of the propagation delay. The symbol ↑ represents a rising transition, ↓ represents a falling transition and – means that there is no transition on the wire. In the best case, when the three wires are switching in the same direction, the delay on the victim wire is the delay without crosstalk (i.e. when Ceff = Cs). However, the bus clock cycle must be adapted exclusively regarding the worst-case delay (i.e. Ceff = Cs+ 4.Cc) to ensure the integrity of the transmitted data. Nevertheless, the power consumption is roughly proportional to the percentage of appearance of the worst-case transition patterns.

The proposed schemes in this paper aim to minimize the switching activity and crosstalk effects of the high data-rate interconnection DSM busses inside modern embedded Systems-on-Chip (SoC) that usually integrate many IPs with long wires. At the same time we keep the complexity of the encoder–decoder circuitry very low so as to make it easily applicable to very large range of on-chip communication links. Furthermore, these new schemes are free of extra overheads due to additional wires that could be used for shielding purposes. Included is a novel scheme that can even reduce the number of required wires exhibiting time overhead. Additionally, in modern Network-on-Chip (NoC) -based designs the benefits come in effect due to reduced cost of wiring and also due to that encoded data need not be transformed when switched by NoC routers.

The organization of this chapter is as follows. Section 19.2 presents briefly related work and discussion. In Sect. 19.3 two encoding schemes are detailed, along with exploring possible extensions and improvements. Section 19.4 presents hardware implementation results which are compared with a few alternative schemes previously reported. Finally, Sect. 19.5 provides a summary and conclusions.

19.2 State of the Art

Static encoding strategies for reducing activity of on-chip busses are based on a priori knowledge of the stream of patterns that will travel on the bus, so as to generate an ad hoc encoder which will minimize the switching activity for that stream [3]. Obviously, these techniques are highly application-dependent, in the sense that their applicability is limited to cases in which it is possible to have detailed information about the traffic on the buses. Opposing to this viewpoint the novel scheme proposed in this work is independent from traffic characteristics.

Often, low power and crosstalk immune schemes may use extra bit-lanes, i.e. the proposed method by Khan in [4] uses a 4-to-6 codebook for a total of 16 additional signals if a 32-bit bus is used. Another option to shield data transfers is proposed by increasing a 16 bit bus to 27 bits, or adding separation bits [5]. Other similar bus encoding techniques such as the variable cycle transmission with temporal redundancy (i.e. the VCTR technique presented in [6]), uses three clock cycles in cases of crosstalk classes 4 and 5, whereas the approach described in [7] uses three and four clock cycles. However, with the proposed scheme only one transition occurs per lane providing clear energy benefits. Philippe in [8] and Najeeb in [9] also propose another temporal coding method which purges crosstalk effects to improve delay but does not reduce mean switching activity. Shin et al. [10] proposes encoding methods for reducing power consumption by reducing the number of bit transitions on data buses; these techniques are variants of the classical bit-invert method [11], where bus data is inverted prior to transmission if the distance between the current data and the previously transmitted data is greater than half the width of the bus.

Serialized low energy transmission coding for on-chip interconnect networks (SILENT) presented in [12] aims at reducing the switching activity in the serial link by employing differential encoding. This is achieved by XORing the data to be serialized with the previous data. However, this scheme is not so effective when the transmitted data have non-uniform statistics.

Besides the static encoding techniques mentioned so far, adaptive encoding schemes have been proposed [13, 14] that reduce switching activity often very efficiently, due to their adaptability to varying statistical parameters. They do not require the a priori knowledge of statistical parameters of the data streams to be encoded, but observe them periodically at system run time and modify the encoding rules accordingly. However, the efficient reduction is achieved at the cost of large hardware requirements.

Kretzschmar et al. argue in [15] that adaptive schemes could obtain up to 40% savings, but the bus length which is required to reduce the overall power consumption is not realistic for on-chip buses. However, crosstalk effects are not considered and additionally in our case of utilizing fewer wires the capacitance is significantly decreased. Further, if P_{Codec} represents the power dissipated by the codec system, then, we claim that in a modern NoC it is not necessary to account P_{Codec} for every intermediate router or switching component:

Switching or routing of encoded information is achieved free of the cost of decoding-switching-and-encoding.

It is easy to design switching elements that take into account the proposed encoding scheme and most other schemes as well. Circuit switching or even packet switching could be easily implemented if headers are not encoded, or easily identified either in encoded format. Thus, the overhead of decoding, switching and encoding is not necessary to be paid for each hop.

An associated issue with coding for energy efficiency is resilience to errors. This concern is very essential for deep submicron technology since circuits are more susceptible to less predictable forms of interference such as noise induced by power grid fluctuations, electromagnetic interference, and alpha particle radiation. Therefore, an encoding scheme is significant to feature properties that favor fault tolerance. Error correcting codes for on-chip busses use algorithms that most times become computationally infeasible for moderate to large bus sizes.

Moreover, retransmission based strategies perform better than those using correction. Actually, techniques have been proposed [16], which work at lower voltage swings because of their error detection capabilities and are more effective from an energy viewpoint. In the same direction the first proposed scheme based on block decoding features clear attributes that offer easy error detection (for instance a nor gate can detect the absence of a "1" in the decoded block, or an XOR gate can detect the existence of a single "1"), but due to space limitations this is not further analyzed.

19.3 Energy-Delay Efficient Bus Encoding

In this work a set of solutions are developed for combined reduction of the switching activity of a bus and of the delay caused by crosstalk effects. Fault tolerant properties of these solutions will be discussed as well.

19.3.1 Temporal Coding Using Block Decoding

One essential property inherent in the first proposed scheme is that the switching activity is constant for each piece of information transmitted; a word will be called hereafter. The term constant is used to describe that it is independent of the type of information moved over the on-chip communication link. It could easily be employed in an address bus, or convey data in a multi-core system on a chip.

Hence, the activity can be calculated accurately and independently of the application we run over the proposed scheme. From one viewpoint it is thus meaningless to benchmark its properties using a variety of applications, as it is necessary for other encoding techniques that are proposed in the literature. In these cases the featured performance and the energy consumed by the interconnection fabric depends on the attributes of the application data transferred. Pushing an encoding scheme to benefit from those attributes may reduce the efficiency of this scheme if an application with a different traffic profile uses the bus. Considering the crosstalk of adjacent wires however, the actual application data conveyed over the bus affect the delay and energy consumed considerably and thus it is an equally important factor compared with the reduction of the switching activity. It is examined in the following paragraphs in detail.

Conceptually, the main idea is that in order to transmit a piece of information of w bits, the optimal way to achieve the most energy efficient transmission is to allow only one signal to make a transition. At the side of the receiver the original word of w bits can be recognized by discovering which individual signal of the bus appears to switch state. The original value of the transmitted word is reconstructed by identifying the relative location of the "1" among the zeros. This idea is essentially realized by a simple pair of decoder–encoder system. While the most intuitive way to think of a decoded value is a bit vector where only one "1" exists, one may consider the result of the decoding operation as a data block n by m of all zeros except one.

Let n be the dimension that defines the available wires a designer will use and m be the time span that the decoded block will complete its transmission. In the sequel it is shown that although high clock frequencies can be realizable, it is not desirable to use more clock cycles than four for the m parameter.

If the data link is w bits wide then we partition it in w/n sub-links with n bits per sub-link. Thus, 2^n bits must be transmitted from the sender to the receiver per sub-link. Only one bit is switching, making it beneficial to use more bits to decode since the percentage of activity is proportional to $1/2^n$. This however, is very costly in wiring resources since it increases exponentially the bus width. Using temporal distribution of the decoded information the overhead of extra signals is avoided. Apparently, the decoded information of 2^n bits must be transmitted over n consecutive clock cycles to avoid the overhead. Unless strict timing restrictions dictate a violation in the operation of the developed system, a balance may be explored between additional wiring and additional clock cycles needed to transfer the decoded block of 2^n bits.

For example, for a 4-bit lane we can use four clock cycles to transfer 16 bits, or two clock cycles to transfer eight bits per cycle. Figure 19.1 shows an example with two 4-bit lanes in which four words are sent in four clock cycles, or in two if double-data-rate technique is used; however, this is beyond the scope of this paper.

Consequently, the percentage of switching activity of a bus which transfers a word of w bits is given by Eq. 19.1.

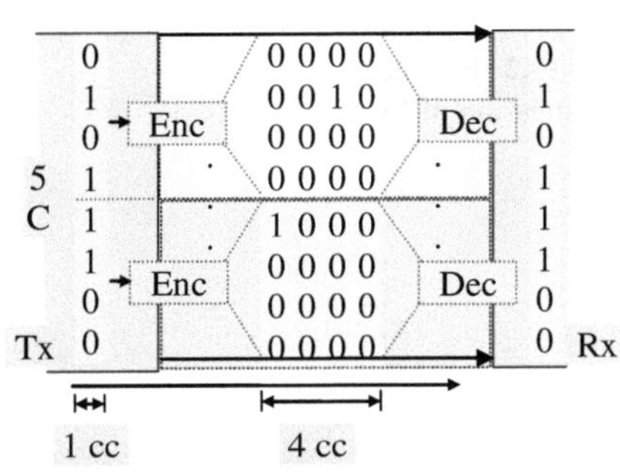

Fig. 19.1 Encoding of 0×5C for transmission over an 8-bit bus

$$w/n \times 1/w \times 10^2 = 1/n \times 10^2 \tag{19.1}$$

A time interval of $t_{dec} = n$ is assumed, where no additional signals are used. Compared with the activity in the original bus that is able to transfer a w-bits word in t_{orig}, or in other words, considering the equivalent rate of information transferred, the switching activity is constant on the condition that the time interval to send one word in both systems is equal, that is $t_{dec} = t_{orig}$. This accounts for a constant $1/n$ activity factor, which is most of the times distributed over n clock cycles. Hence, the proposed scheme excels for traffic patterns that sport activity greater than $1/n$ in un-encoded format or with other encoding techniques. Thus, in the proposed scheme the switching activity is a priori known and constant.

Additionally, the property of just one "1" inside each decoded block of data transferred is very resilient to errors and thus we may apply techniques to further reduce power consumption. By lowering the Vdd supply for instance, the energy savings would be significant, and since the proposed encoding simplifies an error detector, this is another promising direction to be explored.

One can argue that in order to reduce activity, apparently the operation frequency is increased (over-clocking) and consequently, the power consumption which is given by $P = a \times \mathrm{Vdd}^2 \times f \times C$ cannot be improved. Actually, the clock frequency of the encoder and decoder is increased, whereas the factors a and C in the equation are reduced. The principle objective in this scheme is to keep the same data bus frequency as before, and even if we over-clock the encoder and decoder circuits the overall power consumption is reduced as shown by the simulation results in the following sections.

Due to the format of the decoded data that travel over a bus the probability of appearance of the crosstalk classes listed in Table 19.1 is reduced. Type-4 for example, which is one of the worst case patterns, will never appear inside a 16-bit block of zeros with a single one. It may appear between two adjacent lanes. Let $P\alpha$ be the probability that Type-4 transitions occur between neighboring blocks. Since the probability of a "1" at the boundary is ¼ and, excluding all the combinations of transitions that do not cause a Type-4 pattern we have:

$$P_a = \frac{1}{4} \cdot \frac{1}{4} - \sum_{1}^{6} \left(\frac{1}{16} \cdot \frac{1}{16} \right) \tag{19.2}$$

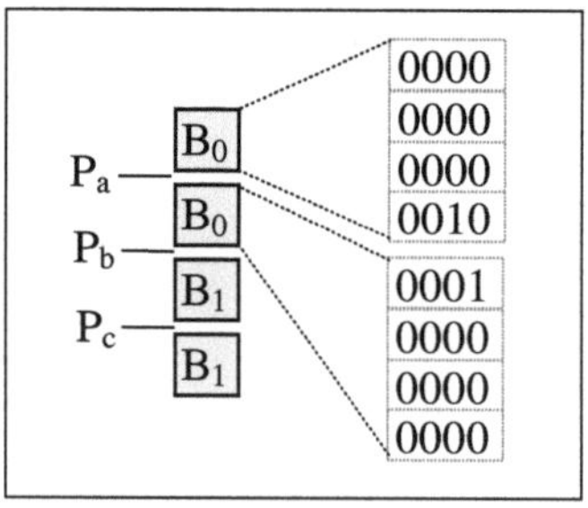

Fig. 19.2 Encoded transmission of four bytes

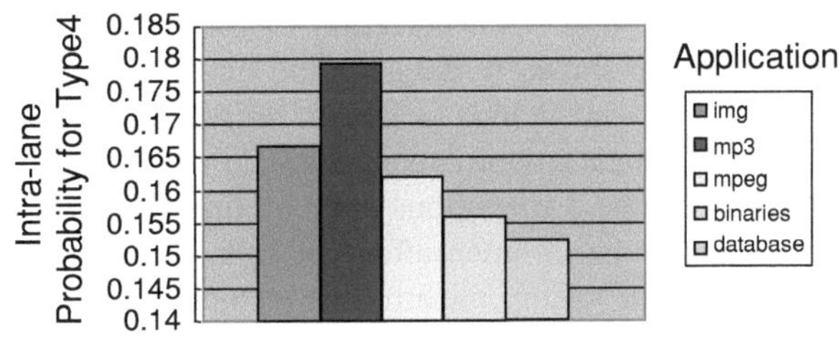

Fig. 19.3 Percentage of Type-4 coupling in encoded 4-byte data words of various applications

If we consider the transmission of two bytes, then, four lanes must be considered. Figure 19.2 outlines this case where each 4-bit word is encoded to a block. Then, the probability of a Type-4 transition is given by:

$$P_{4\text{Lane}} = P_a \cdot P_c + P_a + P_c + P_b(1 - P_a \cdot P_c) \tag{19.3}$$

From Eqs. 19.2 and 19.3 the probability $P_{4\text{Lane}}$ is 0.11865. Similarly, given the transmission of four bytes the probability is 0.29. A tool is developed to analyze data encoded with the proposed scheme and calculates the occurrence of Type-4 crosstalk patterns. Figure 19.3 shows that real data from images, mp3, mpeg, binary code samples and database transaction traces exhibit less probability of Type-4 transitions, with MP3 data being the most susceptible to it.

Next, the analysis is focused on the worst case of crosstalk effect (Type-5) and how it is possible to eliminate this situation.

The worst transition pattern is Type-5, which can occur between two blocks of bits, one sent immediately after the other, so that the last bit vector of one block and the first bit vector of the following block may cause this case. Figure 19.4 shows all the possible cases where the tail and the head of each block inside neighboring lanes are depicted. Symbol *B-1* stands for the previous block and *B* for the subsequent block in time, transferred top to bottom or upside-down, for a total of four possible cases. Unless another redundant vector of zeros is sent so as to eliminate this case, this situation can be alleviated with minor changes of logic.

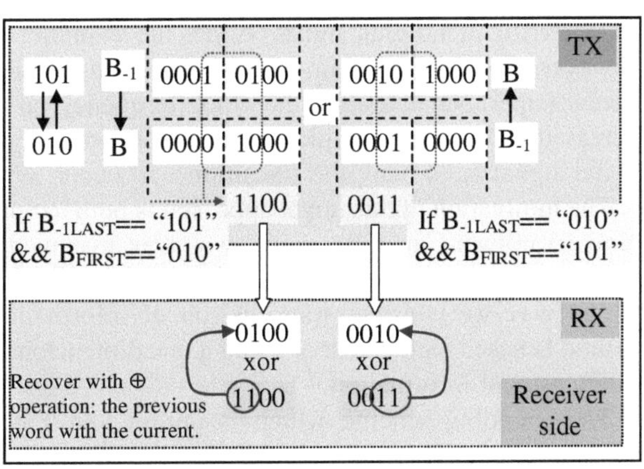

Fig. 19.4 Worst case transitions (Type-5) when transmitting successive blocks. The encoder (TX) can intentionally flip one bit to eliminate this case, while the decoder (RX) can easily detect and fix the original data

The transmitter upon detection of the two worst case transition patterns (the conditions are shown in detail in the figure), can add one more "1" in the transmitted 4-bit word, which is in fact an illegal encoding word. When the receiver detects this invalid word fixes it by doing an XOR operation with the previous word. The Type-5 transitions can be eliminated with this simple logic.

However, the counter effect of this mechanism is that in situations where reliability is of major concern, this technique suffers as it is not easy to recognize at the receiver side whether the flipped bit is intentional or due to coupling noise.

19.3.2 Temporal Coding Using Narrow Links

The initial approach presented in this section so as to manage power consumption of transfers over long on-chip interconnects is to reduce transition activity. After that, crosstalk induced issues are discussed.

The additional benefit of this novel scheme is that power consumption is reduced due to:

i. the hardware complexity to construct routers that can manipulate directly the encoded information is negligible,
ii. the overall size and consequently the cost in silicon of the NoC is significantly reduced, and,
iii. the simple arithmetic performed while using this encoding scheme without the cost in area, time and energy to perform decoding.

The principle idea of this innovative encoding scheme is that a transition of a signal basically indicates a change of state from $0 \rightarrow 1$ (or $1 \rightarrow 0$) to communicate with the receiver part so as to transfer one bit of useful piece of information. The ultimate benefit from the energy perspective would be to increase the amount of information transmitted without wasting any more energy than this of a single transition of one wire. The apparent question is how the receiver will understand the transfer of more bits of information from a single wire transition. An additional characteristic of a digital signal besides the swinging from $0 \rightarrow 1$ is the amplitude. However, this requires analog signal processing, which is beyond the scope of the current paper. Manipulating the clock frequency on the other hand, would cause increased activity and complexity. The most prominent and attractive property of a digital signal is "when" the transition will occur, or phase shift of the signal.

The motivation in this novel approach is both to reduce the number of necessary wires to transmit an n-bit word, which is managed by a parallel-to-serial conversion and, at the same time to reduce the number of signal swings. Modifying a bit the way we consider transmission of information to a target, the proposed method is based on the idea of using a one dimensional physical medium to encode 2-dimensional information.

The encoding scheme which is outlined in Fig. 19.5 is devised to support transmission of a 4-bit piece of information with a single pulse of varying duration

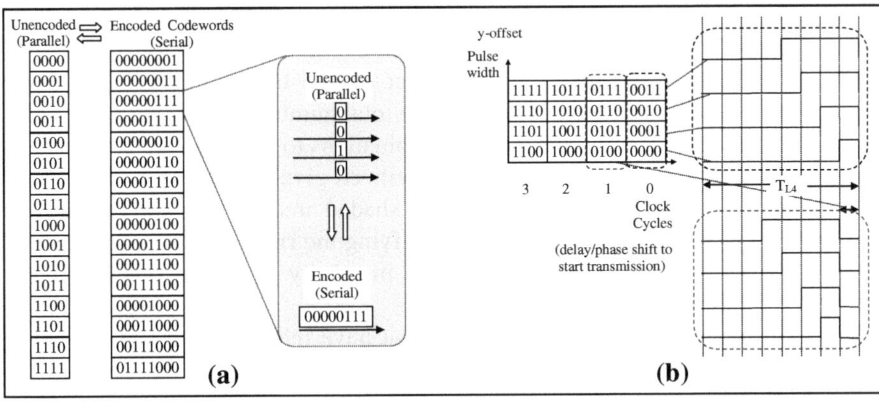

Fig. 19.5 2-D temporal coding of a 4-bit word. **a** Encoding scheme to transform four bits transmitted in parallel in one clock cycle to eight bits serially transmitted, **b** principle of encoding scheme: the two most significant bits indicate when the pulse starts, and the two least significant show the duration of the pulse

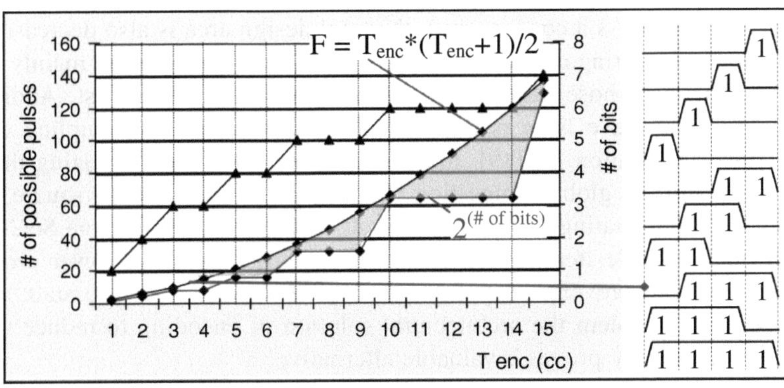

Fig. 19.6 Number of possible pulses to transmit (# of bits) in Tenc clock cycles interval; e.g., to send five bits 7, 8, or 9 clock cycles are required using a 2-D encoding scheme, while the number of possible pulses with two transitions is a lot more (28, 36, 45 accordingly)

that may appear in a time slot window of fixed duration of eight clock cycles. The principle of moving a chunk of n-bit data, called a symbol, is to transmit a single pulse of varying duration and varying offset on a single wire. A symbol is identified and decoded based on its position in a virtual 2-D space. The variable offset of the rising edge identifies the x-coordinate, whilst the variable duration of the pulse represents the y-coordinate of the symbol.

In general using a single wire to transmit a pulse of varying duration from 1 clock cycle to $y = \text{Tenc} - 1$ we can have a total of Tenc $*$ (Tenc $+$ 1)/2 different clock pulses. For example, Fig. 19.6 shows the ten possible pulses assuming an encoding scheme that utilizes four clock cycles. However, in the proposed encoding that illustrates encoding in a 2-D space, the number of utilized clock pulses which abide with scheme described above (see Fig. 19.6) is given by the formula $\lfloor \log_2(\text{Tenc} * (\text{Tenc} + 1)/2) \rfloor$.

Figure 19.6 shows the number of bits that can be transmitted using a single pulse of Tenc duration in clock cycles and the total number of possible pulses. The shaded area between the required binary combinations to send a number of bits and the number of potential pulses that may be utilized gives headroom which allows for developing fault tolerant techniques. This shaded area between the two lines in the plot is very challenging in terms of identifying the right set of energy efficient pulses with two transitions and assisting to immunity to errors; this domain is subject for future research.

Hence, it is obvious that for data types that have inherent high switching rate this proposed scheme is quite effective without even taking into account the crosstalk effects.

Due to the size of modern industrial circuits it is becoming more important factor for logic synthesis and technology mapping to satisfy performance constraints such as timing and power consumption while occupying minimum silicon area. In these directions a few methodologies are proposed in the literature [17–20] to effectively reduce the global wire-length, thus significantly improving congestion across the chip. As a consequence, the total design area is also decreased due to the reduction in wiring demands. Global routing optimizations are mainly based on a cost function whose basic factor is the interconnect wire cost. Additional routing algorithms have been proposed with considerations for coupling effects and crosstalk elimination. In [19], the cost function used for optimizing the tree routing algorithm is a global congestion function and depends on the square of the wire length, demonstrating the significance of wiring to building large SoCs; it is also shown that while iterating on the critical paths to achieve lower crosstalk efforts are made to prevent delay deterioration. However, since crosstalk avoidance is NP-hard problem the architectural solution of encoding to reduce wiring and switching activity proves a valuable alternative.

19.4 Implementation and Evaluation Results

The first proposed encoding scheme based on simple block decoding was implemented using FPGA technology in a Virtex-4 FX100 device, along with Bus Invert (BI), and Coupling driven Bus Invert (CBI) techniques (as presented in [11] and [21, 22] respectively). These two are very efficient coding schemes and various optimizations of these have been proposed in the literature.

All proposed circuits were layout as shown in Fig. 19.7; the transmitter is manually placed at one end of the chip and the receiver at the other end, so as to transfer words over a four mm bus.

The speed grade was set to -11 and thus, the maximum frequency achieved is shown in the first column of Table 19.2. The switching activity is back-annotated in Xpower from XILINX, after simulating the gate-level netlist with uniform

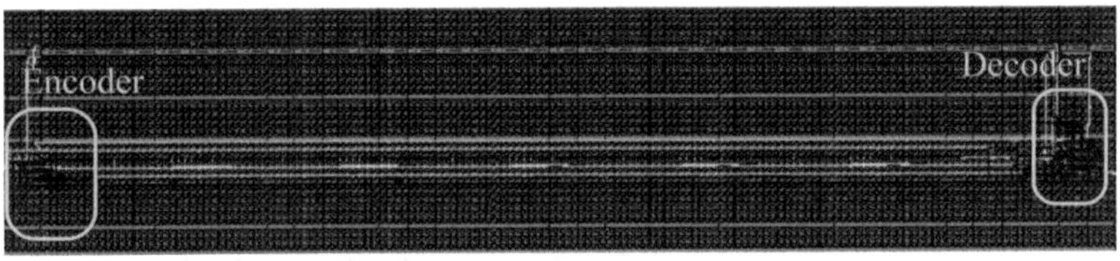

Fig. 19.7 Layout of the block encoding scheme in a Virtex-4 FX100 FPGA; the encoder (Tx) and decoder (Rx) are placed at the edges of the die spanning a 4 mm distance

generated traffic. All simulations were performed at 125 MHz and the data traffic was supplied at full rate, but the same for all the three encoding schemes. The improvement in total power consumption is small; from 4 to 9%, but the improvement regarding the peak power dissipation reaches up to 57%.

Measurements over very large data streams were made with the aid of a custom tool so as to build the activity profile of different types of data. Images, mp3, mpeg and binary traffic were analyzed. The proposed encoding method using block decoding reduces the switching activity exhibiting the following percentage ranges:

- images/mp3 48–49%
- mpeg: 39–48%
- sample traces of binary code: 25–48%

The bus invert, on the other hand, produces an average reduction of about 17% in total switching activity, whereas the CBI scheme resulted in less than 8% energy savings. This was in fact achieved by modifying the majority voter that outputs high when at least six input lines are high out of 15 inputs, instead of eight that was in the original published implementation.

The second class of encoding scheme, which is based on 2-D temporal coding over a narrow bus, was also modeled in VHDL and simulated with different types of data as before. Implementation details can be found in [23]. Figure 19.8 shows the percentage of savings per coupling type considering the un-encoded transfer of 32-bit words and the encoded transfer over 8 wires. Although the time interval is enlarged by eight times the reduction in Type-5 and Type-4 couplings is remarkable. Type-3 couplings are increased, but less than 20% in average.

Although it is also verified in this case of coding with narrow links that it exhibits significant reductions in crosstalk couplings, the actual benefit of the encoding scheme is that all cross-couplings can be eliminated by adding shielding, or by increasing the wire spacing. In complex SoCs with very dense routing there is now enough headroom to make it easily achievable. All the proposed encoding

Table 19.2 Implementation in a Virtex-4 FX100 FPGA

	F_{max}	P_{total} (mW)	P_{peak} (mW)
Idec	208	1,109	1,581
Bi	125	1,223	3,708
Cbi	128	1,152	3,417

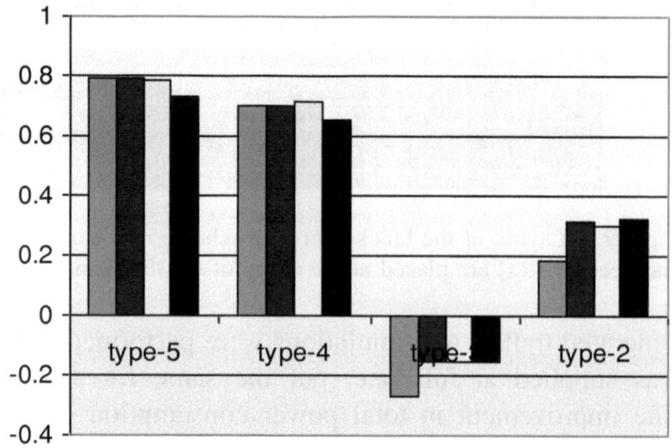

Fig. 19.8 Reduction per crosstalk type transferring images, mpeg, mp3 and executable files respectively

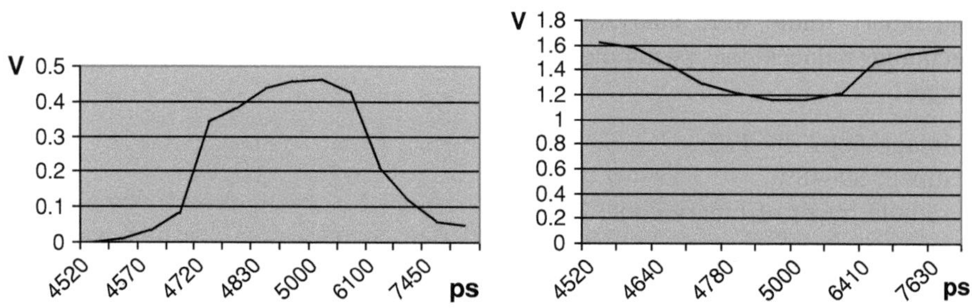

Fig. 19.9 Noise effects in a 0.13 μm technology

schemes were placed and routed in a 0.13 CMOS technology by UMC. The floorplanning was done as in the FPGA implementation, for a maximum bus length of 5 mm. Although the Celtic tool by Cadence [24] was used to reduce crosstalk effects, such as the noise effects shown below in Fig. 19.9, this is quite hard, especially for the schemes BI, CBI, and the block decoding, considering the large number of 32 wires. The achieved clock period is almost 4 ns.

19.5 Conclusions

Coupling effects of on-chip interconnects have become significant for Systems-on-Chip designs in ultra deep submicron technology. Temporal redundancy is used by a set of new proposed encoding schemes to minimize crosstalk in coupled switching which dominate the on-chip bus power consumption. The experimental results show significant energy savings for on-chip interconnects that can reach up to 48%. The used encoder/decoder circuitry is very simple allowing very fast implementations compared with other more complex methods. The temporal redundancy is compensated for, by the high throughput that can be achieved by increasing clock frequency if required. Finally, all the new schemes do not necessitate wiring overhead, but on the contrary, even less number of wires can be used.

References

1. Sotiriadis PP, Chandrakasan A (2000) Bus energy minimization by transition pattern coding in deep sub-micron technologies. In: International conference on computer-aided design (ICCAD), San Jose, CA, November, pp 322–328
2. Benini L, De Micheli G, Macii E, Sciuto D, Silvano C (1998) Address bus encoding techniques for system-level power optimization. In: Design automation and test in Europe (DATE), Paris, France, February, pp 861–866
3. Wong S-K, Tsui C-Y (2004) Re-configurable bus encoding scheme for reducing power consumption of the cross coupling capacitance for deep sub-micron instruction bus. In: Design automation and test in Europe (DATE)
4. Khan Z, Arslan T, Erdogan AT (2005) A novel bus encoding scheme from energy and crosstalk efficiency perspective for AMBA based generic SoC systems. In: Proceedings of international conference on VLSI design—4th international conference on embedded systems design (VLSID'05)
5. Hsieh W-W, Chen P-Y, Hwang TT (2006) A bus architecture for crosstalk elimination in high performance processor design. In: Proceedings of the 4th international conference on hardware/software codesign and system synthesis, pp 247–252
6. Mutyam M et al (2006) Delay and energy efficient data transmission for on-chip buses. In: Annual symposium on emerging VLSI technologies and architectures, April, 2006
7. Li L, Vijaykrishnan N, Kandemir M, Irwin MJ (2004) A crosstalk aware interconnect with variable cycle transmission. In: Design automation and test in Europe, pp 102–107
8. Philippe JM, Pillement S, Sentieys O (2006) Area efficient temporal coding schemes for reducing crosstalk effects. In: Proceedings of the 7th international symposium on quality electronic design (ISQED'06)
9. Najeeb K, Gupta V, Kamakoti V, Mutyam M (2006) Delay and peak power minimization for on-chip buses using temporal redundancy. In: GLSVLSI'06, April 30–May 2, 2006
10. Shin Y, Choi K, Chang YH (2001) Narrow bus encoding for low-power DSP systems. IEEE Trans Very Large Scale Integration (VLSI) Syst 9(5):656–660
11. Stan MR, Burleson WP (1995) Bus-invert coding for low-power I/O. IEEE Trans VLSI Syst 3(1):49–58
12. Lee K, Lee SJ, You HJ (2004) SILENT: serialized low energy transmission coding for on-chip interconnection networks. In: Proceedings of ICCAD, pp 448–451
13. Kretzschmar C, Siegmund R, Muller D (2002) A low overhead auto-optimizing bus encoding scheme for low power data transmission. In: Power and timing modeling and optimization PATMOS, LNCS, vol 2451, pp 342–352
14. Kretzschmar C, Siegmund R, Mueller D (2000) Adaptive bus encoding technique for switching activity reduced data transfer over wide system buses. In: Workshop on power and timing modeling and optimization PATMOS, September, pp 66–75
15. Kretzschmar C, Nieuwland AK, Muller D (2004) Why transition coding for power minimization of on-chip buses does not work. In: Design automation and test in Europe (DATE)
16. Bertozzi D, Benini L, Ricco B (2002) Energy-efficient and reliable low-swing signaling for on-chip buses based on redundant coding. In: IEEE international symposium on circuits and systems, ISCAS

17. Xua J, Honga X, Jinga T, Zhanga L, Gu J (2006) A coupling and crosstalk-considered timing-driven global routing algorithm for high-performance circuit design. Integration VLSI J 39:457–473
18. Pandini D, Pileggi LT, Strojwas AJ (2002) Understanding and addressing the impact of wiring congestion during technology mapping. In: International symposium on physical design (ISPD), US, April, 2002
19. Salek AH, Lou J, Pedram M (1999) An integrated logical and physical design flow for deep submicron circuits. IEEE Trans CAD 18:1305–1315
20. Parakh PN, Brown RB, Sakallah KA (1998) Congestion driven quadratic placement. In: Proceedings of the ACM/IEEE 35th design automation conference, June, pp 275–278
21. Kim K, Baek K, Shanbhag N, Liu C, Kang S (2000) Coupling-driven signal encoding scheme for low-power interface design. In: IEEE/ACM international conference on computer aided design, pp 318–321
22. Lindkvist T, Löfvenberg J, Gustafsson O (2004) Deep submicron bus invert coding. In: The 6th Nordic signal processing symposium, Espoo, Finland, June, 2004
23. Kornaros G (2009) Temporal coding schemes for energy efficient data transmission in systems-on-chip. In: Proceedings of 7th IEEE workshop on intelligent solutions in embedded systems, Ancona, Italy, June 25–26, 2009
24. Cadence Celtic, http://www.cadence.com/products/digital_ic/celtic/index.aspx

Chapter 20
Powersim: Power Estimation with SystemC

Computational Complexity Estimate of a DSR Front-End Compliant to ETSI Standard ES 202 212

Marco Giammarini, Simone Orcioni and Massimo Conti

20.1 Introduction

The International Technology Roadmap for Semiconductors (ITRS) [1] and MEDEA + Roadmap [2] evidence that power and performance analysis are becoming challenging task in current System-on-Chip (SoC) design. In recent years, portable devices has largely spread out implementing more and more complex applications that require a large energy amount from battery. This increasing interest in energy and power consumption of hardware has driven to search new design methodologies and tools to analyze and lower power consumption just in the early stages of the design. To facilitate this analysis, research has focused efforts in developing tools to be used when project is still in system-level phase; this fact gives an estimation of power dissipated before going into other planning levels and then optimize the system to decrease consumption.

SystemC [3, 4], now in its version 2.2.0, add a new library of C++ classes in order to create a language for description of hardware at system-level. In 2005, IEEE has approved a standard, named "IEEE Std. 1666-2005" [5], for SystemC. Although SystemC is now considered one of the more promising languages for system-level design, it does not encompass any functionality for power estimation. However, the object oriented nature of SystemC, can easily be extended to cover this lack.

This chapter extends the work presented in [6] proposing a new framework, called Powersim, that consist of a C++ library added to SystemC, in order to estimate the power consumption of a system described at system-level. Powersim can estimate the power of a system interacting with arithmetic operations, logical functions and mathematical functions of the various modules constituting the

M. Giammarini · S. Orcioni (✉) · M. Conti
DIBET, Università Politecnica delle Marche, Ancona, Italy
e-mail: s.orcioni@univpm.it

M. Conti et al. (eds.), *Solutions on Embedded Systems*,
Lecture Notes in Electrical Engineering, 81, DOI: 10.1007/978-94-007-0638-5_20,

system. Also, thanks to the fact that the Powersim monitors operator and mathematical functions it is able to provide a computational complexity estimate of the system. This feature, combined with accurate power models leads also to a good power consumption estimate.

At the end of this chapter, we presents, an application example of the proposed tool, the computational complexity estimate of a DSR front-end, compliant to ETSI Standard ES 202 212 [7] and implemented at system-level in SystemC.

20.2 Powersim

Powersim [8] is a C++ class library developed to be used inside a SystemC implementation. Its main purpose its the simulation of computational complexity and power consumption of digital system described at system-level. Its main characteristics are:

1. no need to change the source code describing the system;
2. easy configuration via file described in Extended BNF [9] grammar;
3. possibility to assign a different power model for each operation performed on each data type;
4. possibility to add, in a simple way, new power models;
5. possibility to extend, in a simple way, its functionality through a fully object oriented implementation based on C++.

Development of Powersim requested the change of some SystemC classes and the addition of Powersim classes to SystemC library.

The next paragraphs will show in detail the Powersim classes, the changes to SystemC, the creation and use of power models and of the configuration file.

Powersim library consists basically of four classes, in addition to power models and error report classes. Figure 20.1 shows the structure of Powersim and how it interacts with SystemC.

The main class of this library is `ps_kernel`: an object of this class is instantiated inside the `sc_main` function before the modules are created. This class, using the static methods of `ps_configure`, parses the configuration file and stores the read data in a static map. The map contains an object of `ps_module` class for each SystemC module used in the source code. Each object takes care to maintain and update data on computational cost and power dissipation estimation of the module to which it is associated. This is done using the map that each object of `ps_module` class has within. The map contains `ps_power_model` type pointers that point classes derived from `ps_power_model`. Indeed the class `ps_power_model` is an interface which generalises all functions necessary to create a new power model. So the user can implement its own power model by extending `ps_power_model` and implementing all its methods.

Regarding changes to SystemC, we have acted on two types of classes: (i) `sc_module` class and (ii) data-type classes. Changes to the class `sc_module`,

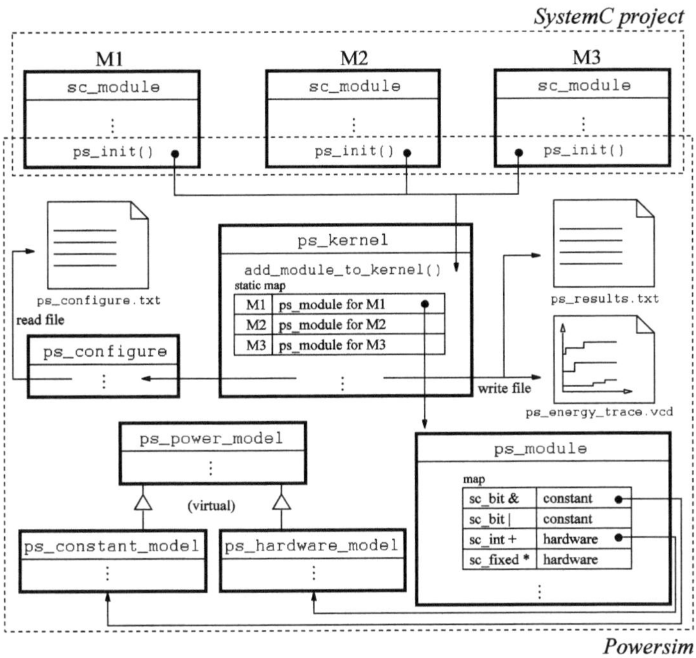

Fig. 20.1 Block scheme of Powersim

the class that let to create a new SystemC module, consist of providing a new method called `ps_init()`. This method, during simulation, calles a static function of `ps_kernel` class. The static method adds the current module under control of the Powersim kernel. Changes to data-type classes consist of providing the necessary code to maintain in each operation the previous value of each variable. Furthermore, the definition of each operator function has been modified to allow the call to `ps_call()`. This function, during simulation, tells to the Powersim kernel that was performed a new operation.

The power models are the instrument through which Powesim estimate the power associated with on operator run on a SystemC data type. As mentioned, the power models are represented by classes that extend `ps_power_model` class and implement also the methods necessary to interact with Powersim kernel. These methods have different purposes: (i) to define the model of the power dissipated by arithmetic or logic operation and mathematical function, (ii) to set the parameters of the power model, using the values that users provide in the configuration file, (iii) to return the total power consumption, (iv) to manage the variables that will be printed in the result tracing file.

Through the configuration file, the user can choose of which modules to monitor the power dissipation. In particular, the user can decide which operators to control the power consumption of and which power model to bind. In order to write a configuration file, the user must follow a grammar defined in Extended

Backus-Naur Form [9], a syntactic meta-language used to describe formal languages syntax. The rules imposed by the grammar are used by `ps_configure` class to parse the input file.

20.3 Case Study

Thanks to the continuous improvement in computer performances and to the development in Digital Signal Processing (DSP), Automatic Speech Recognition (ASR) is spreading in many aspects of everyday life. The application fields of speech recognition go from the help in editing by means of dictation, to command recognition and execution in the automotive or other fields. In general, a voice recognizer can be applied in all situations where the voice may replace hands, including applications to provide support for people with disabilities.

Another area of application of voice recognition is the Distributed Speech Recognition (DSR). Through a client–server based approach in combination with a speech recognizer, DSR can offer a new chance in the field of home automation or mobile communications, for instance, the ability to dictate the notes of a conference directly to your phone immediately after the end of the meeting and to return to office with the text file stored on your PC ready to be edited. This new approach revolutionizes the way of speech recognizing, allowing to replace the communication of the signal samples with a parameterized and compressed representation, which is suitable for the recognition and for the communication over a noisy or limited-capacity channel. In this section we presents a computational complexity estimate, computed by Powersim, of a DSR front-end, compliant to ETSI Standard ES 202 212 [7] and implemented at system-level in SystemC.

20.3.1 The ETSI Front-End

Accordingly to ETSI Standard ES 202 212 a DSR system is composed by a mobile terminal, or front-end, and a server or back-end. The *Front-end* extracts the features of the voice, implementing signal denoising and cepstrum coefficient calculation. Then this features must be compressed and packaged, as show in Fig. 20.2, before being sent to the server, where the recognition take place.

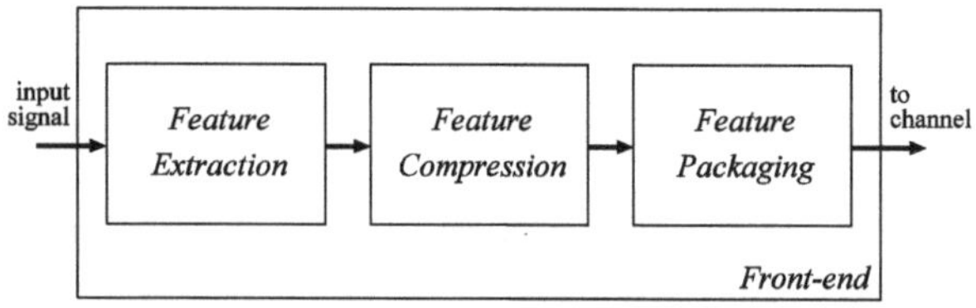

Fig. 20.2 Block scheme the terminal side of ETSI ES 202 212

In this work we have estimated the computational cost of the *Feature Extraction* part, as defined in [7]. Figure 20.3 shows our SystemC implementation of the *Feature Extraction*, divided into five main blocks: the first two blocks make signal denoising, the third makes waveform processing, the fourth performs cepstrum calculation, and the last executes blind equalization of cepstrum coefficients. Before *Feature Extraction*, in *Input Stream* the signal is divided into frames of 80 samples each.

20.3.1.1 First Noise Reduction

The noise reduction consists of two cascaded stages, which, as can be seen in [7], are almost identical. Figure 20.4 shows *First Noise Reduction* SystemC implementation, which consists of eight modules.

In *Buffering* module a four-frame long FIFO (320 samples) is used in order to obtain in each iteration, a 200-sample frame, used to obtain the Wiener-filter coefficients be applied to a single 80-sample long frame. To this end the module applies a 200-sample wide window from the 260th sample to the 61th sample of the buffer, and takes the frame to be denoised from the last but one frame of the FIFO. The *Spectrum Estimation* module performs a power spectrum estimate of its 200-sample long input frame. First the input frame is windowed by Hanning window $s_W(n) = s_{in}(n) \cdot w_{Hann}(n)$ where $0 \le n \le N_{in} - 1$, $N_{in} = 200$, and the Hanning window is

$$w_{Hann}(n) = 0.5 - 0.5\cos\left(\frac{2\pi \cdot (n + 0.5)}{N_{in}}\right). \tag{20.1}$$

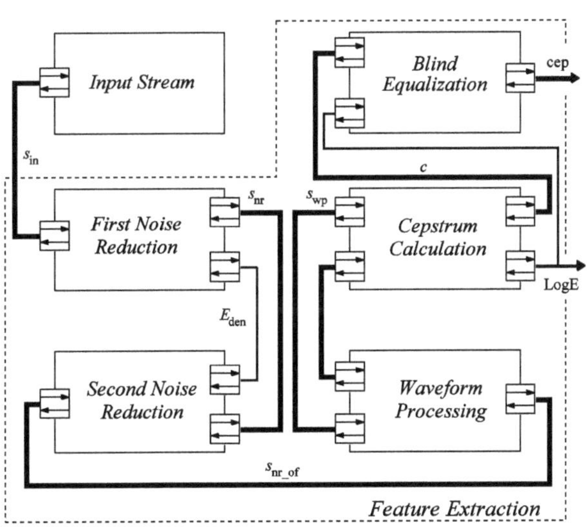

Fig. 20.3 SystemC schematic representation of ETSI ES 202 212 Feature Extraction

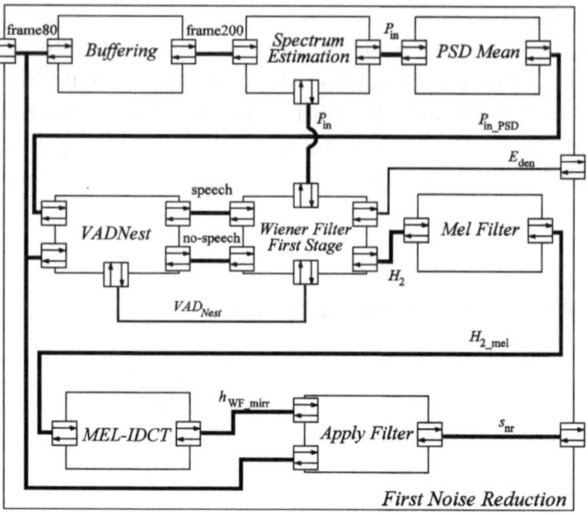

Fig. 20.4 SystemC schematic representation of SystemC First Noise Reduction Module

Then zeros are padded, in order to apply a 256-sample wide Fast Fourier Transform (FFT). The power spectrum is calculated by squaring the module of FFT representation, $X(bin)$, $P(bin) = |X(bin)|^2$, $0 \leq bin \leq N_{\mathrm{FFT}}/2$.

Then the power spectrum is smoothed, as shown in the following $P_{\mathrm{in}}(bin) = (P(2 \cdot bin) + P(2 \cdot bin + 1))/2$, where $0 \leq bin < N_{\mathrm{FFT}}/4$ and $P_{\mathrm{in}}(N_{\mathrm{FFT}}/4) = P(N_{\mathrm{FFT}}/2)$. By means of this last operation, the length of P_{in} is reduced to $N_{\mathrm{SPEC}} = N_{\mathrm{FFT}}/4$.

In the *PSD Mean* module, the mean of power spectral density is performed over the last T_{PSD} frames

$$P_{in\_\mathrm{PSD}}(bin, t) = \frac{1}{T_{\mathrm{PSD}}} \sum_{i=0}^{T_{\mathrm{PSD}}-1} P_{\mathrm{in}}(bin, t - i) \tag{20.2}$$

for $0 \leq bin \leq N_{\mathrm{SPEC}} - 1$, where *bin* is the frequency index ant *t* is the current frame index. *VADNest* module is used to decide if the current frame is speech or not by means of two variables. The former is the logarithmic energy of the last frame of the input signal

$$frameEn = 0.5 + \frac{16}{\ln 2} \cdot \ln\left(\frac{64 + \sum_{n=0}^{M-1} s_{\mathrm{in}}(n)^2}{64}\right). \tag{20.3}$$

This parameter is used to update the second variable *meanEn*. The output of this module is the Boolean variable $flagVAD_{\mathrm{Nest}}$ that indicates if the current frame is speech or not.

Wiener Filter module computes the Wiener filter coefficients that are used to reduce the amount of noise present in a signal by comparison with an estimation of the desired noiseless signal. In the *Wiener Filter First Stage* the noise spectrum estimate $P_{\mathrm{noise}}^{1/2}(bin, t)$ is calculated according to the $flagVAD_{\mathrm{Nest}}$. Then the noiseless signal spectrum $P^{1/2}_{\text{den}}(bin, t)$ is estimated using a "decision-directed" approach and the a priori SNR $\eta(bin, t)$ is computed as $\eta(bin,t) = P_{\text{den}}(bin,t)/P_{\text{noise}}(bin,t)$.

The Wiener filter transfer function $H(bin, t)$ is obtained according to the following equation $H(bin,t) = \sqrt{\eta(bin,t)}/1 + \sqrt{\eta(bin,t)}$ and it is used to improve the estimation of the noiseless signal spectrum $P^{1/2}_{\text{den2}}(bin, t)$. By the new noiseless signal spectrum an improved a priori SNR $\eta_2(bin, t)$ is obtained like

$$\eta_2(bin,t) = \max\left(\frac{P_{\text{den2}}(bin,t)}{P_{\text{noise}}(bin,t)}, \eta^2_{\text{TH}}\right) \tag{20.4}$$

where η_{TH} corresponds to a SNR of −22 dB. Then the improved transfer function $H_2(bin, t)$, that is the module output, is obtained as $H_2(bin,t) = \sqrt{\eta_2(bin,t)}/1 + \sqrt{\eta_2(bin,t)}$, for $0 \le bin \le N_{\text{SPEC}} - 1$. This function is utilized to calculate the new improved noiseless signal spectrum $P^{1/2}_{\text{den3}}(bin, t)$, that will be used to calculate $P^{1/2}_{\text{den}}(bin, t)$ of the next frame. The difference between the *Wiener Filter First Stage* and *Wiener Filter Second Stage* is the method of computation of the noise spectrum estimate $P_{\text{noise}}(bin, t)$, that in the second case does not depend on the *VADNest* module.

The Wiener filter coefficients $H_2(bin)$ are smoothed and transformed to the Mel-frequency scale by *Mel Filter* module. The new coefficients $H_{2\_\text{mel}}(k)$ are calculated by using triangular-shaped, half-overlapped frequency window applied on $H_2(bin)$ as

$$H_{2\_\text{mel}}(k) = \frac{1}{\sum_{i=0}^{N_{\text{SPEC}}-1} W(k,i)} \sum_{i=0}^{N_{\text{SPEC}}-1} W(k,i) H_2(i) \tag{20.5}$$

where k are the transformed frequency, $0 \le k \le K_{\text{FB}} + 1$, with $K_{\text{FB}} = 23$, and $W(k, i)$ is the frequency window.

In the *Mel IDCT* module the time-domain impulse response of Wiener filter is computed from the Mel Wiener filter coefficient $H_{2\_\text{mel}}(k)$ by using Mel-warped inverse DCT.

$$h_{\text{WF}}(n) = \sum_{k=0}^{K_{\text{FB}}+1} H_{2\_\text{mel}}(k) \cdot IDCT_{\text{mel}}(k,n) \tag{20.6}$$

for $0 \le n \le K_{\text{FB}} + 1$, where $IDCT_{\text{mel}}(k, n)$ are Mel-warped inverse DCT, that are obtained as

$$IDCT_{\text{mel}}(k,n) = \cos\left(\frac{2\pi n \cdot f_{\text{centr}}(k)}{f_{\text{samp}}}\right) \cdot df(k) \tag{20.7}$$

for $0 \le k, n \le K_{\text{FB}} + 1$, where $f_{\text{samp}} = 8{,}000$ is the sampling frequency and f_{centr} is the central frequency of each Mel band. The central frequency is computed like $f_{\text{centr}}(k) = 700 \cdot \left(10^{f_{\text{mel}}(k)/2{,}595} - 1\right)$, $0 \le k \le K_{\text{FB}}$ where

$$f_{\text{mel}}(k) = k \cdot \frac{Mel\{f_{\text{lin\_samp}}/2\}}{K_{\text{FB}} + 1} \tag{20.8}$$

where $f_{\text{lin\_samp}}$ is the linear sampling frequency and $Mel\{\cdot\}$ is the function that transform a linear frequency to a Mel scale frequency as

$$Mel\{f_{\text{lin}}\} = 2,595 \cdot \log_{10}\left(1 + {}^{f_{\text{lin}}}/_{700}\right). \tag{20.9}$$

The output of this module is the mirrored impulse response of Wiener filter $h_{\text{WF\_mirr}}(k)$.

In the last module (*Apply Filter*) the noise-reduced signal is produced by means of tree steps. In the former step the causal impulse response is obtained from previous module output. In the second the impulse response is truncated and weighted by a Hanning window. In the latter stage the input signal is filtered like

$$s_{\text{nr}}(n) = \sum_{i=-(\text{FL}-1)/2}^{(\text{FL}-1)/2} h_{\text{WF\_}w}(i + (\text{FL} - 1)/2) s_{\text{in}}(n - i) \tag{20.10}$$

for $0 \leq n \leq M - 1$, where $h_{\text{WF\_}w}$ is the filter impulse response, the filter length FL equals 17 and the frame shift interval M equals 80.

20.3.1.2 Second Noise Reduction

The *Second Noise Reduction*, as show in Fig. 20.5, differs from the former because *VADNest* is not present, instead *Gain Factorization* and *Offset Compensation* have been added.

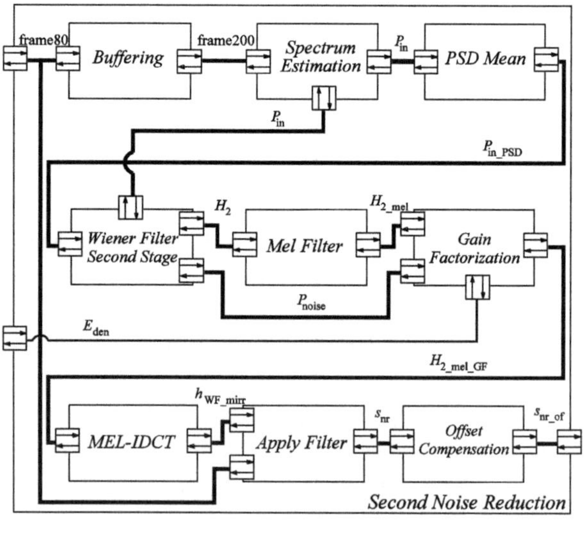

Fig. 20.5 SystemC schematic representation of SystemC Second Noise Reduction Module

Gain Factorization module aims to apply a more aggressive noise reduction to purely noisy frames and less aggressive noise reduction to frames also containing speech. To decide the degree of aggression, SNR value, based on energy values calculated in Wiener filter stages are used. In particular in the *Wiener Filter First Stage*, denoised frame signal energy is calculated by using the denoised power spectrum $P_{\text{den3}}(bin,\, t)$

$$E_{\text{den}}(t) = \sum_{bin=0}^{N_{\text{SPEC}}-1} P_{\text{den3}}^{1/2}(bin, t) \tag{20.11}$$

where t is the current frame index, instead in the *Wiener Filter Second Stage*, the noise energy is computed by using the noise spectrum $P_{\text{noise}}(bin,\, t)$

$$E_{\text{noise}}(t) = \sum_{bin=0}^{N_{\text{SPEC}}-1} P_{\text{noise}}^{1/2}(bin, t). \tag{20.12}$$

Smoothed SNR $SNR_{\text{aver}}(t)$, is evaluated by using tree value of $E_{\text{den}}(t)$ and $E_{\text{noise}}(t)$. At this point, the current SNR estimation is compared to the low SNR tracked value, and the aggression of the second stage Wiener filter is reduced to 10% for speech and noise frames and to 80% for noise frames.

Offset Compensation module removes the DC offset by a notch filtering operation that is applied to the noise-reduced signal like $s_{\text{nr\_of}}(n) = s_{\text{nr}}(n) - s_{\text{nr}}(n-1) + (1 - 1/1{,}024) \cdot s_{\text{nr\_of}}(n-1),\ 0 \le n \le M - 1$, where $s_{\text{nr}}(-1)$ and $s_{\text{nr\_of}}(-1)$ correspond to the last sample of the previous frame.

20.3.1.3 Waveform Processing

After denoising, the *Waveform Processing* part of the standard begins. In this block emphasis on higher energy parts of the signal occurs by means of the action of four blocks, as shown in Fig. 20.6.

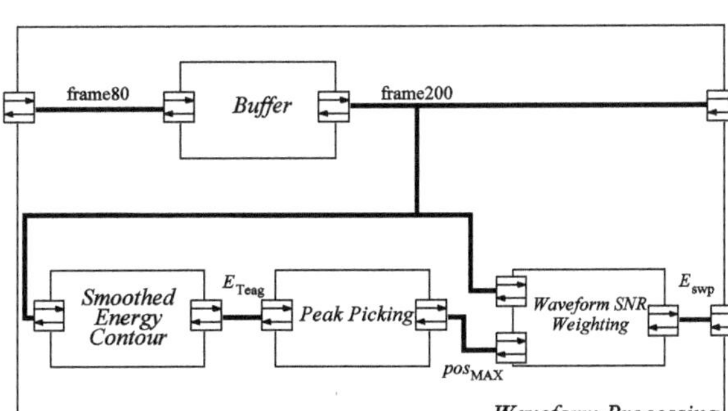

Fig. 20.6 SystemC schematic representation of SystemC Waveform Processing Module

The first block, *Buffer*, stores in a 240-sample buffer the 80-sample long frames given in output by the *Second Noise Reduction*. In this module a 200 (from position 1 to position 200) samples wide window is applied to the buffer.

The second one, *Smoothed Energy Contour*, calculates the Teager-Kaiser energy of the signal and smooth it by means of a FIR filter. The Teager-Kaiser energy is computed for each input frame $E_{\text{Teag}} = \left|s^2_{\text{nr\_of}}(n) - s^2_{\text{nr\_of}}(n-1) \cdot s_{\text{nr\_of}^2(n+1)}\right|$ where $0 \leq n \leq N_{\text{in}} - 1$.

Peak Picking block finds the global maximum in the smoothed energy contour and the maxima on the left and right side of the global maximum, so that maxima related to the fundamental frequency are found.

Such values are used in *Waveform SNR Weighting* block to realize a window function to be applied to the input signal. Indeed having the number of maxima N_{MAX} of the smoothed energy contour and their position pos_{MAX}, a weighting function of length N_{in} is constructed and applied to the input noise-reduced frame like

$$s_{\text{swp}}(n) = 1.2 \cdot w_{\text{swp}}(n) \cdot s_{\text{nr\_of}}(n) + 0.8 \cdot \left(1 - w_{\text{swp}}(n)\right) \cdot s_{\text{nr\_of}}(n) \tag{20.13}$$

where $0 \leq n \leq N_{\text{in}} - 1$ and w_{swp} is a weighting function that equals 1.0 for n belonging to the following interval

$$\begin{aligned}&[(pos_{MAX}(n_{MAX}) - 4), (pos_{MAX}(n_{MAX}) - 4)\\ &\quad +0.8 \cdot (pos_{MAX}(n_{MAX} + 1) - pos_{MAX}(n_{MAX}))]\end{aligned} \tag{20.14}$$

and 0 otherwise.

20.3.1.4 Cepstrum Calculation

The *Cepstrum Calculation* part performs the calculation of cepstrum coefficients and the natural logarithm of the energy of the signal. The our SystemC implementation consist of seven modules, as shown in Fig. 20.7.

First a *Pre-emphasis* filter is applied to the output of *Waveform Processing* block $s_{\text{swp\_pe}}(n) = s_{\text{swp}}(n) - 0.9 \cdot s_{\text{swp}}(n-1)$ followed by a *Windowing*, where the following Hamming window of length $N_{\text{in}} = 200$

$$w_{\text{swp\_w}}(n) = 0.54 - 0.46 \cdot \cos\left(\frac{2\pi \cdot (n + 0.5)}{N_{\text{in}}}\right) \tag{20.15}$$

for $0 \leq n \leq N_{\text{in}} - 1$, is applied to the output of the previous module.

Then a Fast Fourier Transform (*FFT*) is applied. Each frame of N_{in} samples is zero padded to create an extended frame of 256 samples. An FFT is applied to compute the complex spectrum of the denoised signal, then a corresponding power spectrum P_{swp} is calculated.

The next module, *MEL-FB*, recombines the information contained in the FFT according to the Mel band representation. The FFT elements are linearly recombined for each Mel Band. The useful frequency band lies between f_{start} and $f_{\text{sample}}/$

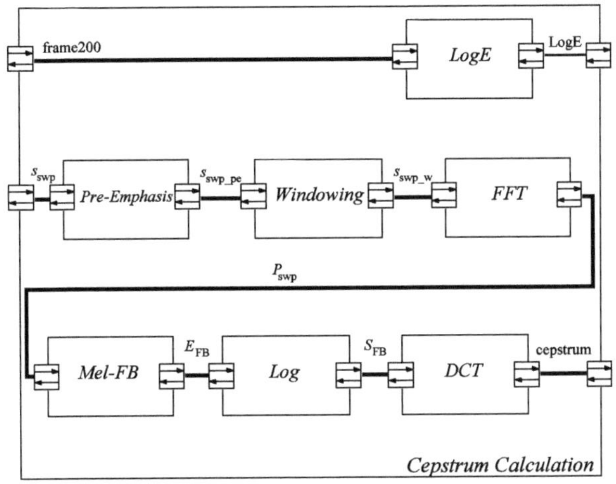

Fig. 20.7 SystemC schematic representation of SystemC Cepstrum Calculation Module

2. This band is divided into K_{FB} equidistant channels in the Mel frequency domain. Each channel has a triangular-shaped frequency window and consecutive channels are half-overlapping. To perform an equidistant distribution of the band in the Mel domain, the central frequency of each filter are calculated from the Mel-function like

$$f_{\text{centr}}(k) = Mel^{-1}\left\{Mel\{f_{\text{start}}\} + k \cdot \frac{Mel\{f_{\text{sample}}/2\} - Mel\{f_{\text{start}}\}}{K_{\text{FB}} + 1}\right\} \tag{20.16}$$

for $0 \leq k \leq K_{\text{FB}}$, where $Mel\{\cdot\}$ is the Mel-function and it is the operator which rescales the frequency domain, likewise Eq. 20.9. Indeed the inverse Mel-function is $Mel^{-1}\{y\} = 700 \cdot \left(e^{y/1{,}127} - 1\right)$.

In terms of FFT index, the central frequency of the band correspond to

$$bin_{\text{centr}}(k) = index\{f_{\text{centr}}(k)\} = round\left\{\frac{f_{\text{centr}}(k)}{f_{\text{samp}}} N_{\text{FFT}}\right\} \tag{20.17}$$

for $0 \leq k \leq K_{\text{FB}}$. For the kth Mel Band, the frequency window $W(i, k)$ is constructed and divided into two parts. The former part accounts for increasing weights, whereas the latter part accounts for decreasing weights. Each frequency window is applied to the denoised power spectrum $P_{\text{swp}}(bin)$ computed in the previous module. The output of each Mel filter is

$$E_{\text{FB}}(k) = \sum_{i=bin_{\text{centr}}(k-1)}^{bin_{\text{centr}}(k)} W_{\text{left}}(i,k) \cdot P_{\text{swp}}(i) + \sum_{i=bin_{\text{centr}}(k)+1}^{bin_{\text{centr}}(k+1)} W_{\text{right}}(i,k) \cdot P_{\text{swp}}(i) \tag{20.18}$$

for $0 \leq k \leq K_{\text{FB}}$.

The *Log* module carries out the logarithmic function on the output of Mel-filtering and finally the thirteen cepstral coefficients are obtained by applying the DCT on the nonlinear transformed FFT by means *DCT* module. The following equation shows how cepstrum coefficients are obtained

$$c(i) = \sum_{k=1}^{K_{\mathrm{FB}}} S_{\mathrm{FB}}(k) \cdot \cos\left(\frac{i \cdot \pi}{K_{\mathrm{FB}}} \cdot (k - 0.5)\right) \tag{20.19}$$

where $0 \leq i \leq 12$. The last module (*LogE*) perform a natural logarithm of the energy of the denoised signal as

$$\log E = \begin{cases} \ln(E_{\mathrm{swp}}) & \textit{if } E_{\mathrm{swp}} \geq E_{\mathrm{THRESH}}, \\ \ln(E_{\mathrm{THRESH}}) & \text{otherwise}, \end{cases} \tag{20.20}$$

where $E_{\mathrm{THRESH}} = e^{-50}$ and E_{swp} is calculated as $E_{\mathrm{swp}} = \sum_{n=0}^{N_{\mathrm{in}}-1} s_{\mathrm{swp}}(n) \cdot s_{\mathrm{swp}}(n)$.

20.3.1.5 Blind Equalization

In the last module of ETSI 202 212 *Feature Extraction* named *Blind Equalization*, twelve cepstral coefficients ($c(1)$, ..., $c(12)$) are equalized according to LMS algorithm. The final feature vector consists of thirteen cepstral coefficient and the log-energy coefficient.

20.3.2 Computational Complexity Estimate

The aim of this section is to show the results of the estimate of the computational complexity of the *Feature Extraction* as computed by Powersim. They will be provided by means of the number values of different mathematical operations performed by each block and by means of the total computational complexity estimate of each block.

Table 20.1 shows the number of operations executed by each SystemC module during a simulation with a registered voice as input. The input signal was sampled at 8 kHz and was 6 s long for a total of 48,000 samples.

To see the computational load of each block, the relative computational complexity of each operation must be estimated. The computational complexity of each operation clearly depends on the hardware where they are executed. As a reference for the relative complexity of operations the Intel® Atom™ N270 [10], largely used in many netbook or embedded systems, has been chosen. The relative cost of each operation has been estimated by simulating ad-hoc programs and calculating the CPU time needed by the execution of each arithmetic operation and mathematical function as implemented in the C ++ *cmath* library [11]. Table 20.2 reports these relative costs.

Table 20.1 Computational cost of SystemC modules, expressed in terms of number of operations

Macro module	Module	Add.	Mul.	Sub.	Div.	Nat. log.
First noise reduction	Spectrum Estimation	2,496,000	3,844,800	0	0	0
	PSD Mean	39,000	39,000	0	0	0
	VADNest	48,700	48,100	1,396	600	600
	Wiener Filter Design	155,740	239,980	45,240	117,000	0
	Mel Filter	0	975,000	0	0	0
	Mel IDCT	375,000	375,000	0	0	0
	Apply Filter	816,000	826,200	0	0	0
Second noise reduction	Spectrum Estimation	2,496,000	3,844,800	0	0	0
	PSD Mean	39,000	39,000	0	0	0
	Wiener Filter Design	273,091	390,741	39,650	320,425	0
	Mel Filter	0	975,000	0	0	0
	Gain Factorization	55,600	18,600	16,200	6,000	0
	Mel IDCT	375,000	375,000	0	0	0
	Apply Filter	816,000	826,200	0	0	0
	Offset Compensation	48,000	48,000	48,000	0	0
Waveform processing	Smoothed En. Contour	960,000	238,800	120,000	120,000	0
	Peak Picking	4,545	0	3,129	0	0
	Wavef. SNR weighting	2,045	2,045	4,090	0	0
Cepstrum calculation	Pre-Emphasis	0	120,000	120,000	0	0
	Windowing	0	120,000	0	0	0
	FFT	2,457,600	3,686,400	0	0	0
	Mel-FB	143,400	104,400	0	0	0
	Log	0	0	0	0	27,600
	DCT	179,400	179,400	0	0	0
	LogE	120,000	120,000	0	0	600
Blind equal.	Blind Equalization	7,200	7,800	15,600	0	0

Table 20.2 Computational cost of algebraic operations and mathematical functions

Operation	Complexity
Addition	1
Multiplication	1
Subtraction	1
Division	5
Cosine	16
Sine	16
Tangent	21
Natural Log	25
Common Log	25

By applying the relative costs shown in Table 20.2 at Table 20.1 the data shown in Table 20.3 has been obtained. The first data column shows the absolute cost of each block while the second one the relative cost. The horizontal lines group together SystemC block belonging to the same ETSI block, respectively, *First Noise Reduction*, *Second Noise Reduction*, *Waveform Processing*, *Cepstrum Calculation* and *Blind Equalization*. Furthermore, since analyzing the algorithms performed the front-end, spectrum calculations performed by means of FFT, were recurrent, the FFT cost has been extracted and shown in last two column of

Table 20.3 Computational cost of single SystemC modules of the "Terminal Front-End"

Macro Module	Module	Comp. Cost	Comp.Cost (% Total)	Comp.Cost of FFT	Comp. Cost FFT (% row)
First Noise Reduction	Spectrum Estimation	6,340,800	19.05	6,144,000	96.90
	PSD Mean	78,000	0.23	0	0.00
	VADNest	116,196	0.35	0	0.00
	Wiener Filter Design	1,025,960	3.08	0	0.00
	Mel Filter	975,000	2.93	0	0.00
	Mel IDCT	750,000	2.25	0	0.00
	Apply Filter	1,642,200	4.93	0	0.00
Second Noise Reduction	Spectrum Estimation	6,340,800	19.05	6,144,000	96.90
	PSD Mean	78,000	0.23	0	0.00
	Wiener Filter Design	2,305,607	6.93	0	0.00
	Mel Filter	975,000	2.93	0	0.00
	Gain Factorization	93,400	0.28	0	0.00
	Mel IDCT	750,000	2.25	0	0.00
	Apply Filter	1,642,200	4.93	0	0.00
	Offset Compensation	144,000	0.43	0	0.00
Waveform Processing	Smoot. En. Contour	1,918,800	5.77	0	0.00
	Peak Picking	7,674	0.02	0	0.00
	Wavef. SNR weight.	8,180	0.02	0	0.00
Cepstrum Calculation	Pre-Emphasis	240,000	0.72	0	0.00
	Windowing	120,000	0.36	0	0.00
	FFT	6,144,000	18.46	6,144,000	100.00
	Mel-FB	247,800	0.74	0	0.00
	Log	690,000	2.07	0	0.00
	DCT	358,800	1.08	0	0.00
	LogE	255,000	0.77	0	0.00
Blind Equal.	Blind Equalization	30,600	0.09	0	0.00
	Total	33,278,017	100.00	18,432,000	55.39

Table 20.3, in absolute and relative values. While the relative computational cost of each block is calculated with respect to the total operations performed by the front-end, the FFT relative cost is relative to each block where it is executed, i.e. relative to each row of the table. So in the last row the cost of FFT is relative to the total cost and it can be noticed that it amounts to the 55.39% of the total. This can suggest the use of specialized hardware for the FFT in a low-power implementation of the front-end because of the relevance of FFT computational cost. In Table 20.4 the computational costs of the main four blocks of the front-end are shown.

The data shown in Table 20.1 can be seen also under another view, grouped by operations performed instead of functional block. Table 20.5 shows this view that reveals that the more expensive operation are respectively multiplication and addition. Table 20.6 shows the same view, but with the FFT cost excluded. In this case also the cost of division becomes relevant.

Table 20.4 Computational cost of the ETSI "Front-End"

Module	Computational cost	Computational cost (%)
First Noise Reduction	10,928,156	32.84
Second Noise Reduction	12,329,007	37.05
Waveform Processing	1,934,654	5.81
Cepstrum Calculation	8,055,600	24.21
Blind Equalization	30,600	0.09
Total	33,278,017	100.00

Table 20.5 Computational cost by operations

Operations	Number of operation	Computational cost	Comput. cost % total
Addition	11,907,321	11,907,321	35.78
Multiplication	17,444,266	17,444,266	52.43
Subtraction	413,305	413,305	1.24
Division	558,625	2,793,125	8.39
Natural log.	28,800	720,000	2.16
Total	30,352,317	33,278,017	100.00

Table 20.6 Computational cost by operation, FFT excluded

Operations	Number of operation	Computational cost	Comput. cost % total
Addition	4,534,521	4,534,521	30.54
Multiplication	6,385,066	6,385,066	43.02
Subtraction	413,305	413,305	2.78
Division	558,625	2,793,125	18.81
Natural Log.	28,800	720,000	4.85
Total	11,920,317	14,846,017	100.00

20.4 Conclusions

In this chapter we presented the Powersim, a new framework for the computational cost and power consumption estimate of a system described in SystemC. Furthermore, we presented, as an application example of Powersim, a computational cost estimation of standard ETSI ES 202 212. This analysis has been carried out at system level by means of a SystemC implementation.

The analysis of computational cost of ETSI 202 212 "Front-End" reveals that the major cost, with more than 55% of the total, can be assigned to the FFT computations performed in the *First Noise Reduction*, *Second Noise Reduction* and *Cepstrum Calculation* functional blocks. So particular care must be taken of this function in a hardware implementation. If a specialized hardware for FFT is not chosen, the more computational expensive operation are multiplication and addition with respectively a 52.43% and 35.78% of the total cost.

References

1. ITRS, International Technology Roadmap for Semiconductors (2005) 2005 edn. Design, December. http://public.itrs.net
2. MEDEA+ (2005) MEDEA Electronic Design Automation (EDA) Roadmap, 5th release, September. http://www.medea.org
3. The Open SystemC Initiative—OSCI, SystemC documentation. http://www.systemc.org
4. Grötker T, Liao S, Martin G, Swan S (2002) System design with SystemC. Kluwer Academic Publishers, New York
5. SystemC Language Reference Manual (2006) IEEE-Std 1666-2005, March
6. Giammarini M, Orcioni S, Conti M (2009) Computational complexity estimate of a DSR front-end compliant to ETSI Standard ES 202 212. In: 2009 seventh workshop on intelligent solutions in embedded systems, WISES09, June, pp 171–177
7. Speech Processing, Transmission and Quality Aspects (STQ); Distributed speech recognition; Extended advanced front-end feature extraction algorithm; Compression algorithms; Back-end speech reconstruction algorithm, ETSI Std. ES 202 212, Rev. 1.1.2, November 2005
8. Powersim 0.1.0—WebSite, Documentation and Source Code. http://sourceforge.net/projects/powersim
9. Information technology—syntactic metalanguage—extended BNF, ISO/IEC Std.14977, December 1996
10. Intel® Atom™ Processor—WebSite and Documentation. http://www.intel.com/products/processor/atom/index.htm
11. CMATH—C numerics library—math.h

Chapter 21
Power Analysis of Embedded Systems

The PKtool Simulation Environment

Giovanni B. Vece and Massimo Conti

21.1 Introduction

Nowadays, microelectronic systems have reached a relevant level of complexity and performances, and such a growth should continue its course in the future years. In addition to great application benefits, such a progress has also caused the arising of new problems, especially in design matters. In particular, the realization of complex digital systems requires to handle effectively their functionalities and to address suitable hw/sw co-design methodologies. For this purpose, new modeling paradigms have been proposed in the last years, within the which SystemC language [1] has reached a primary role as a consequence of its peculiar potentialities. More precisely, SystemC is provided with advanced descriptive means constituting a framework oriented to high abstraction levels. Moreover, SystemC allows the introduction of hardware descriptive constructs within a typical C/C++ software environment, so to support properly an integrated hw/sw codesign.

Another relevant issue related to the design of embedded systems is given by power dissipation. The impact of this factor is often crucial on the physical reliability, when implementing complex functionalities with the demand of high performances. Furthermore, since many embedded systems are supplied by battery power, especially those that are used in mobile devices, the evaluation of low-power solutions is important to increase the battery lifetime. These motivations have led to define operative approaches able to introduce power analysis inside design process. Such analysis should rely on suitable automation means applicable, if possible, in the first design phases. For this need several CAD tools have appeared in the last years; some of the most well-known examples are referenced in [2–5].

G. B. Vece · M. Conti (✉)
DIBET, Università Politecnica delle Marche, Ancona, Italy
e-mail: m.conti@univpm.it

M. Conti et al. (eds.), *Solutions on Embedded Systems*,
Lecture Notes in Electrical Engineering, 81, DOI: 10.1007/978-94-007-0638-5_21,

From all these considerations, the utility of CAD tools able to perform power analysis in SystemC-based designs is evident, especially in the case of embedded systems. This chapter reports an introductory description of a power analysis tool recently made available, the PKtool environment [6, 7]. PKtool is a simulation environment dedicated to power analysis for digital systems described in SystemC/C++ language. The contents of the chapter cover a comprehensive overview of the related features and the application modalities. The chapter is organized as follows. Section 21.2 outlines the PKtool execution flow, while Sect. 21.3 explains how a SystemC module can be interfaced with PKtool environment. Sections 21.4 and 21.5 take into consideration some relevant entities involved in power estimations, i.e. power models and augmented signals. Finally, Sect. 21.6 reports an example of analysis concerning the power performances of Bluetooth networks.

21.2 Simulative Approach and Relations with the SystemC Kernel

The core analysis performed by PKtool is given by the power estimation of a system described in SystemC/C++. PKtool can be directly applied on the single modules composing the system, which are modelled in SystemC by means of specific objects called sc_modules [8]. As final result, a PKtool simulation provides the separate estimations of the power dissipated by the single sc_modules, on the basis of user-selected power models.

The PKtool functionality is realized by several components, among the which the simulation engine, named Power Kernel, covers a primary role. In fact, Power Kernel manages all the tasks involved in a PKtool simulation attending to their synchronization and execution. A PKtool simulation takes place during an ordinary simulation of the monitored system through the SystemC execution kernel. More precisely, the PKtool simulation runs simultaneously with the SystemC simulation, in the form of a complementary appendix dedicated to power analysis. Power Kernel operates in a hidden and non intrusive way, without affecting the correct behaviour of the SystemC kernel. With respect to an ordinary SystemC simulation, the only visible effect might be represented by a variable increase in CPU time. In this way, in a same simulation session, it is possible to address two different targets: the reproduction of the system behaviour as modelled by the SystemC description; the estimation of the power dissipation in the simulated operative conditions.

The orthogonal running between PKtool and SystemC simulations can bring important benefits in an articulated hw/sw co-design. In fact, it is possible to preserve the original simulation workflow without forced modifications or adaptations for integrating the run-time functionality of PKtool. In this way, the introduction of a power analysis does not lead to a worsening in the re-design risks and result reliability.

21.3 Power_Modules

In order to include an sc_module in PKtool analysis, the user has to define and instance a corresponding *power_module*. A power_module is a PKtool entity that allows the interaction between the sc_module and Power Kernel, extending the default capabilities of the sc_module in compliance with PKtool analysis. Such enhancement mainly consists in the linkage to a power model and in additional capabilities related to power estimation tasks. From an external point of view (in particular with regard to I/O ports), a power_module retains the original sc_module structure. Figure 21.1 shows the functional architecture of a power_module, underlining the connections with a traditional sc_module.

The effective monitoring of an sc_module during PKtool simulations is enabled by instancing a corresponding power_module. This step can be realized in flexible and selective way, which means it can be applied to all the system sc_modules as well as a restricted subset chosen by the user. Within complex system descriptions, a monitorable sc_module can be a simple atomic module or a more articulated unit, in turn structured in a submodule hierarchy. During a PKtool simulation, all the monitored sc_modules are handled in independent way; each of them is associated to an its own power model and is simulated as a stand-alone entity. At the end of the simulation, the power estimation of each sc_module is reported in a distinct text file.

These selection modalities allow to introduce a space granularity in PKtool analysis, which can represent an important opportunity in the design of embedded systems constituted by the aggregation of heterogeneous parts. In fact, by instancing specific power_modules, it is possible to make a distinction among different elaboration units and to carry out customized power analysis for each of them. In particular, it is possible to distinguish between hardware and software components and to examine them according to their intrinsic features.

As an example of power_module instance, let us consider a SystemC description that models the interaction between master and slave sc_modules;

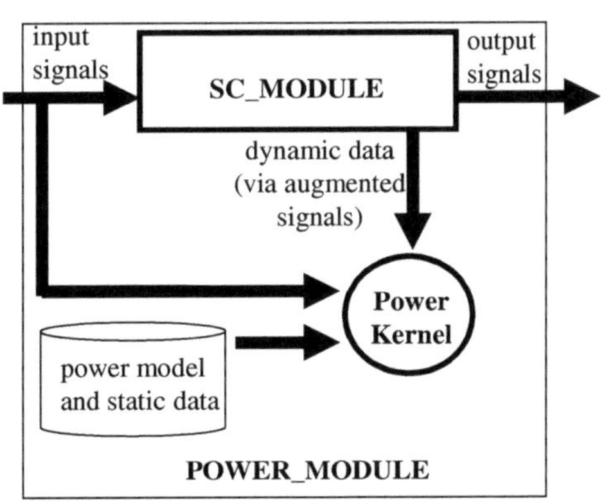

Fig. 21.1 Functional architecture of a power_module

these latter are associated to two distinct classes called respectively *mast_dev* and *sl_dev*. For explanatory reasons, let us assume to apply PKtool only on the master sc_modules. To this end, first we must define a power_module class related to the mast_dev class. After that, we are able to select master sc_modules for PKtool simulations by instancing corresponding power_modules. This step is achieved by simply replacing the instance instructions of the sc_modules with equivalent instructions for the related power_modules.

The following code reports two versions of a SystemC main function (sc_main). The left version is a traditional sc_main that models a top-level architecture constituted by master and slave devices. On the right, there is the same sc_main with the instance of power_module counterparts for the master sc_modules.

```
int sc_main ()
{
// sc_modules instance
mast_dev  m1, m2;

 sl_dev  s1, s2;
// connection signals
 sc_signal<sc_int<64> > data;
 sc_signal<bool >       ready;
 ...
// connection instructions
 m1.out_data( data );
 m1.ready( ready );
 ...
};
```

```
int sc_main()
{
// sc_modules instance
POWER_MODULE( mast_dev )
                  m1, m2;
 sl_dev   s1, s2;
// connection signals
 sc_signal<sc_int<64> > data;
sc_signal<bool >  ready;
 ...
// connection instructions
 m1.out_data( data );
 m1.ready( ready );
 ...
};
```

By comparing the two functions, it is easy to see how the power_modules have been instanced through a simple modification in the original type of the master sc_modules. This represents the only intervention to be done in the original code; due to the external equivalence between sc_modules and power_modules, all the connection instructions do not have to be modified.

21.4 Power Models and Their Characterization

A power model is a formal definition of the power dissipated by a digital system, commonly represented by an analytical/algorithmic formulation. Typically, a power model is not aimed at providing an exact value of the power dissipation, but rather an acceptable estimation. The estimation accuracy can rely on several power model features, such as the reference abstraction level, the computational complexity, the amount of data required. In the technical literature many examples of power models can be found, based on different approaches and accuracy levels. A wide overview is reported in [9].

The PKtool environment is not related to a particular power model, but is linked to an its own power model library that makes available several power models. Such library is integrated in the PKtool implementation and represents an extensible framework where new power models can be defined without strict limitations.

During a PKtool simulation, each instanced power_module has to be associated to a specific power model, that will be applied for computing the related power estimation. This association is carried out at the beginning of the simulation and is based on an interactive routine, where the user is asked to select the power model among the ones provided by the PKtool model library.

The availability of several power models and their independent applicability constitute important features for power analysis on embedded systems. In fact, in synergy with the space separation realized by the instance of power_modules, it is possible to analyze different system components by applying the power models more suitable to their specific features. This chance is further emphasized considering that hardware-based and software-based power modeling are typically referred to different approaches and solutions [9].

The evaluation of a power model is usually based on specific data required in its formulation (model data). Such data are often related to information characterizing the module which the power model is applied to, such as technology, architectural and functional features. From an operative point of view, we can subdivide model data into two distinct categories:

- static data: known a priori and available before the beginning of a simulation;
- dynamic data: available only during the simulation, on the basis of the run-time evolution of the system.

Typical examples of static data are technology parameters; typical examples of dynamic data are signal information, e.g. switching activity and bit probability.

PKtool implements different solutions for the acquisition and the handling of static and dynamic data. Static data must be communicated by the user at the beginning of a simulation, through the same interactive modalities used for the association between power_modules and power models. As concerns dynamic data, PKtool makes available suitable means addressed to their handling during the simulation running. The most important of these means is represented by dedicated components called *augmented signals*.

21.5 Augmented Signals

Augmented signals are aimed at evaluating useful signal data commonly required by many power models as dynamic data. An augmented signal can be regarded as a smart signal, able to show a traditional behaviour with additional capabilities to compute and make available information such as bit size, number of transitions, transition density [10]. Some of these quantities, for example number of transitions, can be computed during a simulation, on the basis of the run-time signal evolution. Augmented signals are realized through a framework of types alternative to the traditional signal types used in SystemC/C++ descriptions.

The usage of augmented signals takes place within a monitored sc_module, by instancing objects referred to specific augmented types. As in the case of power_module instance, this task consists in the replacement of pre-existent signals with augmented counterparts, as shown in the following example:

```
SC_MODULE( mast_dev)
{
// input ports
 sc_in<sc_uint<32> >
           in_1, in_2;
 ...
//output port
 sc_out<sc_uint<32> >
               out_data;
 sc_out<bool>   ready;
 ...
 // internal signal
 sc_uint<3> ctr_1;
 sc_uint<2> ctr_2;
 ...
}
```

```
SC_MODULE( mast_dev)
{
// input ports
 sc_in<sc_uint<32> >
           in_1, in_2;
 ...
//output port
 sc_out_aug<sc_uint<32> >
                    out_data;
 sc_out<bool>       ready;
 ...
 // internal signal
 sc_uint<3> ctr_1;
 sc_uint_aug<2> ctr_2;
 ...
}
```

Considering again the mast_dev sc_module introduced in Sect. 21.3, the code reports two different versions of the related class. The left version represents an original definition, whereas the right one is the same class with the modifications required for instancing two augmented signals. These latter are the output port *out_data* and the internal signal *ctr_2*, whose original types have been replaced by the corresponding augmented types. The conversion simply consists in adding the ending word *_aug* to the name of the original types; this is a general rule for instancing augmented signals.

Once switched into the augmented format, the signals out_data and ctr_2 become able to compute and provide useful data for power estimation, as previously explained. During a PKtool simulation involving mast_dev power_modules, such data are used in application with the selected power models to calculate the output power estimations. All these operations are handled by Power Kernel, which is able to interact with the augmented signals via a nested communication interface.

As shown by the example, augmented signals can be instanced in selective way, with the possibility to consider an arbitrary subset of all the sc_module signals. Actually, this choice is strictly dependent on the signal data required by the applied power model. If, for example, such data were restricted to the input ports, only these signal should be augmented in the sc_module class.

21.6 Application Example on Bluetooth Communication Performances

This section presents a system level power analysis concerning the Bluetooth communication tasks [11, 12] and realized by applying PKtool on a compliant SystemC model. We have considered a simulation scenario constituted by one master and three slave devices, connected through a virtual channel emulating the wireless communication with the presence of noise. Each device has been basically realized through an sc_module called "link_controller", which implements the Bluetooth baseband layer. Each link_controller module is composed by several sub-modules; the power_module conversion has been applied to all the sub-modules of each bluetooth devices.

The power model defined by Eq. (21.1) has been applied for estimating the power dissipation:

$$E = N_{\mathrm{g}}\left[\frac{1}{2}C_{\mathrm{eq}}V_{\mathrm{dd}}^{2}\bar{D}\right] \tag{21.1}$$

This model is an extension of the expression of the dynamic power dissipation of a single CMOS gate. V_{dd} is the supply voltage, $\bar{D}$ is the average number of commutations of the input, output and internal signals of a module, N_{g} is the number of gates of the module, C_{eq} is the average capacitance per gate.

The link_controller unit has been modeled also in VHDL; the synthesizable VHDL code has allowed an estimation of the parameters N_{g} and C_{eq} required by the power model, as illustrated in [7]. Figure 21.2 reports the complexity of each sub-module in the link_controller, in terms of number of logic gates.

The Bluetooth standard allows the optimization of the piconet in terms of throughput or power dissipation, by fixing some parameters of the device and the piconet. Mainly, two types of transmissions are used in the Bluetooth standard: the asynchronous connection-oriented (ACL) logical transport is used to carry control signals and asynchronous user data; the synchronous connection-oriented (SCO) logical transport is a symmetric point-to-point channel between the master and a specific slave, typically used for encoded voice stream [11].

The ACL packet formats are reported in Fig. 21.3. The packet consists in the following fields:

- Access Code (72 bits): it is used for synchronization and identifies all the packets exchanged on piconet channels;

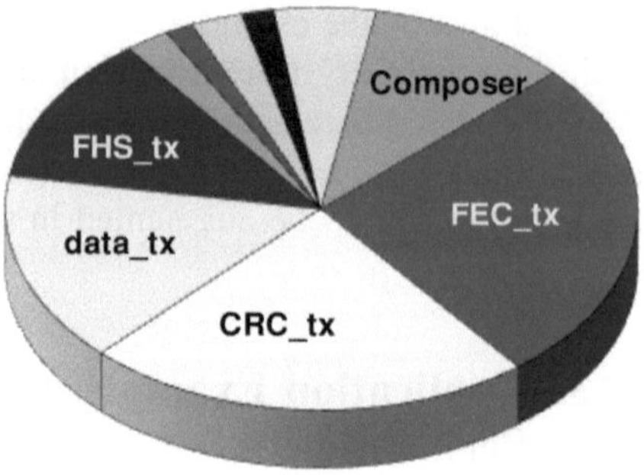

Fig. 21.2 Complexity in terms of logic gates of each module of the link_controller

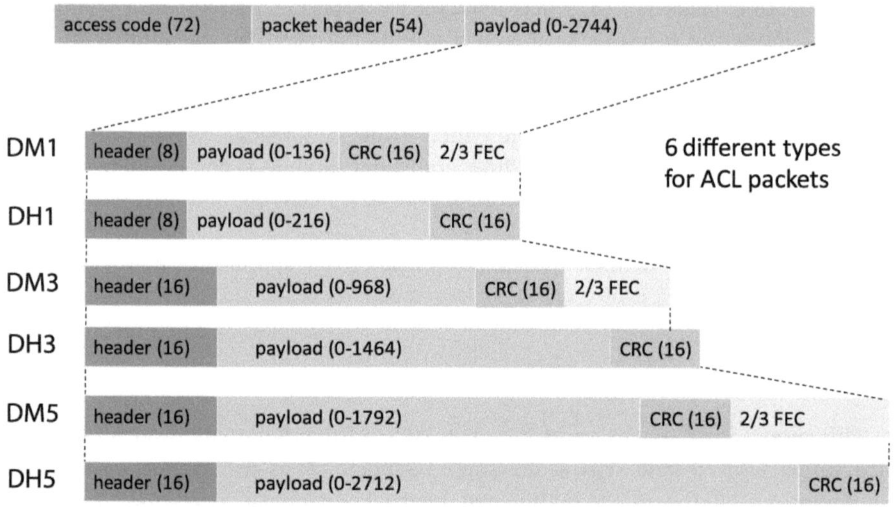

Fig. 21.3 ACL packet formats

- Packet Header (64 bits): it contains the slave address, a code specifying which type of packet is used, the information on transmission errors, buffer status and packet sequence;
- Payload Header (8–16 bits): it specifies the logical channel and the effective length of the transmitted data;
- User Payload (0–2,712 bits): contains the effective user data;
- CRC: 16-bits cyclic redundancy check code is generated from the payload header and user payload segments.

Six principal ACL types of packets exist (DH1, DM1, DH3, DM3, DH5, DM5), offering a balance between robustness and bandwidth. Both DH and DM packets use CRC to detect errors during transmission. DM packets use 2/3 FEC to correct errors on payload header, user payload and CRC segments during transmission, therefore they are more robust against noise but, as a consequence, the effective

Table 21.1 Characteristics of ACL packet types

	Time slots	Payload bytes/packet	Payload bits/packet	Total bits/packet	Min time slots to send 10 kB payload	Total sent bits to send 10 kB payload	Total sent bits/payload bits
DM1	1	17	136	366	603	220,570	2.69
DH1	1	27	216	366	380	138,920	1.70
DM3	3	121	968	1,626	254	137,670	1.68
DH3	3	183	1,464	1,622	168	90,768	1.11
DM5	5	224	1,792	2,865	229	131,023	1.60
DH5	5	339	2,712	2,870	152	86,818	1.06

bandwidth is reduced. Each block of ten information bits is encoded into a 15-bit codeword. This code can correct all single errors and detect all double errors in each codeword.

Table 21.1 reports some characteristics of the ACL packets: the number of time slots used, the user payload bytes and bits per packet, the total bit sent per packet (including access code, packet header, payload header, user payload, CRC and 2/3 FEC).

The SystemC code of the Bluetooth baseband has been used to analyse the performances of a piconet with the master transmitting 10 kB of data to the slaves in presence of noise. Table 21.1 reports the minimum number of time slots required and the total bits necessary to send 10 kB of user data using the different packets. The ratio between the total bits sent and user payload bits is also specified in Table 21.1. The overhead is relevant for the DM packets and for the packets that use 1 time slot, therefore in absence of noise the advantage of DH3 and DH5 packets with respect to the other types of packets is evident.

We have evaluated the power dissipated during the transmission of different types of ACL data packets, by carrying out PKtool simulations onto the piconet models. As simulation scenario, we have considered the transmission of the maximum data allowed by each type of packet. The output results have been normalized by the energy dissipated by the transmission of the DM1 packet.

Table 21.2 reports the mean value of the normalized energy dissipated by the master baseband when transmitting the different ACL packets, the normalized energy per transmitted bit, and the normalized energy per user payload bit.

Table 21.2 Normalized energy dissipated by the baseband layer for different packet types

	(Energy per packet)/ (energy_DM1 per packet)	(Energy/total bits)/ (energy_DM1/total bits)	(Energy/payload bits)/ (energy_DM1/payload bits)
DM1	1.00	1.00	1.00
DH1	0.46	0.46	0.29
DM3	5.25	1.18	0.74
DH3	1.29	0.29	0.12
DM5	9.92	1.27	0.75
DH5	2.09	0.27	0.10

The baseband energy dissipation while transmitting DH packets is considerably lower than DM packets, especially if normalized by the user payload bit; this is due to the fact that the FEC_tx block is inactive. Furthermore, packets using three or five time slots are more energy efficient.

The Bluetooth system allows a high flexibility in the choice of the packet type. This flexibility can be used to optimize the performances of the transmission in presence of noise in the channel. In [13] it has been shown how the packet type can affect the throughput in a noisy channel. In our analysis we have evaluated the effects on the power dissipation; this kind of analysis is useful to evaluate the power dissipated by the baseband in the optimization of the total amount of power dissipated by the device.

The PKtool simulations have been realized with different levels of channel noise, with the aim of identifying the best choice of the type of packet in terms of energy dissipation. The noise has been simulated inserting the module error_gen, which introduces randomly an error in the channel with a probability depending on the fixed BER.

Without loss in generality, in the simulations it has been considered:

- a symmetric user asynchronous data transmission between master and slave, with the piconet already built up;
- a signal/noise ratio constant during time and uniform in the Bluetooth transmission band (2,400–2483.5 MHz);
- a constant type of information (10 kB) for each transmission independently on the type of packet used, in order to make the results comparable.

For each packet type and for each BER many simulations have been performed (40 for DM1, DM3 and DM5 packets and 120 for DH1, DH3 and DH5). A total of 43,000 transmissions have been examined in PKtool analysis.

Figures 21.4 and 21.5 show the mean value and the standard deviation of the number of time slots required to send 10 kB of information in the channel as a function of 1/BER. The standard deviation is indicated through the error bars. The DM packets are protected by 2/3 FEC, therefore they show better performances in terms of channel occupancy when the BER is high. On the other hand DM-based transmissions require more packets to send the same amount of data in absence of noise, as shown in Table 21.1. Therefore, the best packet in terms of channel occupation depends on the level of noise in the channel.

Figures 21.6 and 21.7 show the mean value and the standard deviation of the energy required to send 10 kB of information normalized to the energy required by DM1 packets as a function of 1/BER. The energy per time slot required by the DH packets is about half the energy of the DM packets, as shown in Table 21.2, therefore even when the time slots required by the DH packets are higher with respect to the DM packets, the energy required is less. As a consequence, the best packet in terms of baseband energy dissipation depends on the level of noise in the channel, but in general it does not correspond to the best packet in terms of channel occupation, as it can be seen comparing Figs. 21.4, 21.5, 21.6 and 21.7

Finally, Figs. 21.8 and 21.9 show the mean value and the standard deviation of the energy per transmitted bit normalized to the energy per transmitted bit using DM1 packets as a function of 1/BER. Figures 21.8 and 21.9, compared to Figs. 21.6 and 21.7, evidences that the packets that use three and five time slots dissipate less energy per bit with respect to the corresponding packets that use only one time slot.

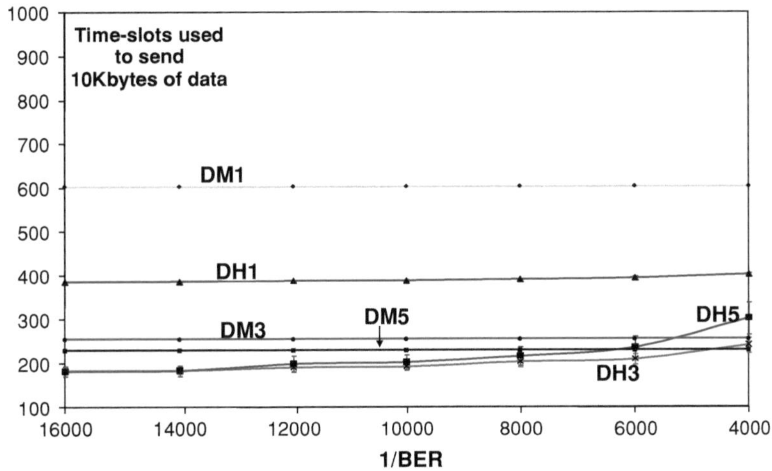

Fig. 21.4 Mean value and standard deviation of the number of time slots used to send 10 kB of data as a function of 1/BER. The error bars indicate the standard deviation

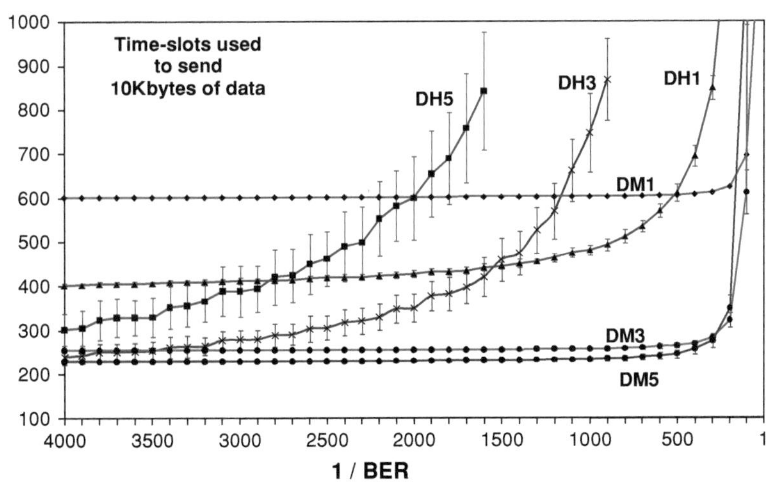

Fig. 21.5 Mean value and standard deviation of the number of time slots used to send 10 kB of data as a function of 1/BER. The error bars indicate the standard deviation

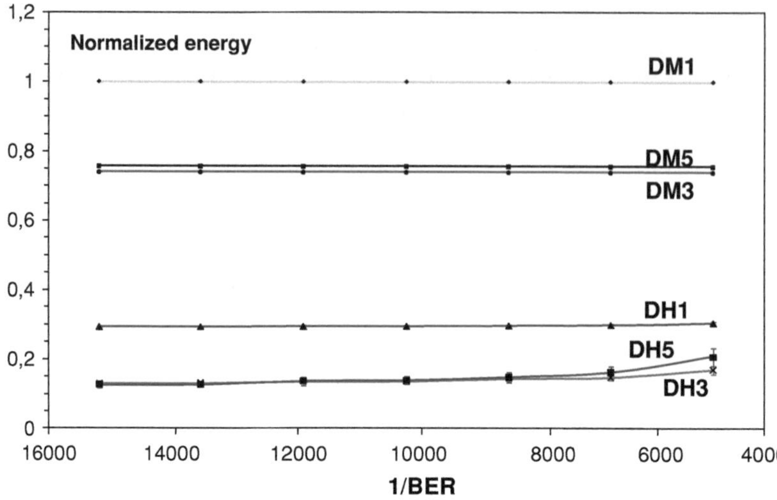

Fig. 21.6 Mean value and standard deviation of the normalized energy spent to send 10 kB of information in the channel as a function of 1/BER. The standard deviation is indicated through the error bars

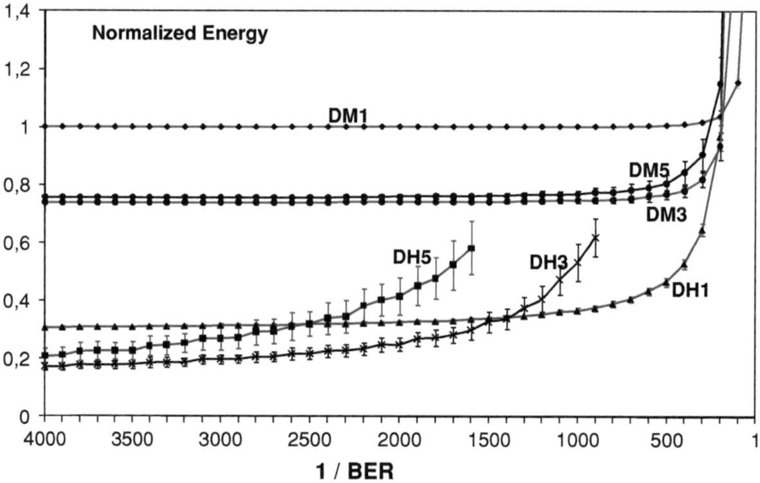

Fig. 21.7 Mean value and standard deviation of the normalized energy spent to send 10 kB of information in the channel as a function of 1/BER. The standard deviation is indicated through the error bars

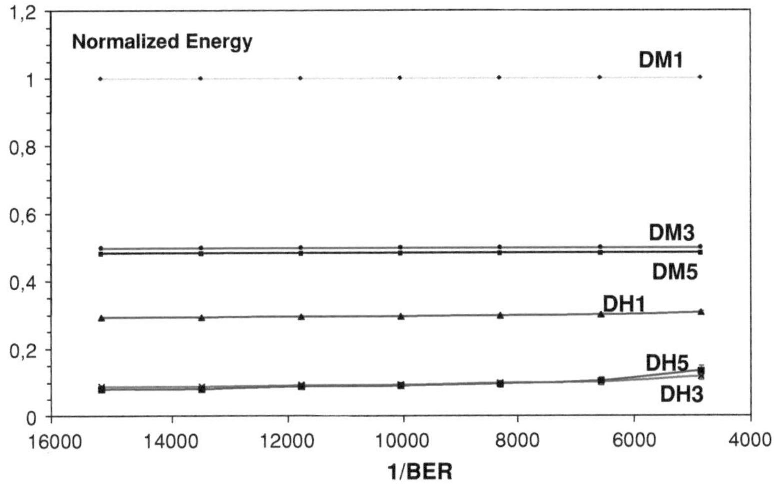

Fig. 21.8 Mean value and standard deviation of the normalized energy per transmitted bit spent to send 10 kB of information in the channel as a function of 1/BER. The error bars indicate the standard deviation

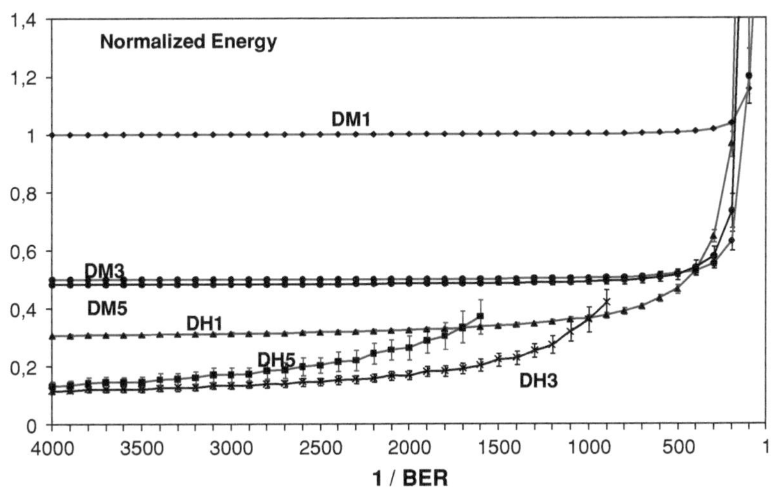

Fig. 21.9 Mean value and standard deviation of the normalized energy per transmitted bit spent to send 10 kB of information in the channel as a function of 1/BER. The error bars indicate the standard deviation

21.7 Conclusions

This work has presented the fundamental features and the application modalities of the PKtool environment, underlining its potentialities within a SystemC-based design context. We have provided a functional description of the main PKtool components, supported by usage examples. As case study, PKtool has been applied to evaluate specific power performances of Bluetooth networks. In particular, analysis of channel throughput and baseband energy dissipation have been carried out.

References

1. SystemC official web-site, http://www.systemc.org
2. PowerTheater, Sequence Design Inc., http://www.sequencedesign.com
3. ORINOCO, Chip Vision Design Systems, http://www.chipvision.com
4. Power Compiler, Synopsis, http://www.synopsys.com
5. PowerChecker, BullDAST s.r.l., http://www.bulldast.com
6. PKtool official web-site, http://www.deit.univpm.it/PKtool
7. Vece G, Conti M (2009) Power estimation in embedded systems within a SystemC-based design context: the PKtool environment. In: Proceedings of the IEEE Seventh International Workshop on Intelligent Solutions in Embedded Systems WISES09, pp 179–184, Ancona, Italy, June 25–26 2009
8. Grotker T, Liao S, Martin G, Swan S (2002) System design with SystemC. Kluwer Academic Publishers, Dordrecht, The Netherlands
9. Piguet C (2006) Low-power CMOS circuit (technology, logic design and CAD tools). Taylor & Francis Group, London
10. Najm F (1993) Transition density: a new measure of activity in digital circuits. In: IEEE Trans on computer-aided design of integrated circuits and systems, vol 12, no 2, Feb 1993, pp 310–323
11. Bluetooth_Core v2.0, Bluetooth official web-site, http://www.bluetooth.org
12. Bray J, Sturman C (2002) Bluetooth connect without cables, 2nd edition. Prentice Hall PTR, EnglewoodCliffs, NJ
13. Caldari M, Conti M, Crippa P, Marozzi G, Di Gennaro F, Orcioni S, Turchetti C (2003) SystemC modeling of a Bluetooth transceiver: dynamic management of packet type in a noisy channel. Design, automation and test in Europe Conference 2003, Munchen, pp 214–219

MIX
Papier aus verantwortungsvollen Quellen
Paper from responsible sources
FSC® C105338

If you have any concerns about our products,
you can contact us on
ProductSafety@springernature.com

In case Publisher is established outside the EU,
the EU authorized representative is:
Springer Nature Customer Service Center GmbH
Europaplatz 3, 69115 Heidelberg, Germany

Printed by Libri Plureos GmbH
in Hamburg, Germany